AQUACULTURE —

An Introduction

AQUACULTURE —

Jasper S. Lee

Professor
Department of Agricultural
and Extension Education
Mississippi State University

An Introduction

Michael E. Newman
Assistant Professor
Department of Agricultural
and Extension Education
Mississippi State University

IPP
INTERSTATE PUBLISHERS, INC.
Danville, Illinois

AQUACULTURE — An Introduction

Library of Congress Catalog Card No. 91-76127

ISBN 0-8134-2911-0

Order from

INTERSTATE PUBLISHERS, INC.
510 North Vermilion Street
P.O. Box 50
Danville, IL 61834-0050

Phone: (800) 843-4774
FAX: (217) 446-9706

1 2 3
4 5 6
7 8 9

To Obed L. Snowden, retired agricultural education professor and administrator at Mississippi State University. "Doc," as Snowden was affectionately known, always had time for students and encouraged them to apply themselves to scholarly pursuits. His counsel and assistance have helped many young people to set and strive for higher goals. The senior author will forever be grateful for the many ways he was helped and supported by the amazing Dr. Snowden.

■ ■ ■

To Rodney D. Childress, retired vocational agriculture instructor at Water Valley High School in Water Valley, Mississippi. Mr. Childress always offered encouragement and taught the value of working hard and trying to achieve the best. Although it may not have been his intention, the professional example he set and his attitude toward his work caused the junior author to choose agricultural education as a profession. It is the author's hope that his students will look up to him the way he has looked up to Mr. Childress.

Cover photos: Courtesy, Marco Nicovich, Manager of Photography, Mississippi Cooperative Extension Service, Mississippi State University

FOREWORD

Aquaculture — An Introduction focuses on introducing the subject of aquaculture to individuals without training or knowledge in the subject, with emphasis on developing a basic understanding of the diversity of aquaculture and the potential it holds for the U.S.

A practical, science-based approach is used throughout. Research findings have been studied and used in undergirding observations in the production of aquaculture crops. New terms are introduced and explained as they are used in the practice of aquaculture.

The bottom line in any production endeavor is to make a profit. Principles of decision making and management are stressed throughout the book. Planning is a step that must be included in any thoughts about beginning an aquabusiness.

Many areas of aquaculture are new and untried in the U.S. As supplies of wildfish and other aquatic foods continue to be depleted, emphasis on aquaculture will increase. Educational programs to prepare people for this new industry will no doubt be popular. Such programs must always be founded on basic principles and the application of the principles in real-world settings. The Council for Agricultural Education in Alexandria, Virginia, is taking an aggressive leadership role in developing and implementing education in schools throughout the nation. Many young people will be able to find satisfying careers in the broad field of aquaculture.

The authors hope that this book will make a contribution to the emergence of the aquaculture industry.

ACKNOWLEDGMENTS

Many individuals are thanked for their assistance in the preparation of this book. The contributions of businesses and agencies involved in aquaculture made it possible for the book to be written. Thus, the authors have also acknowledged the various contributions of businesses and agencies throughout the book. In addition, several individuals are recognized here.

Jeanette R. Evans is acknowledged by the senior author for her untiring effort and dedicated support of all of the author's professional endeavors for over 13 years. Her involvement with this book was most helpful. Reviewing manuscript material and helping critique figures were greatly appreciated. Jeanette, your assistance is always professional, efficient, effective, and considerate. Thank you!

The junior author acknowledges the sacrifices made by his family so that the manuscript could be completed. Thank you, Tricia and Erin, for your understanding and support. Also, thank you, Adrienne Washington, for your help in getting film processed and taking care of other details.

Other individuals to be acknowledged are Jeanne Hartley for assistance with line art and Eddie Harris for his help with photographs. Tom McIlwain and other staff members of the Gulf Coast Research Laboratory provided opportunities for the authors to get information on aquaculture. Gary Fornshell and Mack Fondren of the Mississippi Agricultural and Forestry Experiment Station were most helpful, as were Penney McHann, Chad Howell, Elizabeth Morgan, and Ann Duncan.

CONTENTS

PART THREE — PRODUCTION ALTERNATIVES IN AQUACULTURE

PART ONE

An Overview of Aquaculture

Chapter 1

AQUACULTURE TODAY

Aquaculture is a young and growing industry in the U.S. It is emerging as an alternative to traditional crops for some farmers. And there are important reasons why! Consumption of fish and related foods is increasing each year. The demand for all types of aquacrops is greater than the ability of the earth's oceans and streams to provide them. In fact, some natural bodies of water no longer have the fish that were so abundant just a few years ago. The demand for aquacrops and the development of technology to produce them have resulted in the rapid expansion of aquaculture.

OBJECTIVES

This chapter centers on helping students understand today's aquaculture. Upon completion, the student will be able to:

- Explain aquaculture and related terms.
- Distinguish between cultured and captured aquacrops.
- Describe the types of aquaculture production systems.
- Explain the historical view of aquaculture.
- Explain the importance of aquaculture.
- Explain the role of animal welfare.

AQUACULTURE AND ASSOCIATED AREAS

Aquaculture is the cultural production of plants and animals in water. More specifically, it is the culture and harvest of aquatic animals and plants usually for food and fiber. Aquaculture is much like traditional farming on land except that it occurs in water. Another way of defining aquaculture is to say that it is "water farming."

Many new words are being used in aquaculture. "Aqua" (meaning "water") is often used as a prefix ("in front of") with other words, such as "aquafarm" and "aquabusiness." When this is done, we know that the word means that water is involved.

Fish Predominate

Fish are the predominant aquacrop in the U.S. As such they are produced for one or more of five purposes, as follows:

1. *Food* — These are fish produced almost exclusively for human food. Catfish, trout, and salmon are examples.

Figure 1–1. An aquafarmer examining a catfish at harvest. (Note the seine and filled basket lifting food-size fish.) (Courtesy, Delta Pride Catfish, Inc.)

2. *Bait* — These are small fish used by sport enthusiasts as bait. The fathead minnow is a good example.

3. *Sport fish* — These are fish grown for fee-lakes, for releasing into streams, or for stocking hobby ponds. Examples include trout, catfish, and bream.

4. *Ornamentals (or pets)* — These are fish kept for their aesthetic or personal appeal. They are often found in aquaria or similar water containers. Examples are goldfish, guppies, and angel fish.

5. *Feed* — These fish serve as feed or feed ingredients. Some fish meal is needed in rations for most fish. Examples include goldfish and menhaden, with the latter usually being wild marine fish.

Other Water Crops

The notion that aquaculture involves producing only fish is false. The production of aquaplants, such as watercress, is increasingly emerging as an important area of aquaculture. Aquaculture also includes producing bullfrogs, crawfish (also called crayfish), alligators, turtles, and squid. Regardless of the large variety of aquacrops, four aquacrops (crops grown in water) account for over 90% of the production in the U.S.: trout, salmon, catfish, and crawfish.

Aquafarmers Have Choices

Aquafarmers cultivate plants and animals in water. Many farmers view this as an alternative to traditional agriculture. They hope to find growing aquacrops to be more profitable than what they have been growing.

There are several areas in aquaculture, just as there are many areas in agriculture. Although some of these areas are not yet distinct, others are becoming more identifiable.

Aquacrops may be grown in freshwater, saltwater, or brackish water. Saltwater aquaculture is sometimes referred to as marine aquaculture, or *mariculture*. Brackish water is the water where freshwater and saltwater run together. Those crops adapted to one kind of water likely will not grow well in another. Some aquacrops require both freshwater and saltwater, depending on their stage of growth. For example, freshwater prawns (a kind of shrimp) may grow well in freshwater but must be placed in saltwater to reproduce. Aquacrops are frequently grown in ponds and tanks and in cages in oceans, lakes, and streams.

Seafood includes edible marine fish, shellfish, and other lesser important aquatic plants and animals, which may be wild or cultured. Most often they are

Figure 1–2. Algae is an aquacrop thought to hold potential. (This algae was grown at the Oceanic Institute in Hawaii.)

wild, captured in the seas, oceans, and lakes, which are all bodies of water that are usually salty.

Freshwater aquaculture involves growing aquacrops in streams, ponds, and lakes not containing saltwater. It also includes super-intensive tank culture where spring water is used. Trout, tilapia, catfish, and watercress, a plant crop, are examples of freshwater aquacrops. Current concern over the conservation of freshwater supplies has had little impact on aquaculture, though it might in the future, primarily because most aquaculture has been carried out where there has been plenty of water. Some time is required for the demand for water to have an effect on the supply available. The supply of water below the earth's surface is diminishing in areas where wells are used to provide water for aquaculture. Restrictions on drilling wells for water are being established, and permits are usually required.

Aquaculture encompasses a huge number of crops, terms, and subjects. Much as the scope of agriculture is very broad, so is that of aquaculture. Examples of aquacrops are presented in Table 1-1.

Table 1–1

Examples of Aquacrops
(All aquacrops listed have some potential for culture in the U.S.)

Common Name (scientific name)	Notes on Culture and Importance
Alligator (*Alligator mississippiensis*)	Grown in southern U.S., primarily Louisiana and Texas; flesh is used for food and the skin is used for leather.
Bullfrog (*Rana catesbiana*)	Cultured for food and laboratory use; mostly imported to U.S.
Carp (common) (*Cyprinus carpio*)	Many nations produce and consume; not particularly popular in the U.S.
Carp (grass or white amur) (*Ctenopharyngodon idellus*)	Consumes huge amounts of vegetation; produced in China.
Catfish (channel) (*Ictalurus punctatus*)	Widely grown in southern U.S.
Chinese waterchestnut (*Eleocharis dulcis*)	Plant grown for corms; production is experimental in the U.S.; popular and successful in China; imported food to U.S.
Crawfish (red swamp) (*Procambarus clarkii*)	Primarily grown in Louisiana and Texas for food.
Eel (Japanese) (*Anguilla japonica*)	Popular food in Italy, Japan, and other countries; cultured in Taiwan, Japan, and other countries.
Goldfish (*Carassius auratus*)	Popular ornamental fish in U.S.; grown in Arkansas and Missouri.
Freshwater prawn (*Macrobrachium rosenbergii*)	Experimental in U.S.; successful production in Asian and other countries.
Minnow (fathead) (*Pimephales promelas*)	Popular baitfish in U.S.; grown in Arkansas and other areas.
Oyster (American) (*Crassostrea virginica*)	Limited culture success in Chesapeake Bay area and Eastern Shore of U.S.; successful culture in Japan, Taiwan, and other areas.

(Continued)

Table 1–1 (Continued)

Common Name (scientific name)	Notes on Culture and Importance
Salmon (chinook) *(Oncorhynchus tshawytscha)*	Popular food fish; harvested in several locations.
Shrimp (Chinese white) *(Penaeus chinensis)*	China is world leader in shrimp production; other species produced elsewhere.
Striped bass *(Morone saxatilis)*	Thought to hold much potential for culture, particularly its hybrids; still experimental.
Tilapia (blue) *(Tilapia aurea)*	Grown in warm water or intensive systems; sometimes considered a pest.
Trout (rainbow) *(Salmo gairdneri)*	Popular; produced in northern U.S.
Watercress *(Nasturtium officinale)*	Grown in temperate and subtropical climates, such as in Hawaii; the only aquatic foliage consumed on a regular basis in U.S. in raw salads or cooked.

Sources: Compiled from J. E. Bardach, J. H. Ryther, and W. O. McLarney, *Aquaculture: The Farming and Husbandry of Freshwater and Marine Organisms.* New York: John Wiley & Sons, Inc., 1972; H. K. Dupree and J. V. Huner, *Third Report to the Fish Farmers.* Washington, D.C.: U.S. Fish and Wildlife Service, 1984; B. Rosenberry, *World Shrimp Farming.* San Diego, *Aquaculture Digest,* 1990; and S. W. Waite, B. C. Kinnett, and A. J. Roberts, *The Illinois Aquaculture Industry: Its Status and Potential.* Springfield, State of Illinois, Department of Agriculture, 1988.

Competition with Streams and Lakes

Cultured fish and other aquacrops compete for markets with the wild crops caught in streams, lakes, and oceans. As aquaculture expands, persons in the seafood industry may feel threatened. Some of the companies involved with wild aquafoods are becoming more interested in aquaculture.

The seafood industry has found the capture of wild fish to be a profitable venture. One reason is that there is no investment in feed and other production inputs. Wild fish get their food from what is naturally available in streams, oceans, and lakes.

Aquaculture increases the quantity of good-quality water crops that are on the market and, therefore, creates competition with the wild harvest. Further, some authorities say that cultured aquafoods may be of better quality than those harvested in the wild. This is because the growers can control the quality of the water in which the crops grow and the kind of feed the crops receive.

CULTURED AND CAPTURED AQUAFOODS

Aquaculture emerged over the years so that certain aquafoods would be more readily available. Capturing wild fish in streams and oceans was not always a dependable source of food.

Cultured aquacrops often need expert care in order for production to occur. This husbandry of plants and animals requires the application of scientific and management principles. Aquafarmers must have knowledge in a wide range of subjects, including water management; fish breeding, feeding, and disease control; and harvesting and marketing. Traditional "land" farmers who go into aquaculture have found huge differences between growing row crops and livestock and producing crops in water.

Captured aquafoods grow wild in the oceans and streams much as squirrel and deer grow wild in the forest. Commercial fishers catch the desired species by using nets, hooks, traps, and other means. Of the some 21,000 kinds of fish recognized by scientists, only a few of these are widely known as sources of food.

As contrasted with wild fish, considerable investment may be required to culture fish. The aquafarmer must construct ponds, buy feed for fish, control diseases, and operate hatcheries. Wild fish do not usually necessitate these investments. Therefore, the investment is less for wild fish, but the availability of a supply of fish is less certain.

Some kinds of fish are produced in both wild and cultured forms. This situation often occurs when the supply of wild fish does not meet consumer demand. Of course, the proper equipment and knowledge must be available to culture the fish. Some popular kinds of food fish are virtually extinct due to over-fishing the oceans and the inability to readily culture them. Research programs are trying to develop ways to grow aquafoods that previously could not be cultured. A good example is the popular redfish of the Gulf of Mexico.

Advantages and Disadvantages

The key factor in aquafarming is whether or not the farmer will be able to make a profit. Aquaculture must be economically feasible. In the past, many individuals invested in aquaculture only to fail and lose their money. Thorough knowledge, good planning, and sound management are keys to success.

Beyond making a profit, there are several important advantages and disadvantages of aquaculture. These are listed in Table 1-2.

Figure 1–3. Streams no longer can provide the quantity of wild fish needed to satisfy consumer demand. (Pearl River in central Mississippi, shown here, once produced a large quantity of wild fish.)

Figure 1–4. Oceans have been over-fished, resulting in some areas where catches are small or none. (This is the Pacific Ocean from the Hawaiian Island of Oahu.)

Table 1–2

An Assessment of Aquaculture

Advantages of aquaculture

- Helps insure a supply of desired aquacrops. (Wild supplies may be very limited. Aquafarming helps insure availability.)
- Can control environment of production. (Feeding fish nutritional, wholesome feeds eliminates possibility of contamination with harmful substances, such as pesticides.)
- Helps provide nutrients essential in the human diet. (Fish have been shown to be good sources of protein and other nutrients.)

Disadvantages of aquaculture

- Technology not available for many possible crops. (In order to successfully water farm, the aquafarmer must have information on how to do it. Such information is often not available, though considerable research is underway.)
- Lack of knowledge about aquaculture. (With aquacrops that have been adapted to culture, farmers need education in how to be successful in aquafarming. Farmers who go into aquaculture without good information may lose money.)
- Competition from wild fish and seafood. (Since money is required to produce aquacrops, the aquafarmer must be able to sell the crop at a price to recover costs and earn a profit. Wild fish usually have no production costs.)

TYPES OF AQUACULTURE PRODUCTION SYSTEMS

There are several types of aquaculture production systems. These types can be distinguished by the level of intensity, kind of water facility, type of system (closed or open), water temperature, and kind of species cultured.

Level of Intensity

Intensity refers to the density (crowding together) of the aquacrop being produced. It may be extensive, intensive, or intermediate.

Extensive aquaculture systems involve low population density, minimal control over the system, and little intervention by the farmer. Extensive systems may be used in small natural or artificial ponds or lakes. The number of fish or plants is at a level low enough for the natural food in the water to be sufficient. The farmer makes little or no investment in feed, water management, and other areas. Of course, the volume of the crop harvested is low.

Intensive systems involve producing aquacrops at a high rate of stocking. Considerable skill in management is needed. Tanks and raceways are commonly

Figure 1–5. An aerial view of a commercial aquafarm. (Courtesy, Delta Pride Catfish, Inc.)

used with continuously flowing water. The farmer must be very knowledgeable of feeding, disease control, water management, and other factors. Considerable financial investment is often required. The potential for returns is greater than with extensive systems. Examples of fish produced with intensive systems are tilapia and hybrid striped bass.

Intermediate systems involve some principles of both extensive and intensive aquaculture. The intermediate systems are between the two in level of intensity. Artificial ponds are primarily used. Fish are stocked at a rate requiring feed and water aeration but not in flowing water. Most fish aquaculture in the U.S. has relied on intermediate systems, though there is a trend toward more intensity in these systems. Most catfish farming uses intermediate systems.

Kind of Water Facility

The kind of water facility refers to the design of the water enclosure and the water movement.

Water enclosures include ponds, cages, raceways, and tanks. In whole pond culture, the aquacrop is placed in the pond without restriction by cages or pens. Cages are used to confine fish to a certain area of the water. Raceways are long, narrow structures that use flowing water. Sometimes a series of ponds is linked together so that water flows from one to another much as with a raceway. Tanks may be round or rectangular and use flowing water. In recent years, the trend with tanks has been toward round tanks 20 feet or so in diameter with a considerable volume of flowing water.

Water movement refers to whether or not the water in the enclosure flows. Ponds use static systems but are often artificially aerated with water that is sprayed into the air or with air or oxygen that is injected into the water. Aeration is necessary because fish at the population levels used in aquaculture need more oxygen than static water can provide. Tanks and raceways use flowing water. The density at which fish can be stocked depends on the rate of water flow.

Type of System

Systems may be classified on the basis of the use and re-use of water. The systems may range from closed to open.

An open system is one in which the water is pumped in at one place and removed at another. The water is used once and discarded. This may involve running the water into a holding pond or into a creek or river. Environmental regulations may not allow certain water to be run directly into streams. As freshwater supplies have diminished, interest has increased in ways of reusing water.

Water may be reused by flowing from one tank or other water facility to another in a series. The water is not filtered to remove uneaten suspended feed or other waste matter.

Recirculation involves pumping the water back through the same water facility. In recirculation, the water is often reconditioned. Reconditioning may involve filtering, aerating, or using other procedures to keep the water of high quality for aquaculture.

A closed system is one in which no new water is added. Closed systems involve filtering and recycling the water. The water from closed systems must be treated to remove uneaten feed, fish excrement, and other substances before it is recycled through a tank or a raceway. Some aquafarms use a combination of recycled and "new" water, which is known as a semi-closed system. In practice, very few systems are completely closed. Most require the addition of some new water.

Reconditioning water used in a closed system is a difficult procedure. Engineers are working to develop better systems. Most of the systems that are available today are not fully satisfactory. In addition to problems in making them operate properly, they are costly to set up and maintain.

Water Temperature

Different aquacrops require varying water temperatures. Water requirements may be cold, cool, or warm. Those fish adapted to warm water will not thrive in cool or cold water; they may die. For example, tilapia thrive in water with a temperature of 80–90°F, but will die at water temperatures below 50°F. It is important for the potential aquafarmer to select an aquacrop suitable to the environment. This requires a study of the habitat preferences of the species being considered.

Kind of Species

Some species of aquacrops are compatible with other species and some are not. In addition, cultural practices may require that only one species be grown in a production system. For example, growing two different species of fish together in a pond requires sorting at the time of harvest. Processors of fish want to receive only uniform batches of fish. *Monoculture* is a production system in which only one species is grown at a time.

Figure 1–6. Polycultured salmon and steelhead trout (a type of rainbow trout raised in saltwater) are being dipped from pens at Eastport, Maine, for marketing. (Courtesy, Maine Agricultural Experiment Station)

Polyculture involves growing two or more species together. The species need to be compatible in behavior and environmental requirements. In a good polyculture, the species do not prey on each other. They also do not directly compete with each other for food.

A good example of polyculture is the common farm pond. Bass and bluegill are frequently stocked together. Research into aquafarming is showing that some combinations of species of fish and of fish and other aquacrops may be satisfactory.

A BRIEF HISTORY

Although emphasis on aquaculture is relatively new in the U.S., various forms of aquafarming have been carried out for centuries. The earliest fish farming occurred in China, beginning about 2000 B.C. with the common carp. Early on the Chinese discovered that the productivity of a pond could be enhanced with fertilizer. They used organic fertilizers, primarily manure. From China, fish culture spread to Korea and Japan around the year 200 A.D. By 1700, the Japanese had initiated mariculture (sea farming).

Aquaculture reached Europe during the Middle Ages. Fish were kept in the moats around castles. These moats became known as stew ponds. The principal early use was apparently to store fish caught in the wild until they were needed as food. Some of the fish remained longer than a few days and reproduced. People soon found out about a simple form of fish farming.

In the U.S., interest in aquaculture began to emerge in the mid-1800s. This was largely an outgrowth of depleted stocks of game fish in some natural waters. The early efforts focused on trout and carp. Research in the 1930s in Illinois and Alabama centered on how to alter a body of water to get larger crops of fish. This was followed by the farm pond program of the U.S. government in the 1950s, which encouraged the construction and stocking of farm ponds. These 2 million ponds, however, have been of little importance as a source of food.

In the 1960s, catfish culture began to emerge in the South (primarily in Mississippi, Arkansas, Louisiana, and Alabama). Huge increases in acreage in catfish farming occurred in the 1970s and 1980s. Research and technology developments have resulted in the catfish industry becoming a major aquaindustry. Catfish farming has become a sizeable aquacrop in some areas.

Further stimulation for aquaculture production in the U.S. was provided by the National Aquaculture Act of 1980. This legislation established aquaculture as a national priority. It further stated that various government agencies should work together in planning aquaculture development in the nation. Regional aquaculture centers have been set up in various locations around the country to promote aquaculture research and development work.

Aquaculture appears to have a bright future. Current trends would certainly point in that direction, with a 20% shortfall in the worldwide supply of fish by the year 2,000. Demand for fish is predicted to increase 30% in the next decade in the U.S.

IMPORTANCE OF AQUACULTURE

Aquaculture is important for several reasons. These are briefly discussed here.

Serves as Source of Human Food

Foremost among the important reasons for aquaculture is that it supplies a quality source of nutrition for an expanding population. Some medical authorities support the benefits of fish and seafood in the human diet. With the U.S. population in 1990 a little over 249 million people, aquafoods provide protein and other essential nutrients.

In the U.S., per capita consumption of fish and seafood increased 13% during the decade of the 1980s. Annual consumption was nearing 20 pounds per capita in the early 1990s, with over 90% of this being seafood.

Provides Valuable Non-food Products

Some aquaculture crops provide valuable products and by-products. Fine leather products are manufactured from eel skins and alligator hides. Cultured pearls from oysters are used in making valuable jewelry. Shells and skeleton parts may be incorporated into home and business decorations. Educational programs may use frogs and other aquatic animals in the instructional process. Lower-quality, less-valuable shells may be used in paving roads or in ornamental landscaping.

Contributes to Human Health Research

Aquaculture contributes to human health through research, dietary supplements, and other ways. In research, certain aquaculture animals are studied in an attempt to understand the human body and how it functions. For example, the squid has been used in medical research since the early 1930s. The nervous system of the squid has particular applications that are similar to certain functions of the human nervous system. The University of Texas medical school has been a leader in the use of squid in human research.

Figure 1–7. Wild fish and seafoods are the major source of some foods. Here a commercial fisher is shown baiting traps for lobster in the Atlantic Ocean. (Courtesy, Bob Bayer and Maine Agricultural Experiment Station)

Creates Demand for Grain Crops

The rations used to feed fish and other aquaculture crops often require grain and by-products of grain processing. Soybeans and corn are widely used in feed. The manufacturing of feed creates a demand for agricultural commodities. This should help farmers who produce these ingredients of fish feed to find alternative markets for their crops.

Creates Jobs and Economic Activity

Another way to assess the importance of aquaculture is from the economic activity it generates. Jobs are created for thousands of individuals in producing and marketing aquacrops. For example, one job is created in processing for every 20 acres of catfish ponds. Some 30,000 people are employed in the catfish industry alone. U.S. trout farmers received $72.6 million in 1989, while that same year catfish farmers received $273.1 million for their aquacrop. When the value from processing, merchandising, and restaurant sales is added, the economic impact is tremendous.

International trade involves several aquacrops as well as captured fish and shellfish. Currently, 80% of the shrimp and 65% of the fish consumed in the U.S. are imported. Lobster, tuna, and frogs are examples of other imports. Some exporting also occurs. Crab and salmon are among the exports from the U.S. Some work is being done on developing an export market for catfish.

Revitalizes Rural Areas

Aquaculture has the potential to revitalize rural areas. Many rural areas have experienced a decline in agricultural profits. Some farmers have quit farming traditional crops. A profitable aquacrop could stimulate the economy of an area.

Provides Recreation

The recreational aspects of aquaculture should also be considered. The sport fishing industry is increasingly using fee-lakes stocked with cultured fish, particularly trout and catfish. The success of recreation fish-out ponds depends upon the accessibility of the facilities to towns and cities and the level of management exerted to make the facilities attractive.

Improves Scenic Beauty

Scenic beauty can be enhanced with aquaculture. "Aquascaping" is the term applied to using aquaculture to add aesthetic beauty to an area. Work with wetlands in Florida has helped reclaim swampy areas and convert the land into prime property. Underwater aquascaping has created tourist attractions. Aquatic ornamental plants as well as other features have been used to make attractive designs in successful aquascaping.

ROLE OF ANIMAL WELFARE

Concern for the welfare of animals has arisen in recent years. The focus has primarily been on livestock (specifically hogs and cattle) and pets. Aquaculture producers are adopting certain principles in their operations to insure animal welfare. Processors are following appropriate procedures in the production of quality products.

Some controversy has also arisen about the introduction of exotic species into streams and lakes in the U.S. These species sometimes have negative impacts on the natural populations. Some states have passed laws making it illegal to release certain species. For example, tilapia and grass carp were originally introduced into the U.S. to control aquatic vegetation. After initially posing a threat to native species, they are now being viewed as fish with a potential for aquaculture.

Another concern of animal welfarists has to do with the release of species into environments where they are not well adopted for growth. Placing trout in warm water streams is lethal, as they are cold water fish. Other species which thrive in warm water would not survive in cold water.

Figure 1–8. Aquaculture includes ornamental species that are raised in fish bowls and other containers. Here goldfish are receiving a feeding.

Fish kept as pets or as ornamentals also need to be in environments that are conducive to their growth. The welfare of pet fish in an aquarium is just as important as the welfare of fish on a large commercial fish farm.

SUMMARY

This chapter has focused on the meaning and importance of aquaculture.

Various production systems are used. The species selected for culture must be suited to the climate and production system. The demand for aquacrops is increasing annually; the future looks bright for anyone who may be considering establishing an aquafarm. For success, an aquafarmer must learn as much as possible about the subject, make thorough plans, and follow sound management practices for the aquacrop selected. Assistance from authorities on aquaculture should be sought.

QUESTIONS AND PROBLEMS FOR DISCUSSION

1. What is aquaculture?
2. What are the five purposes for which fish may be produced?
3. What is an aquacrop? Aquafarmer?
4. Distinguish between cultured and captured fish.
5. Describe three advantages and three disadvantages of aquaculture.
6. Identify and describe four ways aquaculture production systems can be classified.
7. Where did aquaculture begin? How did it advance from its beginning?
8. Identify and explain seven ways aquaculture is important.
9. How is animal welfare a part of aquaculture?

Chapter 2

THE IMPORTANCE OF AQUACULTURE

Aquaculture is important. The aquaculture industry contributes to the nation's economy, provides income for agriculture producers and suppliers, and serves as a source of high-protein food for individuals. Several factors point to the significant growth of the aquaculture industry and an increase in its importance to agriculture in the U.S. These factors include (1) improved government regulations of aquaculture as agriculture, (2) seafood inspection regulations that show the quality of aquacrops, (3) an exhausted ocean supply of seafood, (4) concern about the pollution of oceans and other sources of commercial fish, and (5) more consumers who are aware of the nutritional benefits of fish and other aquacrops.

OBJECTIVES

To help students understand the importance of aquaculture in the U.S., several objectives are covered in this chapter. Upon completion, the student will be able to:

- Describe aquaculture as a source of nutrients.
- Describe the scope of aquaculture and its potential for the future.
- Describe consumption trends of aquaculture products.

- Explain the economic perspectives of aquaculture.
- Describe the role of organizations in aquaculture.
- Identify research and educational agencies concerned with aquaculture.
- Identify regulatory agencies concerned with aquaculture.

AQUACULTURE AS A SOURCE OF NUTRIENTS

In the U.S., consumer attitudes toward sources of animal proteins changed dramatically from 1970 to 1990. Seafoods and other white meats have increasingly been recommended by health experts and nutritionists because of their low fat content as compared to other types of meat.

The media has played a large role in this change by publishing articles promoting sources of protein that are lower in fat, including aquacrops. Also, national publications have featured many aquacrops in advertisements that emphasize their nutritional value.

Furthermore, food scientists have determined that fish and some other types of seafood contain high levels of Omega-3 fatty acids; thus, as a result, they have encouraged consumers to include more fish in their diets. Omega-3 fatty acids reportedly raise the levels of high-density lipoprotein (HDL), or "good" cholesterol, in the blood, while at the same time reducing the levels of low-density lipoprotein,

Figure 2–1. By-products of catfish processing are loaded onto a trailer for hauling to a rendering plant. These by-products will be used as a protein source in animal feeds, including catfish feeds.

or "bad" cholesterol. Low-fat alternatives such as poultry with the skin not eaten and veal do not possess this characteristic.

For these reasons, aquacrops are seen as an important part of diets designed to fight two major dietary problems of Americans: high-cholesterol levels that lead to blocked arteries and obesity. As such, aquacrops should increase as a basic staple of American diets as a major source of animal protein. (This increase in the consumption of fish and seafood will be discussed in more detail later in this chapter.)

The role of aquacrops in providing nutrients for both human and animal consumption should increase. The by-products of processing many types of aquacrops provide excellent sources of proteins for use in commercially prepared animal feeds.

SCOPE OF AQUACULTURE

Aquaculture has been the fastest growing agricultural industry in the U.S. in the last 20 years, growing at a rate of about 20% per year. Much of this growth has been due to production increases in the catfish, trout, and salmon industries over this period. Aquaculture in the U.S., however, is very young when compared to aquaculture in other countries and when compared to other types of agriculture in this country. The U.S. is the fourth largest aquaculture-producing nation in the world, behind China, Japan, and South Korea, all of which have been involved in aquaculture for hundreds of years.

The scope of U.S. aquaculture includes many areas. Aquacrops may be produced as fingerlings to sell to other producers, as food fish or other food aquacrops for processing or on-farm retail sales, as baitfish, as ornamental fish for aquaria (plural of *aquarium*), or as stocker fish for fee-lakes or public lakes. Many plants also are produced through the use of aquaculture, although not as much in the U.S. as in other countries. (The role of plant aquaculture is discussed in Chapter 14.)

Several factors have led to the growth of the aquaculture industry in the U.S. and should contribute to its increased growth. Aquafarmers have learned to make their operations more profitable by vertically integrating the industry. Vertical integration means that the same farmer or group of farmers has control over all aspects of the industry — from the time the fish lays the eggs until the finished product is marketed to the retailer or consumer. Also, the U.S. is advanced in reproductive physiology, a necessary component of early advances in the industry as species suitable for intensive production were discovered and developed. Another factor has been the pioneering of intensive or high-density stocking, which means crowding many more animals into a growing facility than would normally be found in nature.

Figure 2-2. Catfish are electrically stunned as they move into a processing plant. Catfish leads aquacrops in pounds produced each year.

Predominant Species in U.S. Aquaculture

The four predominant food species produced by aquaculture in the U.S. are catfish, crawfish, trout, and salmon. These four aquacrops account for about 90% of U.S. aquaculture production, which is about 860 million pounds per year. The production of ornamental fish and baitfish follows these four in volume of production. Table 2-1 gives the statistics for the major aquaculture species produced in the U.S. in 1990.

Over 1,800 aquafarms with more than 140,000 acres in the U.S. produce catfish. These farms produce around 460 million pounds of catfish per year. The largest producing states are in the Southeast. Mississippi is the largest catfish producing state by far, with about 78% of the production.

Over 90 million pounds of crawfish are produced each year, with most of the production concentrated in Louisiana. The value of crawfish production is around $50 million per year.

Pacific salmon production from aquaculture accounts for about 92 million pounds per year, but this figure is misleading. Of the 600 million pounds of Pacific salmon caught each year by commercial fishing (commercial landings), over half

Table 2-1

U.S. Private Aquaculture Production and Value — 1990

Species	1,000 Pounds	1,000 Dollars
Catfish	460,000	370,000
Crawfish	90,000	55,000
Pacific salmon	92,000	40,000
Trout	67,000	81,000
Baitfish	32,000	55,000
Oysters	25,000	50,000
Clams	4,000	14,000
Shrimp	3,000	7,000
Mussels	2,500	3,500
Freshwater prawns	250	1,000
Other species	85,000	85,000
Total	860,750	856,000

Source: Specialty Agriculture Branch, Economic Research Service, U.S. Department of Agriculture.

came from public and private hatcheries, which take advantage of the natural homing instinct of the salmon. This homing instinct leads the salmon back to their place of birth when they get ready to spawn. (The culture of salmon will be further discussed in Chapter 11.) As a result of these commercial landings, salmon is the major seafood export product of the U.S.

Trout production in the U.S. is about 67 million pounds per year. Much of the production is concentrated in Idaho, with smaller concentrations in the Smoky Mountains, the Ozark Mountains, and the Rocky Mountains. Trout are sold as food fish and for stocking in streams and lakes as replacements for fish caught by recreational fishers.

The production of baitfish is prevalent in Arkansas, but it also occurs across the country. Ornamental fish production is largely concentrated in Florida.

Other aquacrops produced in the U.S. include tilapia, freshwater prawns, crabs, lobsters, molluscs (oysters, clams, mussels, and scallops), and gastropods. The production of these crops is discussed in later chapters.

The scope of aquaculture is expected to expand as ongoing research brings about better production practices for aquacrops such as hybrid striped bass, redfish, and freshwater prawns. Other factors that will influence the amount of aquafarming include public acceptance of aquacrops as viable food alternatives;

government regulations that allow commercial aquaculture production; and increased government support for research and education concerning aquaculture production practices.

CONSUMPTION TRENDS

The term "consumption" refers to the amount of a food that people eat. Compared to the rest of the world, people in the U.S. consume little fish and seafood, although the amount is increasing. Of the 240 pounds of animal protein the average person eats each year, about 20 pounds, or less than 9%, comes from fish and seafood. As a comparison, the world average is about 25%. People in many Asian countries receive over 50% of their animal protein from fish and seafood. Table 2-2 shows some of the averages in Asian countries, as well as the U.S.

Several historical factors have contributed to the small percentage of fish and seafood in the diet of the average person in the U.S. These factors include the

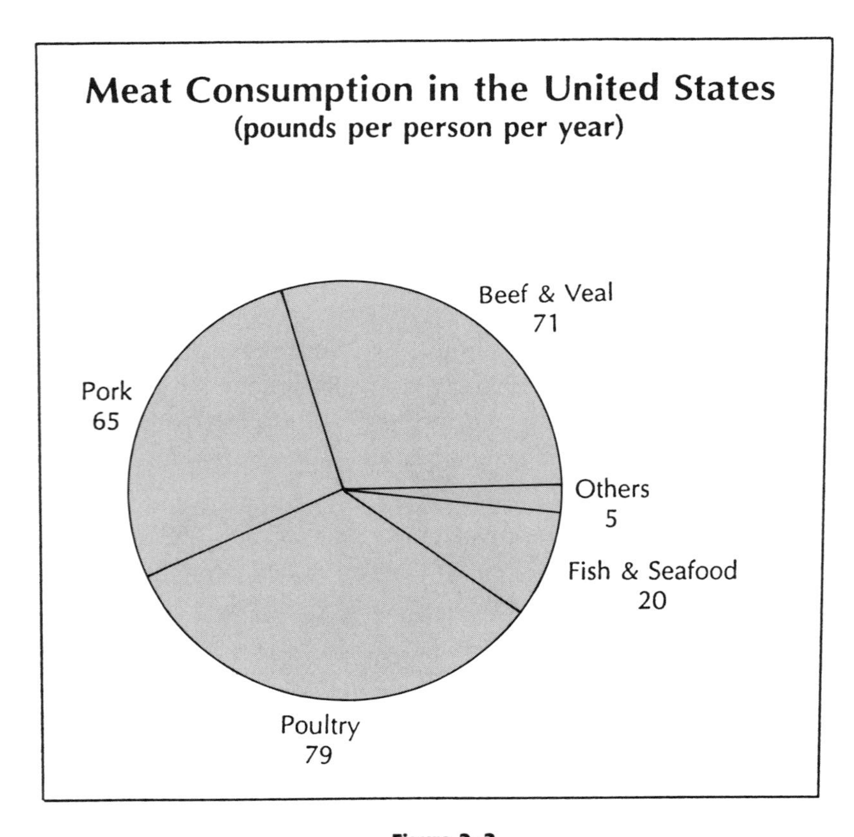

Figure 2–3.

Table 2–2

Population and Percentage of Animal Protein from Fish and Seafood for Selected Countries

Country	1987 Population (millions)	Fish as a Percentage of Animal Protein
Indonesia	174.9	67.9
Philippines	61.5	56.7
Bangladesh	107.1	52.2
Thailand	53.6	51.5
Malaysia	16.1	41.6
China	1,062.0	33.0
India	800.3	15.1
U.S.	240.0	8.4

Source: U.N. Food and Agriculture Organization, Rome, Italy, 1987.

size of the country, a lack of knowledge about cooking fish, and the availability of other sources of animal protein.

Proximity to Fresh Fish and Seafood

People close to sources of fresh fish have always included it as a relatively large part of their diets. Because the U.S. is such a large country, many areas are long distances from coastal waters or other sources of fresh fish or seafood. Before refrigeration, many people did not have access to fresh fish and seafood and therefore did not eat very much of it. The improved methods of keeping fish fresh or frozen and of distributing it to the inland areas more quickly have increased the amount of fish eaten in these areas.

Some of the old habits have been hard to break, however. Because people in some of these areas did not have an abundant supply of fish or seafood, they never learned how to cook fish correctly. As a result, many of them never developed a taste for fish, either because the fish they cooked was not fresh or because they did not do a good job of cooking it. As a result of this shortage and the romance of the old West, for many people beef became the meat of choice.

Alternatives to Fish and Seafood Consumption

Another reason some people did not eat very much fish or seafood was because they had an ample supply of beef or pork, which they preferred when they had the disposable income. This is especially true in the midwestern U.S., where beef and pork have historically been in abundant supply. People in this area often produced their own beef and pork and did not see fish and seafood as a logical part of their diets.

However, with doctors and others recommending fish because of the nutritional benefits discussed earlier in this chapter and with the improved availability of fresh or frozen fish and seafood that is high in quality, people's perceptions concerning taste and their preferences are changing. As more people include fish and seafood in their diets, their ability to prepare these foods will also improve. When the fish and seafood dishes they prepare taste better to them, they will continue to increase consumption. Of course, this will in turn increase the demand for aquacrops.

Promotional Efforts

Another reason that consumption of aquacrops is expected to increase is the promotional efforts of the various producer groups associated with aquafarming. A good example is of the catfish, once thought of as a trash fish by many people in the U.S. The Catfish Institute, the promotional arm of the Catfish Farmers of America, has widely promoted catfish as a delicacy in national publications across the country. This promotional campaign has led to the acceptance of catfish in many of the finest restaurants in the country and surely has increased the market in which producers can sell their fish.

ECONOMIC PERSPECTIVE

The future for aquaculture in the U.S. looks very promising. Because Americans will be eating more fish and seafood as part of a healthier diet, the demand for these products will grow. As a result, the supply of fish and seafood must increase to meet the demand. The increase in supply might come from an increase in commercial landings, from an increase in imports from other countries, or from an increase in aquaculture production in this country.

At present, aquaculture provides about 18% of the total U.S. supply of fish and seafood. Several factors indicate that aquaculture will have to play a much larger role if the demand for fish and seafood by Americans continues to grow as expected. The increased consumption of fish and seafood, the depletion of ocean reserves of fish and seafood, the growing world population, the U.S. trade deficit, and improved aquaculture techniques all point toward a bright future for aquaculture in the U.S. Other benefits of aquaculture, such as better use of processing by-products, the ability of aquafarmers to provide a quality product year-round, and the excellent feed-conversion ratios of many aquacrops, also indicate that aquaculture production in the U.S. will increase over the next few years. Two other factors that point toward aquaculture expansion are the preparation of aquacrops, particularly catfish, in fast food restaurants and the chef training programs sponsored by aquaculture organizations to teach chefs different ways to utilize aquacrops in their menu offerings.

Imported Fish and Seafood

Since the mid-1960s, the U.S. has relied on other countries for over 50% of its edible seafood. During 1988, the U.S. Department of Agriculture (USDA) reported that the U.S. imported almost 3 billion pounds of fish and seafood, while exporting just over 1 billion pounds. The fish and seafood industry had a trade deficit of $6.6 billion. An increase in aquaculture production in the U.S. could substantially lower the existing overall federal trade deficit.

Shrimp leads the list of imports, with almost $2 billion of shrimp imported in 1988. The demand for shrimp by restaurants and food processors has increased while U.S. landings have decreased. As one example, General Mills Restaurant, the parent company for the Red Lobster and Olive Garden restaurant chains, imported shrimp from 38 different countries in 1988. The potential for aquaculture

Figure 2–4. The freshwater prawn could replace some imported shrimp in seafood markets, if production techniques improve to make production more profitable. Note the size of the prawn as compared to the quarter beside it.

of shrimp and freshwater prawns is excellent, if production techniques can be improved to make the enterprise profitable.

Role of Aquaculture in Meeting the Increased Demand

Economists at the Food and Agriculture Organization in Rome, Italy, foresee a world shortage of 20% of the fish and seafood needed to meet demand by the year 2000. Aquaculture promoters can use this anticipated shortage as an incentive to pioneer new techniques and produce new aquacrops to meet the shortage. Some experts believe that aquaculture has the potential to meet at least 25% of the U.S. demand for shrimp and salmon, with other aquacrops replacing some of the major types of imported seafood in the U.S.

Reducing Fishery Imports

Aquaculture should help reduce the federal trade deficit as export markets for trout, salmon, and catfish continue to grow. Michael Dicks, a leading aquaculture economist from Oklahoma State University, believes the increasing world population, the growing demand for fishery products, and advancing income levels overseas will increase the export potential of aquaculture products.

Another way aquaculture should reduce the trade deficit is by greater shrimp production. This increase will decrease the amount of shrimp imported by processors and retailers into the U.S.

The share of U.S. seafood consumption by domestic fish landings consistently dropped 2% each year during the 1980s. The ocean supply of fish and seafood for most of the edible species is at or very near its maximum. This means that the increased supply of fish and seafood to meet the demand will not come from commercial landings. This factor also points to a need for increased aquaculture production. If people want more fish and seafood and the ocean cannot provide it, aquafarmers should take advantage of the opportunity to fill the gap.

Benefits of Aquaculture

The ability of aquaculture to provide a quality product on a year-round basis will continue to make aquacrops strong in the fish and seafood market. As oceans and streams become more polluted, the public becomes more suspicious of fish and seafood from commercial landings. Many people will turn toward aquacrops grown in a clean, controlled environment. In fact, in some areas, an "aquaculture product" label already means that consumers will pay a premium price for the product.

As aquaculture production techniques improve, aquafarmers can plan production in conjunction with processors so that their aquacrops are ready for market throughout the year. The availability of aquacrops year-round is very important to restaurants, which like to keep an item on the menu all of the time, not just "in season." Grocery stores and supermarkets also like the stability of being able to consistently offer a quality product to their customers.

Rural Economic Development Through Aquaculture

Aquaculture can also serve to stimulate the stagnant economies of rural areas. While many aquaculture operations are small, they often provide a needed product in a "niche" market, which is discussed in Chapter 6. Often the success of a few aquafarmers causes an industry to grow substantially as others get involved and markets increase.

An excellent example of how an aquaculture industry "grows up" is the catfish industry in the Mississippi Delta. When the catfish industry began growing before 1970, this was one of the poorest regions of the country. The catfish industry has improved the economy greatly in the counties which are involved heavily in catfish production.

Mississippi produces over 280 million pounds of edible catfish each year. For every 10 million pounds of fish produced, over 200 new jobs are created in production and processing, and over 1,000 new jobs are created in support industries. These support industries include providing inputs such as feed and equipment, services, transporting and marketing products, and providing goods and services to new employees of the production and processing industries.

Other Benefits of Aquaculture

In addition to producing an aquacrop at a profit, many farmers have discovered other benefits of including aquaculture as part of their operations. Farmers who use the waste water from fish operations to irrigate and fertilize their row crops and pastures realize even greater benefits from aquaculture than normal. In a state such as Arizona, for example, where farmers must irrigate, they have found that aquaculture can provide excellent benefits. They can use the water for growing fish in raceways or flow-through systems first and then irrigate their crops with the nutrient-rich water afterwards.

Limitations to Growth

Although the future of aquaculture looks promising, there are some limitations to the growth of the aquaculture industry. The main limitations are the high initial investment required in many types of aquaculture production, the lack of available water, and government regulations that may stifle production of certain species in certain areas.

Another major limitation is the as yet undeveloped technology for producing many aquacrops. Production techniques for many possible aquacrops are still in the development stage. As a result, the production of many crops is very risky. Many people are apprehensive about investing money before these techniques are perfected. Limitations for producing particular species will be further discussed in later chapters.

ORGANIZATIONS

In the young and very diverse aquaculture industry, member organizations will play an important role in the continued growth of U.S. aquaculture. It is very important that aquafarmers present a unified voice to the public — both consumers and policy makers.

The best way to present this voice is by involvement in producer associations. The purpose of these associations is to unify leadership and present a strong voice

Figure 2–5. Several members of a producer association tour a processing plant.

for the members of the associations. If an association has the support of a large number of members, then legislators and regulatory officials will have to listen to the concerns of its members.

Roles and Purposes of Organizations

Organizations have four roles that are generally expected by their members. These roles are (1) promoting the products their members produce, (2) educating their members concerning regulations and practices, (3) effecting policy related to their members and their enterprises, and (4) providing opportunities for their members to interact socially with other producers.

Figure 2–6. Some producer associations sponsor trade shows where manufacturers can demonstrate advances in technology available to the producers.

Effective organizations usually have an elected board and officers made up of members. The board develops the policies of the organization, with approval of the members, and then develops a plan of work to carry out the policies. An administrative staff is often hired to oversee the everyday operations and to carry out the policies of the organization.

The policies formulated by an association must address many issues. An organization must have a membership policy. Membership may be restricted to producers only or may include processors, retailers, or educators in addition to producers. The organization may also limit its membership to producers of a particular aquacrop, such as trout farmers.

Another policy determines the primary audience or audiences to which the association will focus its communication efforts. Some examples include legislators, scientists, and environmental groups. The primary audience may be the general public, especially when the major focus of the organization is the promotion of a particular aquacrop.

The association must also decide which issues to address. Some possible issues include promotion of aquacrops, international trade, market development, food safety, and government regulations.

Examples of Aquaculture Organizations

Aquafarmers have a choice of numerous associations representing general aquaculture or particular aquacrops. Associations may be found at the international, national, and state levels.

The World Aquaculture Society is an international association for the promotion of aquaculture. Its main function is as a source of information and a representative of aquaculture interests around the world.

The National Aquaculture Association represents all types of aquaculture. Leaders of other organizations that represent particular aquacrops formed the association to provide a unified voice for aquaculture in the U.S. The American Fish Farmers Association is also a national organization that represents many aquafarmers whose primary aquacrop is fish.

Several species-specific associations also exist at the national level. These include the United States Trout Farmers Association, the Catfish Farmers of America, and the American Tilapia Association.

In states where a particular aquacrop is prevalent, a state association for that particular aquacrop is usually found. Examples are the Catfish Farmers of Mississippi and the New York Trout Farmers Association. These state associations are usually affiliated with their national counterparts.

Several states also have general aquaculture associations which promote all forms of aquaculture and may be associated with the National Aquaculture Association and / or the American Fish Farmers Association.

RESEARCH AND EDUCATIONAL AGENCIES

In order for aquaculture to continue to grow and become a viable agricultural alternative, state and federal governments must invest in research and education for aquaculture. Research must be conducted concerning methods of disease control, stocking densities, nutritional needs, genetic improvements, reproduction of various species, off-flavor, optimum use of facilities, and many other areas. Education is necessary to make sure the results of research are disseminated to the people who can use them, the aquafarmers themselves.

Research Agencies

The two primary sources of new knowledge concerning aquaculture are

Figure 2–7. A researcher shows blue crab he is using in his research program. Government-sponsored research has played a major role in the growth of U.S. aquaculture.

universities and government-sponsored research stations. The research conducted in universities is usually through the agricultural experiment stations in cooperation with colleges of agriculture, veterinary medicine, and / or natural resources.

Other government-sponsored research is conducted through the U.S. Department of the Interior and the U.S. Department of Agriculture. The Department of the Interior sponsors the Fish Farming Experimental Laboratory in Stuttgart, Arkansas, through the Fish and Wildlife Service. The Department of Agriculture has five regional aquaculture centers that coordinate research in their particular region. Sea Grant programs, which are operated in many coastal states, are sometimes concerned with aquaculture, especially of marine and estuarine species (saltwater and brackish water aquaculture).

Educational Agencies

The two primary government-sponsored educational programs are the cooperative extension service and agricultural education in vocational education. Many states have aquaculture specialists on the staff of their cooperative extension service. These specialists, along with the county agents, provide informal education concerning new aquaculture information.

Figure 2–8. This education center has an aquarium and offers programs for children and adults year-round, including a summer camp.

Agricultural education programs usually focus on formal classroom instruction in high schools, vocational centers, and community colleges, as well as providing adult instruction within the program. Aquaculture is usually a part of a production agriculture curriculum; however, new programs specific to aquaculture have been developed. The Council for Agricultural Education has implemented a project to develop a national curriculum in aquaculture and began field testing the curriculum in 1991.

REGULATORY AGENCIES

The government agencies that regulate aquaculture at the national, state, and local levels play a large role in determining the success or failure of aquaculture. Because aquaculture businesses are often small, with only one or two owners, they do not have large amounts of capital and labor at their disposal with which to battle government rules and regulations. Regulations that require too much time on the part of the aquafarmer or cost too much to comply with can lead to the failure of these small businesses very quickly. Some of these regulations are further discussed in Chapter 4.

Federal Regulatory Agencies

At the federal level, five agencies have the primary responsibility for regulating aquaculture. These five are the U.S. Department of Agriculture (USDA), the U.S. Fish and Wildlife Service (FWS), which is part of the Department of the Interior, the Food and Drug Administration (FDA), the Environmental Protection Agency (EPA), the U.S. Army Corps of Engineers (COE), and the U.S. Department of Labor (DOL).

The USDA is concerned with several aspects of aquaculture, as was previously mentioned in this chapter. From a regulatory standpoint, the USDA is mainly concerned with fish and seafood inspection. Although there is not a mandatory fish and seafood act from the federal government at present, one will probably be passed in the near future. The USDA is the logical choice of a federal agency to oversee fish and seafood inspection, as it already has the experience of overseeing the inspection of other meats such as beef, pork, and poultry.

The FWS is concerned with the natural populations of fish in U.S. lakes, streams, and coastal areas. The FWS operates some hatcheries, although not as many as in the past. Private hatcheries have taken the place of the federal fish hatcheries for the most part.

The FDA is concerned with the use of drugs to prevent diseases in aquacrops and how these drugs affect the aquacrops. Before a drug can be legally sold and applied to fish, it must be approved by the FDA.

The EPA regulates the use of pesticides and approves pesticides for use on certain aquacrops. The main problem with using pesticides in aquaculture is the concern over runoff and how the runoff will affect streams and lakes in the area.

The COE is concerned with wetlands protection. In order to construct an aquaculture pond or raceway, the owner must obtain a wetlands permit from the COE. The COE may work with state agencies to regulate the development of wetlands.

The DOL regulates the use of hired labor on aquafarms, just as it does jobs in other industries. Aquaculture is considered agriculture by the DOL, so aquafarms must follow the same regulations as other farms.

State and Local Regulatory Agencies

The agencies regulating aquaculture at the state level vary, depending on the state, the nature of the aquaculture business, and the scope of the business (number of employees and amount of sales). Often state departments of agriculture, commerce, health, and wildlife or fisheries are involved in some capacity.

State agencies generally focus their efforts on the regulation of ground and surface water use, pollution of discharges from the operation (effluent waters and dead fish, for example), building and construction, and control of non-native species.

Often these state agencies have to balance the interests of other groups with those of aquafarmers. Environmental groups are concerned with the effects of aquaculture on the environment and on endangered species, and recreational boaters and fishers often do not want public lakes, streams, and coastal waters to be used for aquaculture.

An aquafarmer will also have to check with local agencies before engaging in aquaculture production. Most county and city governments have a planning commission and / or a zoning commission and may have local regulations on the use of water from wells or lakes.

An aquafarmer must determine the requirements to meet federal, state, and local regulations and then work to meet these requirements. This task can often be formidable.

An Example of Government Regulations

Suppose Mary Jones wants to begin aquaculture production in Mississippi by building some ponds and drilling a well to provide water for the ponds. In Mississippi (and most other states are very similar), she must get permits from numerous agencies before she can begin producing her aquacrop.

Mary will need a zoning / building permit from the county planning commission. This requirement is common in most counties of most states.

She will need an NPDES (National Pollutant Discharge Elimination System) permit to discharge wastes into the waters of the state. To obtain the permit, Mary has to apply to the Bureau of Pollution Control, part of the state Department of Natural Resources.

The Bureau of Land and Water Resources, also part of the state Department of Natural Resources, issues surface and ground water permits. Because she will probably need a well with a pipe diameter of 6 inches or more, Mary will need a ground water use permit. Because she will probably have to drain ponds at some point for harvesting or disease control or to rework the levees, Mary will also need a surface water permit. Even though a state agency issues these permits, she may also need local approval to drill a well, especially in an area where water is in short supply.

In order to build ponds, Mary may have to obtain a dam construction permit, also from the Bureau of Land and Water Resources. Most aquaculture ponds are not deep enough and do not have enough surface area to make obtaining a dam construction permit necessary.

Mary may also have to obtain a wetlands permit from the Bureau of Marine Resources (part of the state Department of Wildlife Conservation) or from the U.S. Army Corps of Engineers. The Bureau of Marine Resources controls coastal wetlands, while the Corps of Engineers controls navigable waters of the U.S. and any disturbance of upland wetlands areas.

If Mary decides to produce non-native species, such as tilapia, she will have to obtain an aquaculture permit from the Bureau of Fisheries and Wildlife (part of the state Department of Wildlife Conservation). If she chooses to operate a pay fishing lake or to raise catfish, baitfish, or tropical fish in a closed system, she will not have to obtain this permit. If she operates a fee-lake, however, she will have to get a retail business permit from the state Department of Agriculture and Commerce.

Aquaculture as Agriculture

As noted earlier, the hassle of meeting regulations can be too much for the small aquaculture producer. The government must present a consistent treatment of aquaculture to promote growth of the industry.

Most aquafarmers believe that aquaculture should be treated like any other type of agriculture. This type of treatment would allow one government agency at the national and state levels to control and coordinate regulation of the industry.

Another example of the benefits of treating aquaculture as a type of agriculture is in research and development. The U.S. agriculture system leads the world because of the money devoted to research and development as an ongoing process. This process maintains a system where new problems faced by agriculture producers can be researched and solved in a timely manner. Aquaculture would certainly benefit from this type of treatment.

Also, for aquaculture to continue to grow as a viable agricultural alternative, cooperation must occur between the federal and state agencies that regulate the industry. Aquafarmers must be given a chance to make a living based on their ability to produce a quality crop and to manage the production of that crop effectively. Government regulations should provide the incentive for aquaculture production, not hinder it with unnecessary regulations.

SUMMARY

The U.S. needs aquaculture to meet the growing demand for fish and seafood brought about by changes in the dietary habits of its people, who are eating more fish and seafood. This increased demand cannot be met by more commercial landings or increased imports from other countries.

Aquaculture is the fastest growing agriculture industry in the U.S. The main food crops grown are catfish, crawfish, salmon, and trout. These four crops account for about 90% of the aquaculture production in the U.S. The baitfish and ornamental fish industries are also large.

A continued increase in aquaculture will lower the federal trade deficit, allowing U.S. money to be used for U.S. - produced goods and services.

In order for aquaculture to continue to grow and thrive, producer associations must be supported by producers and must take an active role in promoting the benefits of aquaculture to the country. Also, government agencies that regulate aquaculture must take a stance that supports the industry, including cooperation between federal, state, and local agencies.

QUESTIONS AND PROBLEMS FOR DISCUSSION

1. What are the nutritional benefits of including more fish and seafood in the diets of Americans?
2. What percentage of edible fish and seafood in the U.S. is supplied by aquaculture?
3. What are the four major aquacrops produced in the U.S.?

4. Is the consumption of aquaculture products increasing? Why or why not?
5. How can aquaculture help to reduce the federal trade deficit?
6. Other than reducing the federal trade deficit, what are some other economic advantages of aquaculture production?
7. What are some of the economic factors that might limit the growth of the aquaculture industry?
8. What are the roles of producer associations in aquaculture?
9. What are some of the organizations that an aquafarmer might become involved in?
10. Name the six federal agencies that have the primary role in regulating aquaculture at the national level.
11. What are some of the concerns of state and local agencies that regulate aquaculture?
12. What permits would you have to obtain before you could legally build a pond to raise fish in your county?

Chapter 3

CAREER OPPORTUNITIES IN AQUACULTURE

Aquaculture involves many types of jobs in many areas. Careers in aquaculture may entail production, management, research, and many other areas. The work may consist of a lot of hands-on, outdoor activity that requires physical labor. Or it may be mostly indoors with the emphasis on thinking and making management decisions. Most careers in aquaculture will include all of these skills to one degree or another.

The types of careers in aquaculture have changed rapidly and will continue to change as the industry becomes firmly established as part of U.S. agriculture. Some basic types of careers in aquaculture, the type of education necessary for these careers, and the nature of the work involved will be discussed in this chapter. This chapter will also emphasize the nature of the career decision-making process and some characteristics that employers look for in all employees, regardless of the type of work performed.

OBJECTIVES

This chapter focuses on helping students develop an understanding of the nature and types of aquaculture careers and of the skills and education needed to perform successfully in these occupations. Upon completion, the student will be able to:

- Describe the career decision-making process.
- Identify characteristics of good employees.
- Identify types of aquaculture careers.
- Describe the nature of work in aquaculture careers.
- Identify educational requirements for aquaculture careers.
- Describe career ladders in aquaculture.
- Select an appropriate career in aquaculture.

MAKING CAREER DECISIONS IN AQUACULTURE

The choice of a career should not be taken lightly. The career a person chooses will affect his / her lifestyle, friends, amount of money earned, and status in the community. A good career choice should take into account personal interests, the stability of the work, the likely places of employment, the education required, and many other factors.

One of the best ways to begin making a choice about a career is to talk to people who work in the profession or professions in which you are interested. Find out if they like what they do for a living. Ask about the disadvantages as well as the advantages of their jobs. Find out if you can work on a part-time basis to see if you are really interested.

Another important step is to talk to people who know you and your abilities. Your agriculture teacher should be able to answer many of your questions if you are interested in a career in aquaculture or some other area of agriculture. The school guidance counselor can also help. Ask him / her about the educational requirements and where you could obtain the necessary education for the careers in which you are interested. The counselor can help you match your talents and abilities with a career and the education required to be successful in that career.

As with any major decision, your parents should be consulted for advice and encouragement about the career you choose. They may be able to finance the necessary education or help you find a way to finance it. They may also provide land and other capital for you to start your own aquafarm. You may need their support to be successful!

Once you have decided on a general area, experience with the different aspects will help you make your decision. Working within the different areas in aquaculture will help you decide what you like and do not like about different jobs. Developing a supervised agricultural education experience (SAE) program in aquaculture will provide the necessary experience to make your decision. More experience may be required. You may obtain this experience by working full time during the summer months when you are not in school or by working for someone else while you save money and gain experience in order to start your own aquafarm or aquabusiness.

An excellent way to become an entrepreneur (someone who owns his / her own business) is by observing problems faced while you are gaining experience. You may find that a needed service is lacking in your area. If so, that may open the door for you to start a business by providing that service. Many small-business loans and entrepreneurship training programs are available for those interested in providing a useful service to an industry. Check with your local chamber of commerce for details if you have an idea and need help getting started.

As mentioned earlier, making a career decision is one of the most important decisions you will ever make. Many questions must be answered before the decision can be made. One of the best things to do is to talk with others and then select an area of study with some flexibility. Develop skills in communications and human relations that will serve you well even if you change your mind. A good career decision sometimes comes after finding out what you do *not* want to spend your life doing.

CHARACTERISTICS OF GOOD EMPLOYEES

Regardless of the type of work performed, good employees have several characteristics in common. Employers soon find out if an employee lacks these characteristics. Some of those things an employer will look for in a good employee are discussed below.

Ability to Get Along Well with Others

Good employees have a positive attitude about their work and the people they work with. This includes showing respect and concern for others. A kind word will go a long way with co-workers. Remember that you rely on others and that they rely on you to get the job done. Not getting along with others is one of the best ways to lose a job. Show some understanding, be friendly, help others when possible, and try to be flexible about work schedules and sharing of tasks.

Appropriate Knowledge and Skills

Try to prepare yourself for your job as well as possible. Know the extent of your abilities. Do not expect to know everything from the beginning, however. Employers would much rather take a little time to explain a task properly than to spend a lot of time making up for mistakes that result from lack of knowledge or skill. Good employees may not have all of the necessary skills, but they will try to learn quickly and thoroughly. Do not be afraid to ask questions!

Motivation

Good employees motivate themselves to come to work on time, to work enthusiastically, and to perform a task as well as they can. They focus on the pleasurable aspects of the job and do not complain about the not so pleasurable parts. Encourage others and they will share the enthusiasm.

Dedication

Work hard to do each task as well as possible. This benefits the employer and may lead to promotions and raises. Employers appreciate employees who contribute to the profitability of the operation. It makes them look good, and they in turn will provide rewards when possible.

Safety Consciousness

Many occupations in aquaculture involve equipment, facilities, and tasks that can be dangerous if the employee is careless. Injuries due to carelessness slow down operations and cost money. An employee whose carelessness puts his / her own personal safety and that of co-workers in danger cannot be tolerated.

Honesty

An employee must be trustworthy. Dishonesty will get an employee fired very quickly. Each time an employee is assigned a task or left in charge of an operation, the employer must be able to trust that the employee will get the job done and will be honest about the work. In many cases, employees are trusted with money or checks to pay for goods. They may be asked to handle sales transactions in the absence of the employer. They are also trusted with the health of an aquacrop and with the operation of expensive equipment. The employer expects this trust will be rewarded with honest behavior.

Freedom from Substance Abuse

Good employees never show up for work under the influence of drugs or alcohol. To do so could endanger others at the work site. Employees are expected to get plenty of rest before coming to work and to refrain from using drugs or alcohol beforehand. Employers expect workers to maintain certain standards of behavior outside work hours as well as on the job. Alcoholism and drug abuse and their side effects are a common cause of dismissal.

TYPES OF AQUACULTURE CAREERS

The most common type of career in aquaculture involves working on an aquafarm, as the owner, the manager, or a hired laborer. Several types of careers may be found in other areas of aquaculture or in the many service and support industries related to aquaculture. The occupations vary greatly in the amount of physical labor involved, the education required to perform the work, and the amount of responsibility of the worker.

One of the most common occupations in aquaculture is that of a *laborer.* This occupation may be one of a farm laborer at an aquafarm such as a catfish farm or trout farm. It may be at a plant aquafarm. Fish hatcheries, processing plants, and aquabusinesses also employ laborers.

Supervisors may be employed on large aquafarms or aquabusinesses. Plants that add value to aquacrops through further processing often have several supervisors. Also, large farms may have night supervisors responsible for supervising crews to check water for dissolved oxygen and to prevent poaching.

Some aquafarms employ *managers,* who are in charge of the everyday operations. On many aquafarms, the *owner* wears several hats, including that of supervisor and manager, as well as laborer. Owners sometimes work off their farms and employ managers.

Service technicians are needed in several areas of businesses that provide goods and service to aquafarmers. Feeding equipment, tractors, water testing equipment, processing equipment, and electronic controls, just to name a few, must be installed, serviced, and repaired to keep the aquafarms in production.

Many of the goods and service businesses that support aquaculture will employ *salespersons.* Processing plants will also need a sales staff to move the processed product to retailers. Examples include sales of feed, chemicals, feeding equipment, general supplies, and specialized clothing. Large aquafarms may have a sales staff who sell to retailers and restaurants, while many large- and small-scale aquafarms will sell directly from the farm.

Hauling fish from location to location, transporting feed to the aquafarms, and moving processed products to market usually require trucking. These necessitate hiring *drivers* to drive the trucks.

Many laborers at aquafarms must have the ability to handle large equipment, at least around the farm or for short distances on the road. Some aquafarms require *large equipment operators* on a nearly full-time basis, but operating tractors, bulldozers, and other large equipment is usually just part of the work involved.

NATURE OF THE WORK AND THE EDUCATIONAL REQUIREMENTS

Most careers in aquaculture require a combination of physical labor and thinking skills. Workers at each level have to work hard physically, but they also have to make decisions regarding the health and well-being of the aquacrop. These decisions may concern water quality, nutrition, diseases, harvesting, and so on.

The majority of jobs in aquaculture are outdoors. One of the attractions of aquaculture, like many other occupations in agriculture, is the satisfaction people

Figure 3–1. Many aquaculture jobs are outdoors. Here an aquafarm worker prepares to weigh a sample of freshwater prawns to see if they are ready to be harvested.

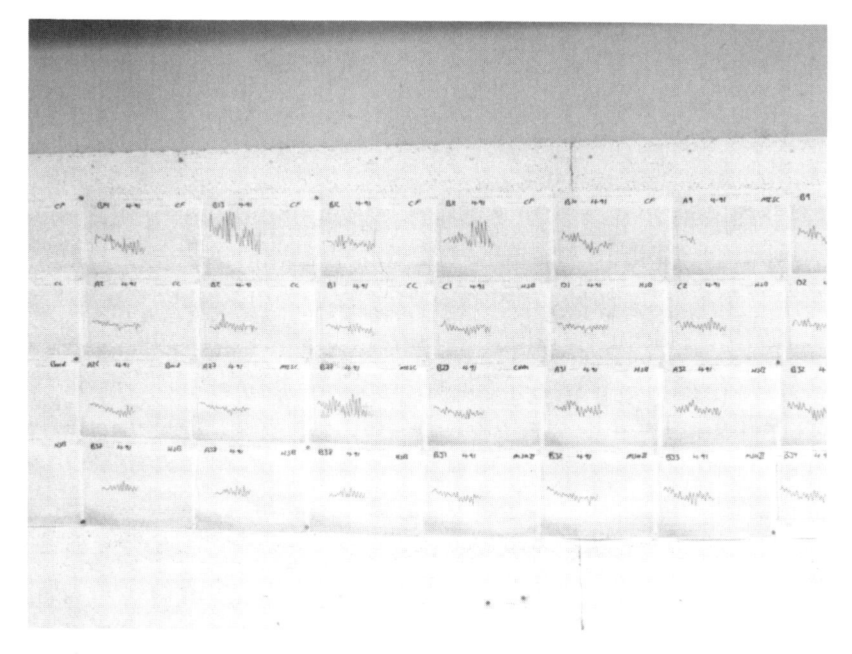

Figure 3–2. Keeping good records is an important part of managing an aquafarm. Each page on this wall represents a single pond with the dissolved oxygen levels plotted.

get from working outside, with their hands, and getting physically tired from their work.

At times, inclement weather makes the outdoors part of aquaculture less appealing. The trend of making a quality aquaculture product available at all times means that processing must occur year-round. This requires harvesting of the aquacrops during all times of the year, including cold winter days. Wading in a pond in 4 feet of water is not as pleasant in the winter as it is in the summer.

As with other types of agriculture, however, the physical labor is just part of the job. Successful aquafarms require sound management. This includes keeping records, analyzing expenses, developing marketing plans, making production decisions, and engaging in other "thinking-type" activities that are not usually accomplished outdoors. The use of computers and consultants may reduce the time spent on these activities, but not their importance.

Some of the common careers in aquaculture and a description of the activities performed as part of those careers are discussed below.

Aquafarmer

An aquafarmer may be the owner of an aquafarm or someone who is hired as the manager. While the responsibility of managers and even owners varies from farm to farm, it is assumed that the person running the aquafarm takes responsibility for all of the processes that determine whether or not the aquafarm makes a profit.

An aquafarmer performs many tasks. Some of the more important of these tasks are making sure adequate facilities for producing the aquacrop are in place and that an abundant supply of clean water is available. The aquafarmer must select a site for facilities and then plan the layout of the different facilities. He / she may also be involved in constructing facilities such as ponds, tanks, raceways, cages, and pens. Of course, if one of the ponds or other facilities becomes damaged, the aquafarmer is responsible for repairing it or hiring someone else to repair it.

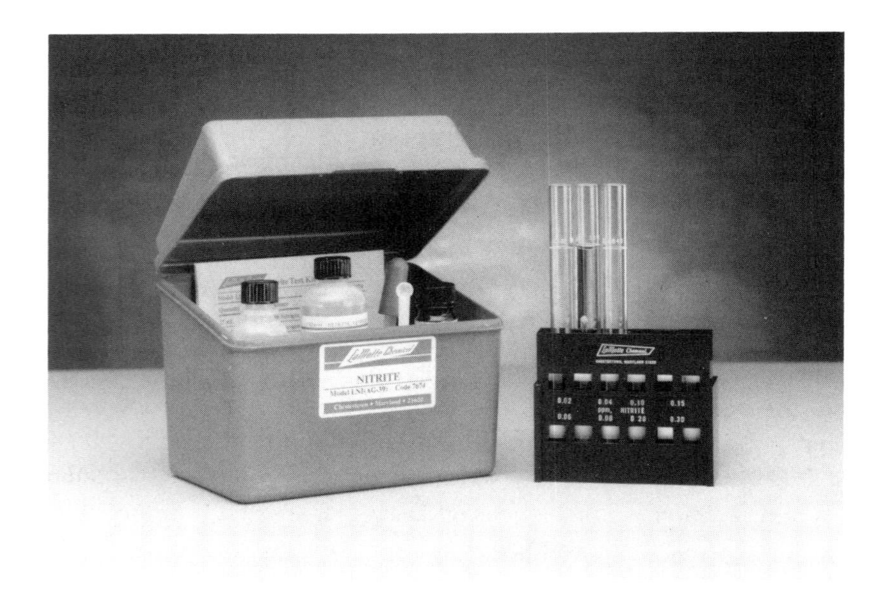

Figure 3–3. This nitrite kit is an example of testing equipment used to monitor water quality. (Courtesy, LaMotte Company, Chestertown, Maryland)

Being able to secure an adequate supply of water will determine the site of facilities. The aquafarmer must determine the source of the water and the amount of water needed and must also be responsible for testing water to make sure it is suitable for the aquacrop. (Chapter 6 presents more information on water quality.)

To make sure the water is adequate, the aquafarmer may have to establish a systematic testing program to monitor dissolved oxygen and to use aerators or other means to correct oxygen problems. This program will usually be required for ponds and tanks that keep the same water for long periods of time. Flow-through systems and raceways that continuously bring in fresh, oxygenated water will not require constant monitoring. Other systems also have to be checked, just not as often. The water must be checked for chemical contamination, nitrates, and excessive weed growth.

As with any farm, the aquafarmer must secure adequate finances to employ laborers, purchase equipment, construct facilities, and maintain an adequate inventory of supplies and aquacrops. Appropriate stocking rates must be determined. High-quality fingerlings must be selected. The management of an aquafarm will require complying with government regulations and budgeting financial resources.

With any aquacrop, nutrition is very important. To maximize profits, the aquafarmer must feed the fish or other aquacrop the right type and the right amount of feed. The aquafarmer must also select which method of feeding is best for a particular aquacrop and facility. Feeding by truck, by demand, and by hand are the common methods used.

To insure that the aquacrop completes the growth process and makes it to a market, pests and diseases must be controlled. The aquafarmer must determine symptoms, select treatments, calculate treatment rates, and treat fish. **It is very important to follow label directions and to use only chemicals approved by the EPA for aquaculture.** Predators such as snakes, birds, and turtles must also be controlled. This control may involve using a number of scare tactics, plus shooting, netting, and employing other means. Careful consideration must be given to predators that are on the endangered species list or that are otherwise protected by state and federal laws.

Figure 3–4. This four-wheeler is equipped with a dissolved oxygen meter and a boom to lower the probe into the water. This allows the worker to measure and record the dissolved oxygen without leaving the machine.

General Worker

The general worker on an aquafarm will often be engaged in the same tasks as the aquafarmer. He / she will often be more involved in the hands-on work on the aquafarm. For example, the general worker will have the responsibility of actually feeding the aquacrop, operating feeding equipment, and checking the aquacrop for behaviors that might signal disease or injury.

The general worker will often be directly involved in the harvesting of fish or other aquacrops. This involvement may include setting the seine, making sure the seine stays on the bottom of the pond or other facility, and removing undesirable species from those collected. The general worker on an aquafarm that grows plants is also heavily involved in harvesting, either operating the equipment or harvesting the aquacrop by hand, depending on the aquacrop grown.

Figure 3–5. A common job on an aquafarm is "riding" the seine at the point where the bottom begins to slope up toward the bank. The worker keeps one foot on the bottom so fish do not escape underneath. (Courtesy, Otterbine Aerators / Barebo, Inc.)

While the aquafarmer or aquafarm manager will usually make the so-called "big" decisions, the general worker who is checking the dissolved oxygen in a pond often has to make decisions that immediately affect the well-being of the aquacrop. The general worker must have some knowledge of the growth process of the aquacrop and of factors that influence this growth process. The general worker's knowledge and reactions to problems often mean the difference between whether or not an aquacrop survives.

An aquafarm general worker will usually get his / her training through experience working on an aquafarm and through a high school or community college agricultural education program. The agricultural education program provides classroom instruction, supervised experience, and leadership training.

Supervisors

Supervisors are usually found in two types of aquaculture operations on the farm. Large aquafarms may have supervisors for crews that perform particular tasks very frequently or exclusive of other tasks. Custom operations may have supervisors for crews that harvest, construct ponds or other facilities, feed aquacrops, apply chemicals, test for water quality, or diagnose causes of death or diseases in aquacrops.

Many aquaculture-related industries, including those that add value to aquacrops through further processing and those that provide inputs to the aquafarmer, employ supervisors.

A supervisor often participates in the same activities as the workers being supervised. However, the supervisor has been given the authority to assign tasks in order to see that the overall goals of the work are accomplished. The supervisor should try to utilize the talents of the members of the group as fully as possible. The supervisor also should try to motivate employees to be more efficient and to complete their work in a conscientious manner. Often the supervisor has some incentive, such as a pay bonus or vacation time, to keep his / her crew producing as much or well as possible. For example, a supervisor of a night crew that monitors dissolved oxygen in catfish ponds may be offered a bonus as an incentive. If no fish are lost due to the crew's performance in a certain period of time, then the supervisor will receive the bonus.

Incentives are common in processing plants. Here supervisors may get bonuses if their crews produce at a rate higher than expected and do this safely.

Supervisors must be trained not only in the processes their workers perform but also in human relations and management. They usually need some college work in these areas to be effective, but extensive experience may serve as a substitute.

Figure 3–6. A processing plant employee uses a vacuum to remove the offal from a fish.

Service Technicians

Service technicians help the aquaculture industry by making sure that the numerous pieces of equipment used on aquafarms or in processing plants work consistently. The types of equipment may include electronic devices that monitor ponds and control aeration or feeding machines, the feeding machines, processing plant equipment, and portable monitoring equipment used to test for water quality.

Service technicians usually help install the equipment or machines and often have the job of servicing the equipment and repairing it when it breaks down. They must have skills in all of these areas. The work is usually outdoors, although this may depend on the type of equipment. The service technician is usually on an aquafarm making service calls. However, some machines and equipment are brought in to a service center for repair and maintenance.

Service technicians must have the ability to get along well with the people who use their products. Once the product is sold, the service technician is often the only person from the company to interact with the aquafarmer. The quality of service often determines if the company will get repeat business.

A service technician may work with an experienced technician in an apprentice-type situation to learn the job. More often, a vocational or technical program will provide the necessary skills. This training will probably be in electronics, mechanics, or other areas and will not be directly related to aquaculture.

Salespersons

Salespersons work in many areas involved in aquaculture, both directly and indirectly. Any large business will have one or more salespersons who service existing customers and try to find new customers. The primary purpose of a salesperson is to find out the customer or potential customer's needs and to provide a product that meets those needs.

Many of the businesses that support aquaculture employ salespersons. These include feed companies, specialized aquaculture equipment manufacturers (such as makers of seines, feeders, and water-monitoring devices), general suppliers, tractor companies, and makers of specialized clothing such as boots and waders.

Salespersons have two types of customers as a general rule. One is the actual aquafarmer who buys products directly from the manufacturer. The items sold this way may include feed, seines and seine reels, and feeders. These items may be sold through on-the-farm visits or through direct marketing such as sending the potential customers a catalog.

The other type of customer is a general aquaculture supplier, who buys goods to resell to aquafarmers. A lot of the specialized equipment is sold this way. In turn, the aquaculture supplier will also employ salespersons to sell directly to aquafarmers. They will usually wait for the aquafarmers to come into the business, but they may also make on-the-farm visits or send the potential customers a catalog.

Salespersons who work for these suppliers must have a general knowledge of many different types of products and how they are used. Aquafarmers will depend on them to provide solutions to problems. Salespersons may have to recommend feeds and feed rations, chemicals to control diseases and pests, types of equipment, and many other products. How well the aquafarmer feels the salesperson met his / her needs will probably determine if he / she will repeat as a customer. The salesperson has a responsibility to help meet the aquafarmer's needs in such a way that is cost-effective to the aquafarmer.

Salespersons may receive training for their work in several different ways. A good salesperson must meet two basic requirements. The first is a knowledge of the products being sold and how they are used. The second is the ability to interact with people, to make the customer feel comfortable with the product and the person selling it.

Some large companies hire college graduates for their sales positions. Smaller companies with local markets will hire people with less education, provided they have extensive experience working with or for the clients and the products. Many salespersons start by working for an aquaculture supplier or on an aquafarm while they are attending school. Regardless of their salespersons' education and experience, many companies require them to go through a company training program and may require a probationary period of employment.

Drivers

Processing plants and feed companies are the two aquaculture-related businesses that employ the most drivers. Processing plants usually have a fleet of live-haul trucks that go to the farms to pick up the aquacrops and haul them back to the plant for processing. Feed companies usually have a fleet of trucks to haul feed to the aquafarms.

The drivers have an important job within the industry. Like service technicians, they are the ones from their company most likely to come in contact with the customers, the aquafarmers. As a result, the drivers must have good interpersonal skills to maintain good relations with the aquafarmers.

Drivers often handle billing and receiving procedures for the company. For example, a driver for a processing plant must weigh the aquacrop as it is loaded in the truck and then provide the aquafarmer with a receipt. Other requirements include a safe driving record and a commercial driver's license.

Drivers usually do not have to have a college degree. Some companies hire drivers from driving schools, while others try to hire drivers who already have experience driving for other companies. A few companies provide their own training programs.

Figure 3–7. Processing plants employ drivers to go to the pond bank and collect fish for processing.

Large Equipment Operators

Only large aquafarms will employ persons to operate large equipment on a regular basis. Large equipment operators usually work for custom service providers, such as people who construct ponds or raceways for aquafarms or help aquafarmers maintain pond levees. The work is outdoors and requires a great deal of skill and experience in using the equipment. Earth movers and bulldozers are the two primary types of large equipment used in aquaculture. Large tractors with special equipment are also used.

Figure 3–8. One example of equipment that must be operated on an aquafarm is a tractor with a seine reel.

Large equipment operators must do more than just sit in the driver's seat and operate the equipment. When constructing a pond, for example, the operator must be able to read the maps, follow the surveyor's stakes, and judge the transformation of the earth very accurately. The operator may also be required to assist with the surveying and layout of the pond. Constructing the pond with the right specifications in the minimum amount of time is the main goal.

The operators usually are responsible for the service of the equipment and often must make minor repairs. They may also have other jobs, such as transporting the equipment from one farm to the other.

The primary skill necessary for the job is to be able to operate the equipment efficiently and safely. This equipment is very expensive and requires care. Carelessness in the operation or maintenance of the equipment should not be tolerated. Of course, the operator must also be concerned with his / her own safety, as well.

Most operators learn to use the equipment as apprentices with experienced operators. Some technical schools provide training in large equipment operation.

Other Occupations

Like any industry, aquaculture creates many jobs that may not be directly related. Engineers design and build special equipment. Chemists develop chemicals to control pests. Veterinarians determine methods of controlling disease and ways to improve health. As discussed in Chapter 2, researchers may study any of the phases of aquaculture production. However, the more common occupations in aquaculture have been discussed.

Figure 3–9. This researcher is studying ways of producing phytoplankton for filter feeders such as oysters.

CAREER LADDERS IN AQUACULTURE

The term "career ladder" can be defined as "the process of moving from a lower-level occupation, usually with lower pay, to a better occupation. Career ladders may be vertical or horizontal.

Moving on a vertical career ladder would involve moving to a better occupation doing similar work. For example, an aquafarm general worker moves to a position as a supervisor of several general workers or a manager of an aquafarm decides to leave the aquafarm and start one of his / her own.

Moving on a horizontal career ladder would consist of moving to a better occupation doing work that is not similar, though it is usually somewhat related. An example would be an aquafarm general worker leaving his / her job to go to work as a feed salesperson.

In aquaculture, moving on a vertical ladder usually involves lower-level workers on aquafarms, in processing plants, and in aquaculture supply companies becoming supervisors, managers, or salespersons.

Moving on a horizontal ladder in aquaculture may necessitate that workers move to and from several different occupations. The most common involve moving from jobs as aquafarm general workers to other jobs with suppliers or processors. Many of the moves on a horizontal career ladder require obtaining more education or training.

SELECTING A CAREER IN AQUACULTURE

Once you decide aquaculture is the career for you, think about the career opportunities discussed earlier in this chapter. Talk to some of the people who work in aquaculture. Find out the advantages and disadvantages of their occupations.

Use the supervised agricultural education experience (SAE) program offered by your school to find out more by getting some experience. With your parent(s) and your agriculture teacher, plan some activities to answer your questions. If you have a specific job picked out, plan on spending some time working at that job after school and in the summer to find out if that is really what you want to do for the rest of your life. If you are not sure, set up an exploratory SAE program to help you learn about as many different jobs as possible. Remember to find out what abilities and education are required to be successful. Do not pick a job that requires more education than you can handle or finance.

If the occupation you select involves owning your own operation, do some investigating. Locate sources by which you can obtain the necessary capital to start the business. Give yourself a chance to be successful as your own boss. Remember that entrepreneurs share in the risks as well as the rewards. This responsibility cannot be handled by everyone.

Choose a career in aquaculture that meets your requirements for a chosen career. Does it allow you to work where you want? Will you enjoy the nature of the work? Can you successfully complete the work required with your present mental and physical abilities? Will you need to go to college or complete a vocational - technical program to perform well? Do you have the necessary experience or the means of getting it? Will you be able to provide a stable, adequate income for yourself and your possible dependents? Will you have to change your lifestyle? These and many other questions will have to be answered before you make your final choice.

SUMMARY

Aquaculture has many exciting careers. Opportunities abound in this fairly young industry. However, many opportunities for success often mean many opportunities for failure. A career decision should be made carefully because it is so important.

The career opportunities in aquaculture range from those as general workers with limited entry skill requirements to owners / managers with high requirements needed in both ability and education. Most young people with an interest can find an occupation in aquaculture that fits their needs, abilities, and education. The nature of the work and the opportunities to move up career ladders must also be considered.

To be successful in an aquaculture career, a person should acquire the necessary experience and training, as well as develop the personal characteristics of successful employees in any occupation.

QUESTIONS AND PROBLEMS FOR DISCUSSION

1. Why is the career decision-making process so important?
2. What are some of the important factors which must be considered when a person is making a career decision?
3. Outline the steps in the career decision-making process.
4. What are the six characteristics of good employees? Briefly discuss each.
5. List eight types of aquaculture careers.
6. Select the occupation in aquaculture that you would most like to work in. Describe the nature of the work and the education required.
7. Interview someone who works in aquaculture in your area to find out what that person likes and does not like about his / her job.
8. What is a career ladder?
9. Identify some career ladders in aquaculture in your area. Identify one person who has moved on a vertical or horizontal career ladder in your area.

PART TWO

The Science of Aquaculture

Chapter 4

DETERMINING AQUACULTURE REQUIREMENTS

Aquaculture requires knowledge, facilities, financing, and other inputs. Aquafarmers need to be familiar with these in order to make sound decisions.

Planning an aquaculture enterprise requires certain essential information. Knowing what to do and how to do it does not come easily. Obtaining needed finances is a key. Avoiding unnecessary risks can help insure success. Knowing about needed industry linkages helps the aquafarmer. Coping with regulations and permits is essential. Studying the requirements in aquaculture can pay big dividends to the beginning aquafarmer.

OBJECTIVES

Several objectives that will help students understand the requirements in aquaculture are covered in this chapter. Upon completion, the student will be able to:

- Define "feasibility" and identify areas to consider.
- Explain ways of doing business in aquaculture.
- Describe general requirements for aquaculture.
- Explain considerations in species selection.

- Explain economic considerations in aquaculture.
- Describe infrastructure relationships needed for aquaculture.
- Explain the role of regulations and permits in aquaculture.

FEASIBILITY

Feasibility refers to what can be done successfully in aquaculture. Various aquabusinesses (aquafarms, aquaculture supply businesses, aquaculture processors, etc.) have been started. Their owners want them to be successful. In studying feasibility, a person is trying to answer the question "What holds the best possibilities for me?"

In assessing the possibilities in aquaculture, a person should look at feasibility from several viewpoints:

1. *The potential of an aquacrop to make a profit is important.* Profit is the money left over after all costs of production have been paid. Unfortunately, some people have gone into aquaculture and not made a profit. Some have even lost huge amounts of money. This area is sometimes referred to as economic feasibility.
2. *The "way" an individual starts an aquabusiness must be determined.* Aquafarming is a business much the same as a business that provides the supplies needed to produce aquacrops. Starting an aquabusiness is a

Figure 4-1. Freshwater prawns are thought to hold much promise as an aquacrop in the U.S. Research is underway determining their feasibility. (Courtesy, Gary Fornshell, Mississippi State University)

serious matter. Money is required to get started. Many people have limited resources and must be creative in how they go about getting started.

3. *The kind of aquacrop to produce must be carefully selected.* Aquacrops usually require certain conditions for best growth. Choosing a crop that is not suitable puts the aquafarmer at a disadvantage. For example, an aquafarmer who has only cold water must select a crop that grows well in cold water.

4. *The resources available in the local area to support the aquacrop must be considered.* Aquafarmers get resources from a variety of providers. Equipment such as pumps, seines, and tractors must be available. Seedstock, feed, medications, and other inputs are needed. There must be buyers and processors for the produced crops. Certainly, the ability to get assistance from experts in aquaculture is helpful. Local communities that do not have the resources pose a significant challenge to the potential aquafarmer.

WAYS OF DOING BUSINESS

People who are going into aquaculture must know more than how to produce an aquacrop. They must figure out how they are going to do business. Few individuals have the finances necessary to start a business. Most people must seek needed finances elsewhere. How aquabusinesses are organized can help in the generation of finances or other resources, such as specialized skills.

The U.S. has a free enterprise economic system, also known as capitalism. In a free enterprise, individuals or groups of individuals may privately own property. The owner may be one person or more. The owners have the freedom to make choices about their business. There is a minimum of government control. Those controls that do exist are intended to provide for the overall welfare of society. For example, regulations on uniform and wholesome aquafood products help protect the consumer from inferior products. Good products help aquaculture to prosper. A few bad products on the market can give it a bad reputation in a hurry, particularly if the information is carried as a news item by the media. Purchases of aquafoods can drop drastically in just a few hours!

There are three ways of doing business: sole proprietorship, partnership, and corporation. (Cooperatives are used in aquaculture; these are special forms of corporations.) It is important to remember that farms are businesses. They are established to produce goods or services that provide monetary gain to the owner(s).

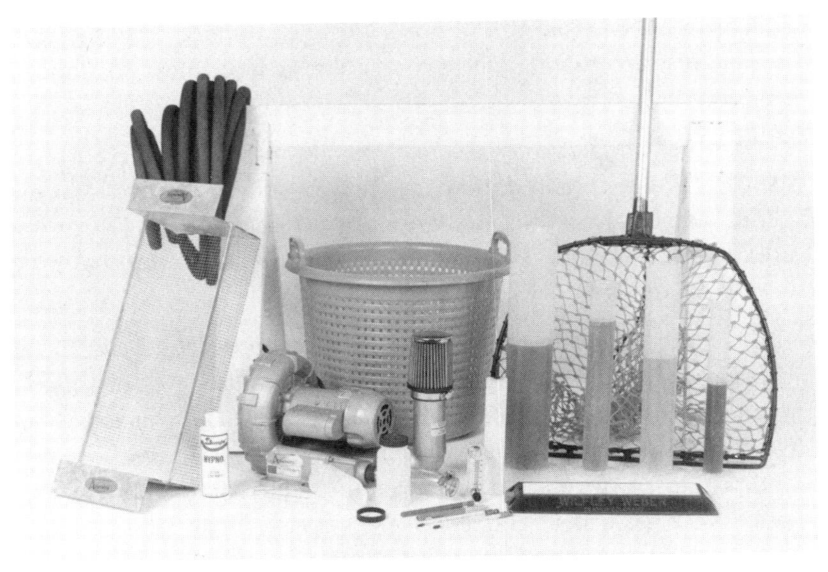

Figure 4–2. The business of aquaculture requires investment in a wide range of small equipment and supplies. (Courtesy, Aquacenter, Inc.)

Sole Proprietorship

A sole proprietorship involves one person owning a business. This person has the legal title to the business, whether it is an aquafarm, a retail market, a processing plant, or a fish feed mill. The owner is responsible for all debts and receives all profits, if any profit is made. The sole proprietorship is the most common form of ownership for small and medium-size aquabusinesses. Most aquafarms are sole proprietorships.

Partnership

Partnerships involve two or more people going into business together. This way of doing business allows people to "pool" their resources, which may be finances, skills, and / or facilities. When resources are pooled, more support is available for the aquabusiness. People with specialized skills can use them in a productive way. They can team up with people who have skills they do not have. An example would be an accountant going into partnership with a person who knows aquaculture. The accountant might know very little about aquaculture but would be very skilled in the accounting areas. The person skilled in aquaculture would provide the expertise with the water farming enterprise and leave the financial details to the accountant.

In a partnership, the profits, if any, are divided among the partners. Likewise, any debts are shared by the partners. Individuals considering a partnership should carefully assess what they want to do. Normally a written document that describes the partnership would be prepared and signed. Some aquafarms involve partnerships. Sometimes non-farm aquabusinesses are partnerships.

Corporations

Setting up a corporation involves creating an "artificial person" or a legal entity. A charter of incorporation has to be established with the state government. All applicable laws, which vary from one state to another, must be met. The owners are known as shareholders because they buy shares in the corporation. A shareholder is eligible to vote on matters about the corporation. Each share entitles the owner to one vote. The more shares, the more votes a person has. Any profit is divided among the shareholders. An individual can lose no more than the amount invested in shares. Corporations are able to get larger amounts of money through the sale of shares than individuals or partnerships are usually able to raise.

A board of directors is elected by the shareholders. The board establishes policy and appoints officers to run the business. The officers normally include a president, vice president, and secretary-treasurer. Larger aquafarms, processing plants, and equipment manufacturers may be corporations. The word "incorporated" or "inc." appears after the name of the corporation, or the word "corporation" may be used in its name.

A special form of corporation is the cooperative. A cooperative is an association formed to provide certain services to its members. For example, aquafarmers may form a cooperative to manufacture feed or to process their aquacrops. None of the farmers could individually establish a feed mill or a processing plant, but by working together they help each other. Cooperatives must be established under applicable state and federal laws. In addition, the federal government has special financial assistance available to certain cooperatives.

Cooperatives are not intended to produce profits but are intended to serve their members. To be a member, an individual must usually buy shares in the cooperative. Each member has only one vote regardless of the number of shares owned. The big advantage of a cooperative is that larger volumes of fish, feed, or other items are involved when individuals join together. Financial advantages usually are found in volume buying and selling. For example, since a greater quantity of feed is needed collectively, it is usually available at a lower price. Very few aquafarms are large enough to use a feed mill of their own efficiently.

Each of the three ways of doing business involves some rules and regulations. There are more of these for corporations than for partnerships and sole proprietorships. Appropriate legal counsel is important. This involves additional cost for lawyers' fees. Certain reports may also be required on an annual basis.

Aquabusinesses must comply with a wide range of regulations that also apply to other businesses. Property taxes, income taxes, privilege taxes, and taxes in other areas must be handled. Social security must be paid for the employees. Most medium- and large-size aquabusinesses regularly use accountant services to help with these details. The largest ones may employ accountants on a full-time basis.

GENERAL REQUIREMENTS

The general requirements are those essentials needed to "make it" in the aquaculture business. And there are many of them! The requirements can be put into seven areas: management, labor, land, water, species, markets, and financial resources.

Management

Management is a human factor. It involves handling a wide range of mental tasks to carry out aquaculture successfully. Commercial aquaculture is a business. Success in a business depends on the ability to operate it in such a manner that a profit is made. In other words, the returns from the marketed aquacrop must be greater than the costs of producing the crop.

Education, the ability to make and carry out decisions, a high energy level, the ability to relate to other people, personal ethics, and personal preferences are important in aquaculture management.

Figure 4–3. A wide range of literature on aquaculture is available to help provide information needed in making decisions.

Education refers to the knowledge and skill an individual has about a particular aquaculture enterprise. An individual needs to be able to both "know" and "do." Formal classes, workshops, seminars, and field days are good ways to learn the basics. The learning to "do" occurs by actually working in the industry of aquaculture. For example, it is very beneficial for an aspiring aquaculture manager to spend time working on an aquafarm. First-hand experience teaches about the "real world." Some important areas for education are:

- Biology of the species under consideration
- Cultural requirements for the species
- Facility and equipment needs
- Water management
- Nutrition requirements and feed ingredients
- Disease prevention and control
- Harvesting and marketing procedures

Making decisions and then following through on them is important. Education and experience help make an aquafarmer a better decision maker. Decision making involves selecting the direction to take with an aquafarm, such as what species to grow, production system to use, and size of farm to have. Any problem usually involves several possible actions. The role of the manager is to select the possible action that is in the best interest of the aquafarm. Several approaches are available in decision making. Routine decisions can be made by following a rather simple process. The process is:

Step 1. ***Become aware of the problem.*** This means that a good manager must keep up with what is going on. "Keeping in touch" requires putting time and energy into whatever is being managed.

Step 2. ***Get enough information on the problem.*** This involves being informed. It is important that the information be accurate and considered in making the decision.

Step 3. ***Determine alternatives.*** This involves identifying the different decisions. What options are available?

Step 4. ***Evaluate alternatives.*** Some alternatives in solving a problem are better than others. Good information about the consequences of the different alternatives is needed for the evaluation. Seek the input of people who have had experience with the different alternatives that are available. In aquaculture, there are "for-hire" consultants who can help in assessing the alternatives.

Step 5. ***Choose the best alternative.*** A good manager selects the alternative that is best based on the situation that exists at the time. Risk is always involved. Every decision maker faces uncertainty. Using a thorough process in making decisions can reduce the risk.

Step 6. ***Implement and follow up.*** Good decisions are no better than how well they are carried out. The manager needs to communicate decisions to others carefully, as appropriate. A decision to take a particular course of action needs to be followed unless a new decision is made that changes the original direction. Following up provides the manager with an opportunity to see how well decisions are being implemented and if they are working out okay. This is the same as evaluation. On small aquafarms, the manager may be the only individual involved. Decisions are still made to be carried out.

Step 7. ***Recycle through the decision-making process (optional).*** If a decision does not work, start over with Step 1 and repeat the process.

Good managers have a high level of energy. This means that they are capable of spending long hours at work on the aquafarm. Managers must use both physical and mental energy. Physical activity involves doing all of the tasks that occur in an aquaculture operation. Mental activity refers to making decisions and recalling information. Any person who is beginning an aquafarm must have a high level of energy.

An aquafarm manager must be able to relate to other people. This includes communicating effectively with employees. Certainly, the ability to relate to the suppliers of feed and other inputs for the farm is essential. Good relationships are also needed in marketing the crop, since the crop must be disposed of in such a way as to yield a profit.

All actions in aquaculture should involve honesty and fairness with all people. Ethical relationships are essential for the prosperity of individuals as well as the aquaculture industry. A few pointers on aquaculture ethics are:

- Managers must follow through on their word. (If they say they are going to do something, they must do it.)
- Aquacrops should be produced to insure a nutritional, healthy product.
- Environmental considerations, such as getting and disposing of water, need to be observed.
- Animal crops should be produced in a manner consistent with animal welfare.
- Business agreements should be carried out as was the original intent unless both parties agree to changes.

- Legal regulations must be observed. (Aquaculture is different from row crop or dairy farming. The beginning aquafarmer should spend time learning about the laws and rules that apply.)
- Dealings should be fair to all concerned.

Personal preference plays a big role in decisions about aquaculture. Individuals need to assess their personal interests and preferences. If they do not like the nature of the work, they should not go into aquaculture. If they prefer to work with one kind of aquacrop over another, they should go into the one they prefer, provided it is economically feasible. Lifestyle is shaped by operating an aquafarm. Factors such as length of work days, repetition of tasks, and risk taking must be dealt with.

Labor

Aquafarming involves a lot of work. Labor needs vary with the size and nature of the farm. Small farms may need only one person — the operator-manager. Larger farms require more labor. The labor requirements for some crops are greater than for others. Some of the work, such as harvesting, can often be

Figure 4–4. Sophisticated new machinery has been developed to reduce the amount of labor required in aquaculture. A good example is the power seine hauler used in harvesting fish. (Courtesy, Master Systems)

handled on a custom basis (hire others to provide equipment and do it). Education and training vary with the nature of the enterprise. Labor must be dependable. For example, some aquacrops require the night-time monitoring of oxygen levels in the water. This duty must be accurately carried out on schedule during the night hours.

Activities requiring labor on a typical food fish farm include:

- Preparing facilities for stocking
- Stocking growing facilities
- Feeding
- Monitoring water
- Maintaining pumps and aerators
- Collecting samples of aquacrops
- Calculating amounts of chemicals, etc., to use
- Administering disease treatments
- Operating trucks, tractors, etc.

Land

Adequate land must be available. The land should have the appropriate physical features for aquaculture. Different aquacrops have different requirements. Some require warm water, while others require cold water; some prefer flowing water and others do not. In general, aquafarmers have found that level land is best for pond construction. Harvesting of fish is also easier in ponds on level land.

In selecting land for an aquafarm, the aquafarmer should consider several factors. These are:

- ***Past history of the land***—If the land has previously been used for row crops where high levels of pesticides were applied, it might be unfit for aquaculture. Pesticide residues might have built up in the soil. These residues could contaminate the water and the crop. A chemical laboratory analysis of the soil can tell if dangerous pesticide residues exist.
- ***Nature of the soil***—Soil high in clay content is preferred. Sandy and loamy soils do not hold water as well.
- ***Flooding***—A site above the floodplain of a creek or river should be selected. Water that overflows into a fish pond can result in the loss of the fish crop. The fish crop could just swim away in the flood water!

Flooding can also restrict access to the pond and make it impossible to feed the fish. If this occurs, the feeding habits of the fish are interrupted, and they may go off feed (stop eating even when fed). Several days may be required to get them back on feed.

- ***Availability of water*** — A good supply of suitable water is essential. Checking on the availability of well-drilling permits is a part of the selection process. Chemical analysis of the water may be needed to determine if it is suitable.

- ***Cost*** — The cost of land should be in line with the capability of the land to produce an aquacrop at a profit. Sometimes land is overpriced. Land near cities or in productive farming areas tends to be higher than land in other locations. Of course, a recreation facility should be nearer population centers in order to obtain adequate business.

- ***Proximity to market*** — Consideration must be given to the proximity of land for an aquafarm to a market outlet. If no market is readily available, the selection of a particular location should be reconsidered.

- ***Size*** — The amount of land required varies with the intended use for aquaculture. For example, a hatchery requires less land than a large food fish operation. Site selection should take into consideration all needs — acreage in ponds as well as for roads, storage buildings, and other uses.

- ***Accessibility*** — A site should be accessible by all-weather roads. Electric power lines should be in the vicinity; otherwise, the aquafarmer may have the additional expense of power line construction.

Water

All aquafarms require water. The water must be available in adequate supply and of appropriate quality and chemistry. Sources of water include wells, springs, rainfall runoff, lakes, oceans, streams, and industrial effluent (water released by an industry, such as an electric power generator). The species of aquacrop to be grown is a definite consideration. For example, warm water fish must have water of a suitable temperature in order for them to thrive. Water from thermal springs or wells is sometimes available. Industrial effluent may have a uniformly warm temperature as it is discharged from the plant. Water should undergo thorough analysis before a final decision is made on the establishment of an aquafarm. (Chapter 6 provides details on water.)

Figure 4–5. An abundance of good water is needed in aquaculture. (Courtesy, Kasco Marine, Inc.)

Species

The species selected for an aquafarm must be suited to the production system. Further, a ready market must exist for the species to be produced. Some fish farmers have been able to establish a market niche. A niche is a specialized outlet for a crop. For example, tilapia grown in the southern U.S. may be shipped without processing to ethnic markets in the large cities of the northern states.

Aquafarmers are advised to produce species for which there is an established market outlet. With aquafoods, processors should be available. In a few cases, farmers operate small roadside markets, but their potential is often small-scale.

Individuals considering aquafarming should learn all they can about the environmental requirements of a species before going into the business. Species should be well suited to a production system. With fish, adequate breeding stock or fingerlings should be available. (More information is presented on species selection later in this chapter.)

Markets

It is impossible to operate an aquafarm profitably without appropriate markets. A superior aquacrop is of little value if no market exists.

In making decisions about which aquacrop to produce, farmers should consider the availability of markets. Transportation costs reduce profit if aquacrops have to be hauled long distances. Discussion with buyers of aquacrops should be held. In some cases, contracts are established between growers and processors. These insure the processors of an adequate supply of product and the aquafarmers of a ready market outlet. (More information on marketing is presented in Chapter 9.)

Financial Resources

Establishing an aquafarm may require a sizeable financial investment. The amount of money needed depends on many factors, including size of aquafarm; cost of equipment, production inputs, harvesting, and marketing; labor; and water availability.

Some lending institutions do not make aquaculture loans; others do. The interest rate and the number of years for repayment are definite considerations. Regardless, adequate finances must be available.

CONSIDERATIONS IN SPECIES SELECTION

Do you want to grow aquaplants or aquaanimals? Which species are appropriate to your farm? These and other questions should be answered. Important considerations in species selection are climate, market, site, finances, and species potential.

Climate

Selecting a species of aquacrop requires knowledge of the characteristics of the crop under consideration. The major climate factor is temperature. Since aquacrops grow in water, the temperature of the water may be more important than the temperature of the atmosphere. Ground water may be too cool for some aquacrops if it is pumped directly into ponds or other growing facilities. In some locations, thermal springs and wells provide warm water. This water can be directly used with warm water fish. Cool well water may need to be held for several days in a large reservoir for natural warming by the sun if it is to be used with warm water aquacrops. Those crops preferring cool and cold water should be selected if the water temperature is satisfactory. Cool spring water may work well with those aquacrops that prefer cool water.

A few examples of temperature requirements for aquacrops are shown in Table 4-1.

Table 4-1

Water Temperature Requirements for Growing Selected Aquacrops

Species	Optimal Water Temperature (°F)
Tilapia	80–90
Striped bass	65–90
Flathead minnow	65–80
Channel catfish	70–85
Alligator	80–90
Bullfrog	60–80
Freshwater prawn	75–95
Crawfish	60–80
Trout	50–68

Sources: Compiled from W. O. McLarney, *The Freshwater Aquaculture Book*. Point Roberts, Washington: Hartley & Marks, Inc., 1987; H. K. Dupree and J. V. Huner, *Third Report to the Fish Farmers*. Washington, D.C.: U.S. Fish and Wildlife Service, 1984; and J. E. Bardach, J. H. Ryther, and W. O. McLarney, *Aquaculture: The Farming and Husbandry of Freshwater and Marine Organisms*. New York: John Wiley & Sons, Inc., 1972.

Market Outlets

An aquacrop is of little value if it cannot be sold for a satisfactory price. Any decision to start producing a particular crop should be based on careful consideration of marketing possibilities. A potential producer should be very cautious in growing a crop that has no established market.

Site

Aquacrops have characteristics that make them better suited to some sites than others. The species grown needs to "match" the site well. The type of production system to be used has a bearing on how appropriate a site may be.

An intensive system involving tanks can be established on land that would not be suited for pond culture.

In addition to the biological aspects of site selection, consideration must be given to government regulations, cost of the property, taxes, zoning regulations, accessibility to electric power and water, and existence of all-weather roads.

Financial Resources

Money is needed for start-up and operating costs. Start-up costs include the land, facilities, and seedstock needed to carry out aquaculture. Operating costs include those for feed, electricity, labor, and other variable production inputs.

Large-scale commercial fish farms require huge financial investments for establishing and operating. Individuals with smaller resources will need to select a species and production system that can be accommodated. For example, ornamental fish can be grown in aquaria in small buildings occupying nothing more than the corner of a lot in a city. (More information is presented on financial matters later in this chapter in the section on economic considerations.)

Species Potential

The aquacrop grown must have the potential of making a profit for the producer. Some species have greater potential than others. Considerations should include adaptability to culture, demand for the species, legal regulations (some species are not legal in some states), personal preferences, and availability of technical assistance when problems arise. Aquaculture specialists with land-grant universities can provide useful information on many aspects of species potential.

ECONOMIC CONSIDERATIONS

The potential to make a profit is the most important consideration in commercial aquaculture. The income derived from the sale of the aquacrop must be greater than the cost of producing the crop. Profit is not a concern to the individual in aquaculture for a hobby, and there are a few people who grow aquacrops for this purpose.

Economic considerations should involve an analysis of both costs and returns potential as well as sources of risk.

Costs

Costs include initial costs, annual fixed costs, and variable production costs. Initial costs include land and facilities. Constructing ponds; drilling water wells; buying or constructing tanks, hatchery equipment, aeration and feeding equipment; and establishing breeding stock are initial costs. The cost for a large catfish operation may be millions of dollars. Good planning is essential in identifying all of the initial costs of setting up an aquafarm.

Annual fixed costs include interest on loans, repayment of loans, property taxes, insurance, and depreciation on facilities and equipment. These costs are often overlooked, but are very important in getting a precise financial picture of a proposed aquaculture enterprise.

Production costs vary with the size of the operation. Feed, seedstock, chemicals, electricity, and fuel are examples. Custom services (such as harvesting) are also variable costs.

Returns

Income from the aquaculture enterprise comes when the aquacrop is sold. Processors typically pay the producer for the volume and quality of the aquafood that is processed. Producers of eggs, fry, or fingerlings receive income when they sell their product to growers. Operators of recreational lakes get income from the sale of fishing privileges and product caught. Income for bait and ornamental producers comes about when they sell their product.

Returns are measured as "gross return" and "net return." Gross return is the total of the money received for an aquacrop. Net return is the amount of money that remains after all expenses for the production of the crop have been subtracted from the gross return. The most important figure to a producer is net return—profit.

Aquafarms that have large gross returns do not necessarily have large net returns. Efficiency involves keeping down costs and maximizing sales. Some aquafarms return sizeable profits each year; others do not.

Financing Aquaculture

Sufficient money must be available to establish and operate an aquaculture enterprise. Some producers have the money they need to establish and operate an aquafarm. Others must rely on credit to provide the needed finances.

Sources of credit include individuals, banks, the Farm Credit System, and aquaculture businesses, such as feed mills and processors. Short-term (sometimes referred to as seasonal or production loans), intermediate-term, and long-term credit may be needed. Intermediate-term credit is usually for more than one year but less than five to seven years. Long-term credit is used to finance the establishment of the aquaculture production facility, with repayment over several years.

Basic credit factors that lenders consider are the management ability of the aquafarmer, repayment ability, financial conditions and operations, economic need, purpose of the loan, and collateral. Certainly the reputation of the borrower is another important factor in the decision of a lender to make a loan.

Risks in Aquaculture

The production of aquacrops involves risk. The producer can keep risk at a minimum by following good management practices. Areas of risk include possible loss of a crop, poor-quality seedstock, failure of water supply, loss of a supplier of inputs, contamination by dangerous chemicals, losses at time of harvest, and instability in the market place.

Carefully observing an aquacrop and applying proper management can minimize the loss of a crop. For example, fish in ponds must be regularly monitored for disease, oxygen depletion, and other problems. Once a problem is suspected, steps must be taken to correct it. An entire crop of fish can be lost quickly due to oxygen depletion in the water. Through proper aeration, these losses can be minimized if not eliminated altogether.

The quality of the seedstock used is largely a function of the reputation of the seedstock producer. Fingerlings bought by a food fish producer to stock a pond must be of high quality. Knowledge of the practices followed by the fingerling producer will help make the decision about which sources are best. Producers of fingerlings need to use good broodfish and follow approved production practices. These affect fingerling quality. Healthy fingerlings free of parasites and diseases should always be used.

Water supply failure can be a catastrophe. It may be due to a power outage, a pump breakdown, a well going bad, or another cause. Standby systems need to be available if at all possible. There should be electrical generators if the electric power fails and the pumps and aerators stop operating. Spare pumps and pump parts should be kept on hand. Personnel who can trouble-shoot and repair pumps, generators, aerators, and other mechanical devices need to be available or on-call for emergencies.

An aquafarmer makes plans on the basis of dependable sources of feed, chemicals, electric power, and other inputs. The failure of any of these providers of inputs can have a major negative impact on the aquafarm. Having contracts

Figure 4–6. An essential investment is an emergency electric generator that operates when the electric power fails.

for supplies from established, reputable dealers can help minimize this area of risk.

Chemical residues in aquacrops can be costly. Residues of some substances make the product unfit for human consumption. Select a site where the soil is free of pesticide contamination. Nearby farm operations should be careful to avoid accidentally getting pesticides into aquaculture facilities. Any feed should be free of residues.

Once a quality aquacrop has been produced, efficient harvesting procedures should assure a minimum of loss. Harvesting equipment should be in good repair and operated in the proper manner. Proper hauling equipment should be used.

In making financial plans for aquaculture, the aquafarmer should consider the amount to be received for the crop to be an important item. Most plans use the current or predicted price level. An increase in price works to the advantage of the producer. A decrease in price can be financial disaster. Instability of price and market outlet is a concern, but some of the risk can be worked out ahead of time by agreements between the producer and the processor or other buyer.

INFRASTRUCTURE RELATIONSHIPS

Infrastructure refers to all of the network of activities required for an aquaculture commodity to become commercially important. Aquafarmers cannot go it alone. They must have finances, equipment, feed, chemicals, processing, and marketing. Without all of these to support the production phase, it is very difficult to establish an aquaculture enterprise. New areas of aquaculture begin slowly, and the infrastructure develops along with the aquacrop. All of what is needed cannot instantly be put in place.

There must be a balance among the infrastructure parts. For example, a large processing plant with only a few fish to process loses efficiency. A large farm without a market outlet has no place to sell the aquacrop. Synchronization of all phases of an aquacrop is essential.

A good example of the importance of a balance in the infrastructure is catfish. Wayne Shell of Auburn University cites the bandsaw in a processing plant. A production standard is that a bandsaw operator should be able to remove the heads of about 45 catfish a minute. This translates into about 19,000 pounds a day and 5 million pounds a 365-day work year. At the current levels of production, 1,250 acres of pond would be needed to keep this one bandsaw going. Some 8.5 million pounds of feed would be needed to raise the fish. Three million pounds of ice-packed fish or 2 million pounds of fillets would be produced. The market must be in place for this quantity of fish.

A first step is to develop a marketing plan. The plan determines the market demand for the aquacrop. It sets a production schedule. Credit can be arranged, seedstock can be obtained, water facilities can be built, feed mills can be put into operation, and processing and marketing facilities can be developed. Advertising can promote the aquacrop to the consumer.

An aquaculture industry can emerge when all of the components are in place. These components must be balanced to accommodate the size of the industry. Growth occurs as each part expands and gets a little out of balance. Good communication in the industry is essential. Agreements among producers, providers, and processors are beneficial. Individuals involved at all steps must be educated. Supermarkets must allocate market space, and restaurants must add

Figure 4–7. A restaurant chef proudly displays the dishes he has prepared using farm-raised fish. (Courtesy, Delta Pride Catfish, Inc.)

items to their menus. Chefs must be trained in how to prepare the food. Interested parties must provide leadership for the industry. Lending agencies must allocate money for aquaculture loans. It takes time and a proven record for a new aquaculture industry to emerge.

REGULATIONS AND PERMITS

All phases of aquaculture are subject to certain regulations and required permits. Some people refer to these as "government redtape." Regardless, the intent of the regulations is to provide for the overall welfare of society. This may be by controlling pollution or assuring that a product is wholesome and up to standard. Information on the regulations in a particular state can be obtained from an aquaculture specialist at a land-grant university or a state or federal agency responsible for aquaculture.

The regulations are typically in four areas: environmental protection, worker health and safety, licensing, and product transportation.

Environmental Protection

Regulations in this area serve to protect the environment from pollution and to conserve natural resources. Water is the major area of concern. Water for aquacrops must be obtained in a manner that complies with federal and state regulations. Aquafarmers cannot put a pump into a creek or river and start taking out water as they please. Wells cannot be drilled in just any way. Permits are needed to drill wells and remove water from streams. Individuals who own property that adjoins streams can usually have access to the water but must not alter the water or interfere with the rights of other property owners. State laws vary, and aquafarmers should become familiar with those of their own states. In areas where aquaculture is important, attorneys are developing legal competence in the matters pertaining to aquafarming.

Figure 4–8. Water disposal may require careful planning by intensive tank production systems, such as the one shown here. (Courtesy, Chore-Time Equipment)

Discharging water from aquaculture facilities is also a matter of regulation. Water from aquacrop facilities often contains feed particles, feces, and other wastes. Sometimes fish may be released with it. The release of some species of fish is illegal even if not intended, as would be the case when fish escape in waste water. Water that changes the nature of the water in a stream cannot usually be released directly into it. Such water would need to be treated or held in a reservoir where it could be made suitable for release. Most states either have laws regarding effluent water or will have them soon.

The disposal of wastes from processing plants requires careful planning. A landfill may need to be constructed according to legal requirements. Both processing plants and farmers must properly dispose of fish wastes, dead fish, or other animals.

Construction of earthen facilities in certain locations is regulated. Altering the banks of streams or shores of lakes, constructing levees on land designated as "wetlands," and altering floodplains require a permit from the U.S. Army Corps of Engineers.

Aquaculture must be carried out so as not to threaten endangered species of other plants or animals. A common problem is predatory water birds that consume hundreds of pounds of fish. For example, cormorants are protected birds. They can consume 1 or 2 pounds of fish a day. Federal law protects them from being killed. However, when they develop into particularly big nuisances, special permits can sometimes be obtained. Creative ways that are legal have been tried in an attempt to run them away.

Worker Health and Safety

Two federal laws shape the work environment in aquaculture. The first is the Fair Labor Standards Act which establishes minimum wage, overtime pay, equal pay, recordkeeping, and child labor standards affecting millions of workers in aquaculture and non-aquaculture occupations. All employees of aquabusinesses engaged in interstate commerce are included. If the only employees of an aquabusiness are the owner's immediate family members, that business is not covered. Very small businesses also are not covered. A few exemptions exist.

The Child Labor Laws provisions of the Act regarding the employment of children under the legal minimum age are specific. The Secretary of Labor determines which occupations are hazardous. Persons under the age of 18 cannot be employed in them; however, 16 has been established as the basic minimum age for employment. Examples of hazardous work include certain jobs in processing plants and operating motor vehicles in certain locations and power-driven machinery. The Act also establishes the number of hours that under-aged youths may work. In some occupations, it is possible for high school students to get a learners permit when they are enrolled in related agricultural education programs preparing them for a career in aquaculture.

The major regulatory agency on worker health and safety is the U.S. Department of Labor through the Occupational Safety and Health Administration (OSHA). The regulations of this agency are designed to promote a safe and healthy work environment. Areas likely to receive attention in aquaculture include repetitive motions by workers in processing plants, noise protection, eye protection, protective clothing, and built-in safety features in buildings and facilities. Many states have established regulations approved by OSHA.

A problem that has received considerable attention in recent years occurs when workers repeatedly perform the same physical movements, such as filleting a fish. This repetitive motion job-place health hazard is also known as carpal tunnel syndrome. It occurs when people quickly perform a single activity over and over, hour after hour, day after day. It can occur in any type of manufacturing or industry involving human repetition. People with the problem have nerve damage and begin to experience pain and tingling in their hands and may lose certain movements. Providing a variety of job tasks in the work setting can help minimize the risk of repetitive motion diseases. Knives with new designs have been introduced in an attempt to change the positions used by the workers when they hold and cut with a knife.

Licensing

Licensing requirements vary. In some cases, they depend on the size and type of aquaculture to be carried out. Retail markets must usually have a license to operate as a retail business. Fee-lake operators may be required to have permits for vending machines or other related sales at the lake site. Operating a retail market or processing facility usually requires a license. Certificates of inspection on sanitation may be needed in areas where food is processed. A license to produce an aquacrop may not be needed in a state where aquaculture has been defined as agriculture.

Product Transportation

Several areas of regulation apply to the transportation of aquacrops. These vary from one state to another. Hauling certain products across state lines may be prohibited. Health certificates may be needed to insure that a shipment of aquacrops is disease-free.

International shipments may be subjected to the laws of different countries. Permits to import goods may be required. In some cases, it is illegal to import certain species of fish.

All motor vehicles must be properly licensed to operate on public roads. Those that operate only within the confines of a farm may not need licensing, depending on state and local regulations. Certainly, the operators of all motor vehicles should be properly licensed. Boats and other marine equipment may need to be registered with the appropriate offices.

Individuals who ship aquaculture materials, such as eggs or fry, by common carrier need to consult with the carrier about the appropriate packaging material. Air freight may be used for tropical fish, high-value food fish, and other items. Certain regulations apply to these shipments.

Figure 4–9. Live fish shipped by air must be packaged to withstand extreme weather conditions while enroute, such as this snowy day at the Minneapolis, Minnesota, airport.

SUMMARY

Aquaculture requires people who know what they are doing and follow good management principles. In order to aquafarm, individuals must be able to obtain certain resources. In some cases, the kind of aquacrop is limited by the resources available. A good match between the requirements of the crop and the resources is essential.

The most important factor in any commercial aquaculture venture is the ability to make a profit. This requires careful consideration of a number of economic matters. Overcoming risks is made easier when good cultural techniques are followed.

Any aquaculture production must have a network of support in providing the supplies needed to produce the crop and the marketing capability to get the commodity from the farm to the consumer. In some cases, consumer demand will have to be developed. Certainly, no aquafarmer would want to violate the regulations that apply to aquaculture. The key is education!

QUESTIONS AND PROBLEMS FOR DISCUSSION

1. What is feasibility? Explain four areas of feasibility.
2. What are the ways of doing business in aquaculture? Briefly distinguish between them.
3. Name seven general requirements in aquaculture.
4. What is management? Why is it important to success in aquaculture?

5. Explain how an aquafarmer might go about making a decision.
6. What are some qualities of good aquafarm managers?
7. What major factors should a person consider in selecting land for an aquafarm? Briefly explain each.
8. Why is species selection important? What are some important considerations in species selection?
9. What costs are involved in aquaculture?
10. How is aquaculture financed? What are some sources of finances?
11. What is infrastructure in aquaculture? Why is it important? How does it develop?
12. What are the areas in which regulations exist for aquaculture? Briefly explain each.

Chapter 5

FUNDAMENTALS OF AQUACULTURE BIOLOGY

In order to raise aquacrops that are healthy and profitable, and that meet the needs of the buyer or consumer, the aquafarmer must first understand the different life processes of these aquacrops. Many different species have potential as aquacrops. Each species has its own biological makeup which determines how it is fed, cared for, reproduced, harvested, and marketed.

This chapter focuses on those biological principles that will help the aquafarmer in producing a vigorous, attractive aquacrop that will bring a suitable price.

OBJECTIVES

The following objectives should help the student better understand the genetic makeup of important aquacrops and their biological features. Upon completion, the student will be able to:

- Describe scientific classification systems.
- Classify important aquaculture species.
- Identify body systems of aquaculture species.
- Describe general habitat considerations of aquaculture species.

- Describe reproductive processes of aquaculture species.
- Describe nutrient requirements of aquaculture species.
- Discuss genetic potentials in aquaculture.

SCIENTIFIC CLASSIFICATION SYSTEMS

Whenever a number of items becomes too large to remember easily, people search out some orderly manner by which to classify the items. Over 1½ million different species of living organisms are known to exist. Therefore, scientists developed a system to classify animals, plants, and other living organisms.

A scientific classification, also called a *taxonomy*, is a system of arranging organisms into groups. The classification is based on the genetic makeup of the organism. Obviously, a pine tree and a catfish have little in common. But what about a catfish and an oyster? Scientific classification allows for the classification of living organisms in such a manner as to tell what they have in common and what is different about them. While the differences are usually noticed first, the likenesses show up more than the differences upon close inspection.

The early scientific classification systems classified all living organisms as either plants or animals. An organism was placed in either the animal kingdom (Animalia) or the plant kingdom (Plantae). Single-celled organisms such as bacteria, paramecia, and fungi were classified into one of these two kingdoms, depending on the scientist doing the classifying.

Modern scientific classification systems have four or five kingdoms. A widely accepted classification used today has five kingdoms: Monera, Protista, Fungi, Plantae, and Animalia. The Monera kingdom contains very primitive organisms—cells without membranes, such as bacteria and blue-green algae. The Protista kingdom contains single-celled and very primitive multi-celled organisms, such as slime molds, protozoa, and primitive fungi. The Fungi kingdom consists of true fungi (plants that do not produce their own food). The Plantae kingdom is composed of plants that produce their own food through photosynthesis. The Animalia kingdom is made up of multi-cellular animals.

This chapter focuses on the Plantae and Animalia kingdoms, although some organisms in the other kingdoms play an important role in aquaculture. Some single-celled organisms make up the primary food source for aquacrops. Others, most notably blue-green algae, play an important part in oxygenating the water in which aquacrops are grown.

Each organism is classified into seven principal areas. These are kingdom, phylum (division for plants), class, order, family, genus, and species. Most classifications are further differentiated by sub or super categories; for example, a subphylum would be between a phylum and a class, a superorder would come between class and order. These classifications may be different, based on the particular system used. For this book, the species are classified into the seven principal areas only.

Most species are referred to in books and reports based on their genus and species name only, along with their common name. This is their taxonomic name.

CLASSIFICATION OF IMPORTANT AQUACULTURE SPECIES

Some of the important species of aquacrops and those deemed to have potential as important species in the future are classified in Figures 5-1 through 5-7. Many that have the same kingdom, phylum, and order are grouped together. In parentheses, a brief phrase identifying the classification area is given where appropriate to help the reader understand the different classifications. When the same genus is referred to consecutively in the text, it is abbreviated to the first letter, as is common scientific practice. For example, the two trout mentioned below are *Salmo gairdneri* and *S. trutta (Salmo trutta)*. The species name is usually lowercase, while the genus and other classifications usually begin with a capital letter.

Fish

All of the freshwater, saltwater, and brackish water fish; ornamental fish; and baitfish that are important to aquaculture belong to the class Osteichthyes. Osteichthyes have skeletons with true bones, a skull with sutures, teeth usually fused to the bones if present, nasal openings on each side, premaxillae and maxillae, and a swim bladder or a functional lung.

Ictalurus punctatus (channel catfish) is the only important species in the order Siluriformes, which means catfish. The channel catfish and its culture are further discussed in Chapter 10.

The order Salmoniformes and the family Salmonidae, which means "salmon-like," contain several important aquaculture species. The trout *Salmo gairdneri* (rainbow trout) and *S. trutta* (brown trout) are classified in this family. Also, the various species of Pacific salmon are included here. These species include *Oncorhynchus tshawytscha* (chinook salmon), *O. kisutch* (coho salmon), *O. nerka* (sockeye salmon), *O. keta* (chum salmon), and *O. gorbuscha* (pink salmon).

The order Cypriniformes is a large and varied order of freshwater fish. The primary family, Cyprinidae, contains the baitfish, some of the ornamental fish, and one important food fish, the *Cyprinus carpio* (common carp). The baitfish species include *Pimephales promelas* (fathead minnows), *Carassius auratus* (goldfish), and *Notemigonus crysoleucas* (golden shiners). Many ornamental species are classified

Scientific Classification of Freshwater Food Fish

Kingdom — Animalia (animals)
Phylum — Chordata (with spinal cord)
Class — Osteichthyes (bony fish)

Channel Catfish:

Order — Siluriformes (catfish)
Family — Ictaluridae (freshwater catfish)
Genus — *Ictalurus* (catfish)
Species — *punctatus* (spotted or channel)

Rainbow Trout:

Order — Salmoniformes (maxilla in gape of mouth)
Family — Salmonidae (salmon-like)
Genus — *Salmo* (trout)
Species — *gairdneri* (rainbow trout)

Brown Trout:

Order — Salmoniformes
Family — Salmonidae
Genus — *Salmo*
Species — *trutta* (brown trout)

Tilapia:

Order — Perciformes (very diverse, scaly fish)
Family — Cichlidae (cichlids)
Genus — *Tilapia*
Species — *aurea* (blue)
mossambica (Java)

Hybrid Striped Bass:

Order — Perciformes
Family — Percichthyidae (temperate bass)
Genus — *Morone* (freshwater bass)
Species — *saxatilis* (striped bass, female)
chrysops (white bass, male)

Bream (sunfish):

Order — Perciformes
Family — Centrarchidae
Genus — *Lepomis* (sunfish)
Species — *microchirus* (bluegill)
cyanellus (green sunfish)
microlophus (redear sunfish)

Common Carp:

Order — Cypriniformes
Family — Cyprinidae
Genus — *Cyprinus* (common carp)
Species — *carpio*

Figure 5-1.

Scientific Classification of Baitfish

Kingdom — Animalia
Phylum — Chordata
Class — Osteichthyes

Fathead Minnow:

Order — Cypriniformes (protractile, toothless mouths)
Family — Cyprinidae (minnows or carp)
Genus — *Pimephales* (North American minnows)
Species — *promelas* (fatheads)

Goldfish:

Order — Cypriniformes
Family — Cyprinidae
Genus — *Carassius* (goldfish)
Species — *auratus*

Golden Shiner:

Order — Cypriniformes
Family — Cyprinidae
Genus — *Notemigonus*
Species — *crysoleucas*

Figure 5–2.

in this family, including the genera *Puntins, Barbodes,* and *Capoeta* (barbs); *Brachydanio* (danios); and *Rasbora* (rasboras).

The order Perciformes is the most diverse of all fish orders. It is the dominant order in saltwater and brackish water fish, and in some tropical areas is the dominant freshwater fish order, as well. The major freshwater fish of this order are tilapia, hybrid striped bass, and bream. Of the ornamental fish, gouramis and cichlids are from the order Perciformes. Several of the salt and brackish aquaculture species are from this order.

The two primary tilapia species raised in the U.S. are *Tilapia aurea* (blue) and *T. mossambica* (Java), both from the family Cichlidae. The culture of several other tilapia species and hybrids (for example, *T. nilotica*) is still experimental. The hybrid striped bass is a cross of two Perciformes species: usually a female *Morone saxatilis* (striped bass) and a male *M. chrysops* (white bass). Bream, usually cultured for stocking in ponds and lakes, include the *Lepomis microchirus* (bluegill), *L. cyanellus* (green sunfish), and *L. microlophus* (redear sunfish).

The ornamental gouramis classified in the order Perciformes include *Trichogaster trichopterus* (three-spot gouramis), *T. leeri* (pearl gouramis), and *Helostoma temminki* (kissing gouramis). The family Cichlidae includes numerous ornamental species from the genera *Pterophyllum* (angel fish), *Symphysodon*

Scientific Classification of Ornamental Fish

Kingdom — Animalia (animals)
Phylum — Chordata
Class — Osteichthyes

Guppies and Mollies:

Order — Cyprinidontiformes
Family — Cyprinidontidae (killifish or toothcarp)
Genus — *Poecilia*
Species — *reticulata* (guppies)
latipinna (sailfin mollies)
sphenops (common mollies)

Swordtails and Platy:

Order — Cyprinidontiformes
Family — Cyprinidontidae
Genus — *Xiphophorus*
Species — *helleri* (swordtails)
maculatus (platy)

Barbs:

Order — Cypriniformes
Family — Cyprinidae
Genus — *Puntins*
Barbodes
Capoeta
Species — (numerous living species)

Danios:

Order — Cypriniformes
Family — Cyprinidae
Genus — *Brachydanio* (danios)
Species — *rerio* (zebra danios)
albolineatus (pearl danios)

Rasboras:

Order — Cypriniformes
Family — Cyprinidae
Genus — *Rasbora*
Species — *heteromorpha* (harlequin fish)
trilineata (scissortails)
einthoveni (brilliant rasboras)

Tetras:

Order — Characiformes (small, extremely colorful)
Family — Characidae (characins)
Genus/Species — *Hyphessobrycon innesi* (neon tetras)
Cheirodon axelrodi (cardinal tetras)
Hemigrammus (several living species)

Figure 5–3.

Cichlids:

Order — Perciformes
Family — Cichlidae
Genus — *Pterophyllum* (angel fish)
Species — *scalare* (angel fish)
Genus — *Symphysodon* (discus fish)
Species — (several living species)
Genus — *Nannacara* (dwarf cichlids)
Species — *anomala* (golden drawf cichlids)
Genus — *Apistogramma* (drawf cichlids)
Species — *agassizi* (Agassizi's drawf cichlids)
rameriz (butterfly cichlids)
Genus — *Pelvichachromis* (kribensis)
Species — *pulcher* (kribensis)
taeniatus (striped kribensis)
Genus — *Cichlasoma* ("normal" cichlids)
Species — *meeki* (firemouths)
festivum (flag cichlids)

Top Minnows:

Order — Cyprinidontiformes
Family — Cyprinidontidae
Genus — *Aphyosemion*
Species — *ahli* (Ahl's lyretails)
australe (lyretail panchax)
Genus — *Aplocheilichthys*
Species — *lineatus* (striped panchax)
Genus — *Epiplatys*
Species — *dageti* (red-chinned panchax)

Gouramis:

Order — Perciformes
Family — Belontiidae (gouramis)
Genus — *Trichogaster*
Species — *trichopterus* (three-spot gouramis)
leeri (pearl gouramis)
Family — Helostomatidae (kissing gouramis)
Genus — *Helostoma*
Species — *temmincki*

Figure 5–3 (continued).

(discus fish), *Nannacara* (dwarf cichlids), *Apistogramma* (also dwarf cichlids), *Pelvichachromis* (kribensis), and *Cichlasoma* ("normal" cichlids).

Several saltwater and brackish water fish are Perciformes. The most important aquaculture species are *Mugil cephalus* (striped mullets), *Micropogon undulatus* (Atlantic croakers), *Sciaenops ocellata* (red drums or redfish), and *Trachinotus carolinus* (pompanos).

Scientific Classification of Saltwater and Brackish Water Fish

Kingdom — Animalia (animals)
Phylum — Chordata (with spinal cord)
Class — Osteichthyes (bony fish)

Striped Mullets:

Order — Perciformes
Family — Mugiladae
Genus — *Mugil*
Species — *cephalus*

Pacific Salmon:

Order — Salmoniformes
Family — Salmonidae
Genus — *Oncorhynchus* (Pacific salmon)
Species — *tshawytscha* (chinook salmon)
kisutch (coho salmon)
nerka (sockeye salmon)
keta (chum salmon)
gorbuscha (pink salmon)

Milkfish:

Order — Gonorynchiformes (swim bladder)
Family — Chanidae (milkfish)
Genus — *Chanos*
Species — *chanos*

Atlantic Croakers:

Order — Perciformes
Family — Sciaenidae (drums, croakers)
Genus — *Micropogon*
Species — *undulatus*

Red Drums:

Order — Perciformes
Family — Sciaenidae (drums)
Genus — *Sciaenops*
Species — *ocellata* (red drums)

Pompanos:

Order — Perciformes
Family — Carangidae (jacks and pompanos)
Genus — *Trachinotus* (pompanos)
Species — *carolinus*

Figure 5–4.

Several species of ornamental fish are classified in the order Cyprinodontiformes and family Cyprinodontidae, also called killifish or toothcarp. These species include the guppies and mollies from the genus *Poecillia*, the *Xiphophorus helleri* (swordtails), and *X. maculatus* (platy). Top minnows from the genera *Aphyosemion, Aplocheilichthys,* and *Epiplatys* also come from this order and family.

The order Characiformes and family Characidae contain several species of tetras, ornamental fish. These species include *Paracheirodon innesi* (neon tetras), *Cheirodon axelrodi* (cardinal tetras), and several species from the genus *Hemigrammus*.

The only important aquaculture species from the Gonorynchiformes order is the *Chanos chanos* (milkfish), a brackish water food fish cultured widely in Southeast Asia, but with much potential for Hawaii and California.

Frogs

The only amphibian species with potential for culture in the U.S. is *Rana catesbiana* (bullfrogs). Experimental bullfrog culture has been tried because the food product (frogs legs) is a very expensive item and because there is a demand for frogs for use as experimental laboratory animals. Most experiments have shown the risks of disease and predators to be great. These factors, along with the relatively low meat production per amount of water area and the frogs' erratic reproduction cycles, keep raising bullfrogs from being more popular. The best method of culturing the bullfrog seems to be in a polyculture with one or more aquaculture species.

Decapods

Several species of decapod crustaceans are important to aquaculture or have potential importance to aquaculture. Examples of decapods include the lobster, shrimp, blue crab, crawfish, and freshwater prawn.

Homarus americanus (lobsters) and the shrimp species, *Penaeus aztecus* (brown shrimp), *P. setiferus* (white shrimp), and *P. duorarum* (pink shrimp), are the most important saltwater and brackish water decapod species. These species are discussed further in Chapter 11.

Of freshwater crustaceans, only *Procambarus clarkii* (red crawfish) and *P. blandingi* (white crawfish) are cultured widely. Crawfish culture is presented in Chapter 10.

The culture of two decapods, the *Callinectes sapidus* (blue crabs), a brackish water species, and *Macrobrachium rosenbergii* (freshwater prawns) is still in the

Scientific Classification of Other Freshwater Animal Aquacrops

Kingdom — Animalia

Crawfish:

Phylum — Arthropoda (jointed legs, hard exoskeleton)
Class — Crustacea (two pairs of antennae, maxillae, one pair of mandibles)
Order — Decapoda (10 legs)
Family — Cambaridae
Genus — *Procambarus* (crawfish)
Species — *clarkii* (red)
blandingi (white)

Freshwater Prawns:

Phylum — Arthropoda
Class — Crustacea
Order — Decapoda
Family — Atyidae
Genus — *Macrobrachium*
Species — *rosenbergii*

Bullfrogs:

Phylum — Chordata
Class — Amphibia
Order — Anura
Family — Ranidae
Genus — *Rana*
Species — *catesbiana*

Figure 5–5.

experimental stages. Both of these species show much potential, as techniques for producing them continue to improve.

Oysters

The *Crassostrea virginica* (oyster) is the important aquaculture species in the Pelecypoda class of the phylum Mollusca. The oyster is cultured all over the world and in most coastal waters of the U.S. Its culture is further described in Chapter 11.

Gastropods

In California, the *Haliotis rufescens* (red abalone) is cultured, the only gastro-

Scientific Classification of Other Saltwater and Brackish Water Aquacrops

Kingdom — Animalia (animals)

Oysters:

Phylum — Mollusca (molluscs)
Class — Pelecypoda (bivalve molluscs)
Order — Isodontida
Family —Ostreidae
Genus — *Crassostrea* (oysters)
Species — *virginica* (American oysters)

Lobsters

Phylum — Arthropoda (jointed legs, hard exoskeleton)
Class — Crustacea
Order — Decapoda
Family — Homaridae
Genus — *Homarus*
Species — *americanus*

Shrimp:

Phylum — Arthropoda
Class — Crustacea
Order — Decapoda
Family — Penaeidae
Genus — *Penaeus* (shrimp)
Species — *aztecus* (brown shrimp)
setiferus (white shrimp)
duorarum (pink shrimp)

Blue Crab:

Phylum — Arthropoda
Class — Crustacea
Order — Decapoda
Family — Geryonidae
Genus — *Callinectes*
Species — *sapidus*

Red Abalone:

Phylum — Mollusca
Class — Gastropoda
Order — Archaeogastropoda
Family — Haliotidae (abalone)
Genus — *Haliotis*
Species — *rufescens* (red abalone)

Figure 5–6.

Scientific Classification of Food Plants

Kingdom — Plantae

Red Algae:

Division — Rhodophyta
Class — Rhodophyceae
Order — Bangiales
Family — Bangiaceae
Genus — *Porphyra*
Species — *angusta*
kuniedai
pseudolinealis
tenera
yezoensis

Green Algae:

Division — Chloraphyta
Class — Chlorophycea
Order — Ulvales
Family — Monostromataceae
Genus — *Monostroma*
Enteromorpha

Brown Algae:

Division — Phaephyta
Class — Phaeophyceae
Order — Laminariales (kelp)
Family — Alariaceae (Japanese kelp)
Genus — *Undaria*
Species — *pinnatifida*
Family — Lessoniaceae (giant kelp)
Genus — *Macrocystis*
Species — *pyrifera*
integrifolia

Watercress:

Division — Anthophyta
Class — Magnoliopsida
Order — Cruciferales
Family — Cruciferae
Genus — *Nasturtium*
Species — *officinale*

Chinese Waterchestnuts:

Division — Anthophyta
Class — Magnoliopsida
Order — Myrtales
Family — Trapaceae
Genus — *Eleocharis*
Species — *dulcis*

Figure 5–7.

pod culture in the U.S. of any consequence. This culture is further presented in Chapter 11.

Plants

Several species of algae, sometimes called seaweed, are cultured around the world, although not very much is done in the U.S. Most seaweed, which is viewed as a health food, is not expected to become a staple of most American diets. The three primary types of algae are classified into three divisions: several species in the genus *Porphyra* (red algae) in the division Rhodophyta; *Monostroma enteromorpha* (green algae), the primary species in the division Chlorophyta; and *Undaria pinnatifida*, *Macrocystis pyrifera*, and *M. integrifolia* (brown algae, also called kelp) in the division Phaephyta. Probably the most widely harvested kelp in the U.S. is *Macrocystis pyrifera* and *M. integrifolia*, two species of giant kelp.

Two freshwater aquatic plants cultured to some extent in the U.S. are *Eleocharis dulcis* (Chinese waterchestnuts) and *Nasturtium officinale* (watercress). Both of these, which are expensive foods for humans, are not widely cultured.

Blue-Green Algae

An important species in aquaculture is blue-green algae, which plays the significant role in freshwater ponds of oxygenating the water as it undergoes photosynthesis. Blue-green algae is usually classified in the Monera kingdom and Cyanophyta division. About 1,800 living species of blue-green algae exist. Five of the most common genera are *Gloeocapsa*, *Microcystis*, *Oscillatoria*, *Nostoc*, and *Scytonema*.

BODY SYSTEMS OF AQUACULTURE SPECIES

As demonstrated in the previous section of this chapter, many species can be identified as aquaculture species. In this section, the body systems of the more common types of aquaculture species are presented. Some general terms are given to help in identifying the body types of different animal aquacrops. (Plant systems are discussed further in Chapter 14.)

Fish

Almost all fish are bilaterally symmetrical; that is, they could be divided into

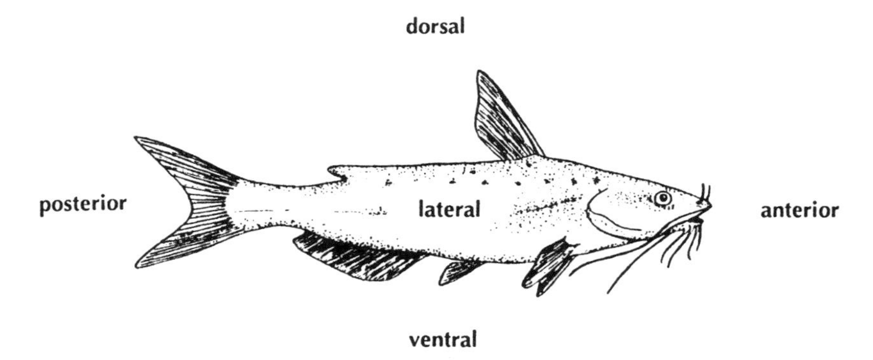

Figure 5–8. Location of common sides of fish.

two mirror-identical halves. Anterior refers to the front of the fish (the end with the mouth and head). Posterior refers to the tail-end of the fish. The belly plane of the fish is ventral. The opposite of ventral or back plane is dorsal. The side of the fish is lateral.

Bodies of fish can be divided into three parts: head, trunk, and tail. The head begins at the nose and ends at the posterior end of the operculum (gill cover). The snout is located between the anterior end of the premaxillary (outer jaw bones) to the eye. The nares, or nostrils, are located on the snout, usually one on each side. The dorsal part of the head is called the nape. The ventral part is called the thorax.

The trunk begins at the posterior end of the operculum and extends to the anus. The humeral area is the side of the trunk. The trunk contains the dorsal fin and a pair of pectoral fins and pelvic fins. The length of the trunk varies widely in the different fish. In the channel catfish, the trunk is very short when compared to that of the rainbow trout, for example.

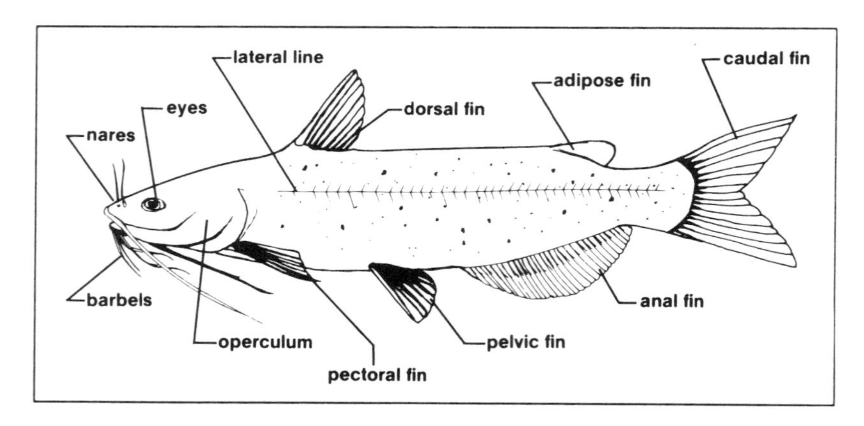

Figure 5–9. External parts of an *Ictalurus punctatus*. (Source: T. Wellborn, *Channel Catfish: Life History and Biology*, Southern Regional Aquaculture Center, Publication No. 180)

The tail extends from the anus to the posterior tip of the caudal fin. The tail contains another fin, the anal fin, which is located just posterior to the anus. The tail may also have an adipose fin, a fleshy fin found on salmon, trout, and catfish.

The fins on a fish are supported by soft rays and sometimes by spines. The spines, if present, will be located anterior to the soft rays of the fin.

The lateral line, visible on the trunk and tail, is a series of pores on a linear series of scales. On fish without scales, the lateral line shows up as a small indention running the length of the trunk and tail. The lateral line is part of the sensory system of the fish.

Nervous system. The nervous system of fish has two major parts: the central nervous system and the peripheral nervous system. The central nervous system contains the brain and the spinal cord. The peripheral nervous system contains two types of paired nerves. Nerves that branch from the spinal cord are referred to as spinal nerves, while nerves that branch from the brain are called cranial nerves.

Circulatory system. Fish have a heart and various blood vessels similar to other higher animals. Both white and red blood cells are present in the system. Arteries carry blood from the heart. The primary arteries in fish are the dorsal aorta, ventral aorta, cartoid artery, and coelicomesenteric artery. Veins carry blood to the heart. The portal system transports blood to the liver, where it is then transported to the heart through the hepatic vein. Other primary veins are the portal vein, the caudal vein, and the anterior cardinals.

Sensory system. The sensory system of fish consists of the eyes, ears, nares (with olfactory sacs), and skin (and its appendages). The eyes will vary, depending on the type of fish. Fish that must see their feed, such as trout, have larger eyes. Fish that find food by their sense of smell, such as catfish, have smaller eyes. Most fish do not have external ears but sense vibrations through the ear bones in their heads. The sense of smell is accomplished through the olfactory sacs, which open through the nares, usually located between the eyes and the snout.

The skin and appendages, such as barbels and cirri, give fish a sense of touch. The skin secretes a substance that makes the fish slimy, or slippery. Also, on some fish, the scales are developed from the embryonic cells of the dermis, the inner layer of the skin. The outer layer of the skin is called the epidermis.

Skeletal system. The skeletal system of fish consists of two major parts: the axial skeleton and the appendicular skeleton. The axial skeleton contains the skull — the cranium, which protects the brain, and the visceral skeleton, which is made up of the jaws and gill arches. The remainder of the axial skeleton is composed of the vertebrae and ribs.

The appendicular skeleton consists of the girdles of the paired fins and the bones supporting the fins.

Muscular system. Fish move through the water by undulating their bodies. The muscular system is composed of a series of *myotomes,* or muscle segments, arranged in the shape of a W lying on its side. The myotomes are separated by connective tissues called *myosepta.*

Respiratory system. In fish, the primary organs that oxygenate the blood and exchange carbon dioxide are the gills. Gills are blood-filled membranes located under the operculum. The lamellae have spaces through which blood moves

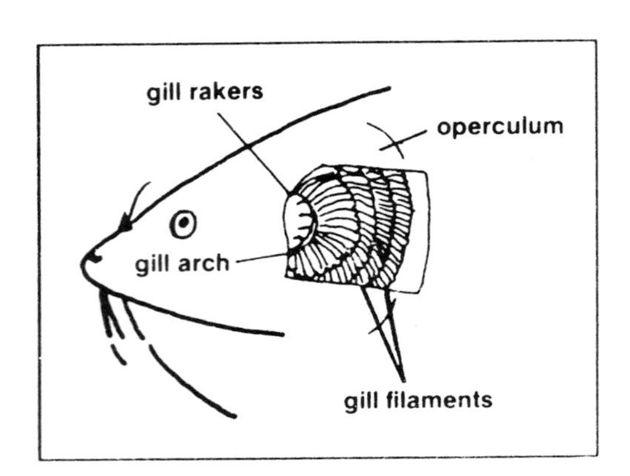

Figure 5–10. Location of gills on a fish. (Source: T. Wellborn, *Channel Catfish: Life History and Biology,* Southern Regional Aquaculture Center Publication No. 180)

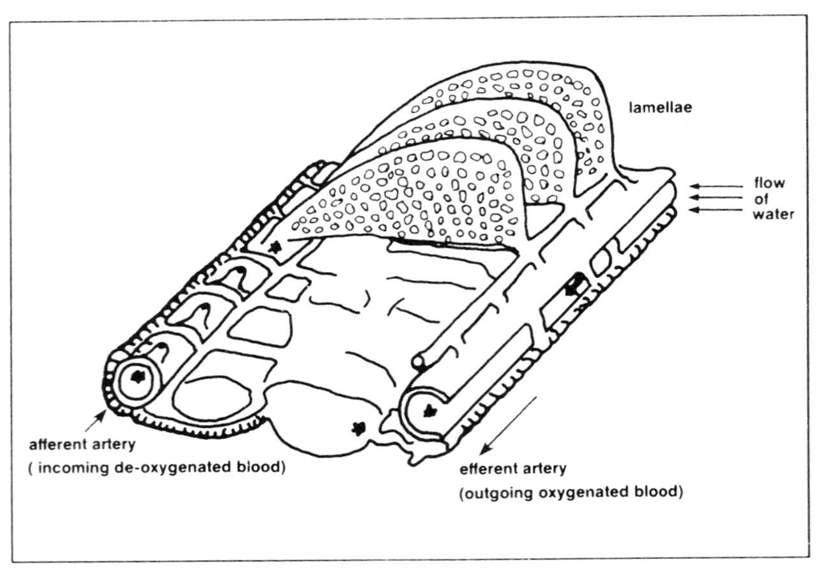

Figure 5–11. Parts of a gill filament. (Source: T. Wellborn, *Channel Catfish: Life History and Biology,* Southern Regional Aquaculture Center Publication No. 180)

rapidly. The afferent artery carries blood to the gills, and the efferent artery carries oxygenated blood back to the heart.

As the fish swims, water is forced through the mouth and passes over the gills, where the oxygen – carbon dioxide transfer occurs. The oxygenated blood is then transferred to the heart and the rest of the body to continue the cycle.

Digestive system. The digestive system of fish is composed of the mouth, esophagus, stomach, intestine, and anus. The food moves through the mouth and esophagus to the stomach, where most of the digestion occurs. The digestive process is completed in the intestine and wastes are excreted by the anus.

The size and shape of the parts will vary according to the type of fish and its natural diet. For example, the catfish, a bottom-feeder in its natural habitat, has an inferior mouth, while the rainbow trout has a terminal, or normal mouth. Fish that are strictly surface feeders usually have a superior mouth. Fish that feed on algae and microorganisms will have a small stomach and a long intestine. Carnivorous fish (those that eat other animals) will have a large stomach and a short intestine.

Decapod Crustaceans

Decapod crustaceans (decapods) are part of the same order as insects, and their body parts may be broken down in much the same way. The three primary body parts of crustaceans are the head, carapace (similar to the thorax in insects), and abdomen.

The head of a decapod contains the antenna, the eyes, and the maxilliped. These parts may look quite different, depending on the species.

The carapace consists of the walking legs and the chelipeds (clawed appendages).

The abdomen is the body part that has most of the "meat" of a crustacean. The swimmerets (short legs that help in swimming) are part of the abdomen. The "tail" of a decapod is made up of the uropods and telson.

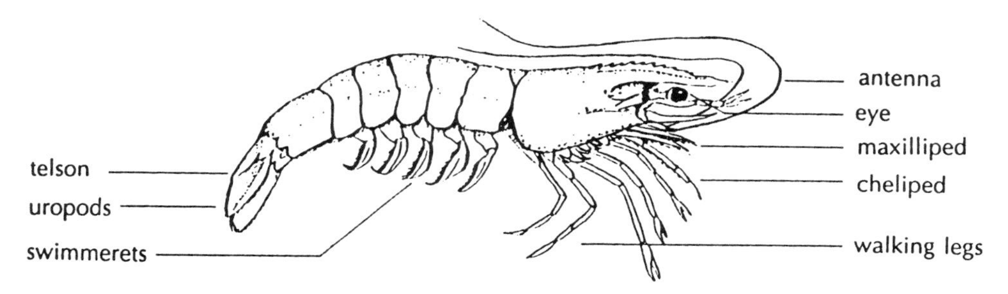

Figure 5–12. External parts of a shrimp.

Nervous system. The decapod's nervous system is similar to that of vertebrates, but not nearly as complex. Decapods have a central nervous system with a brain (or at least a ganglionic mass) and a ventral nerve cord. Several ganglia (masses of nerves) branch from the ventral nerve cord to serve the major appendages.

Sensory system. Decapods are very responsive to light and to movement. The primary sense organs include the eyes, antennae, and tactile hairs. The antennae and the tactile hairs are responsive to the tactile stimuli of contact with other organisms or movement of water. Lobsters and crabs have very good vision, which help them find prey.

Circulatory system. Decapods have an open circulatory system with a single-chambered heart. The blood or hemolymph flows freely through the hemocoelic cavity. The heart pumps the blood, but circulation also occurs as the result of movement and gut contractions. The decapods have arteries to carry blood away from the heart, but they lack veins. Blood returns to the hemocoelic cavity through passages called sinuses.

Skeletal system. All decapods have an exoskeleton, which means that the skeleton is outside the body. The exoskeleton is made of a chitin-protein material that is secreted by the epidermis. The epidermis is a single layer of skin cells.

Decapods such as the lobster, blue crab, shrimp, freshwater prawn, and crawfish all grow by the process of molting. *Molting* refers to the animals' shedding their outer shell (exoskeleton) and increasing in size as they develop a new one.

Figure 5–13. The *Procambarus clarkii* is an example of a decapod crustacean. Note the hard, outer exoskeleton.

Muscular system. Crustaceans have a well-developed system of muscles for various types of locomotion. Decapods have evolved so that special appendages, all the same on lower crustaceans, have specialized functions and muscles to control those functions. The cheliped, the large claw on a lobster, for example, is used for protection and for tearing some foods. The appendages on the abdomen, the swimmerets, are obviously used for swimming. Between these, the walking legs have developed for terrestrial (land) movement.

Respiratory system. When in water, decapods breathe through their gills. The water may be moving from front to back or back to front. The decapods have appropriate openings for either case. In shrimp, the carapace has a negative pressure to draw the water in. Respiration also occurs through cell diffusion in the exoskeleton.

Digestive system. The basic digestive system of decapods consists of a foregut, midgut, and hindgut. The foregut varies, depending on the diet of the particular species. It may be a simple esophagus or a series of straining or filtering mechanisms. The midgut normally contains various outpockets referred to as ceca, midgut glands, and diverticula. The hindgut is usually very short and similar to an intestine. It ends at the anus.

Molluscs

Although clams and mussels are cultured in some areas, the most important mollusc to aquaculture is the oyster *Crassostrea virginica*. The oyster is a bivalve mollusc. This means it has two sides of a shell hinged together by a ligament. The ligament is located at the pointed end of the shell called the beak, or anterior region. The rounded posterior end contains an oval-shaped, pigmented (colored) area called the muscle scar. The muscle scar is from the adductor muscle. The oyster uses the adductor muscle to pull the shell closed when it senses danger.

The adductor muscle must be cut before the shell can be opened to expose the inner parts of the oyster. This is usually done with a specialized oyster knife.

With the shell opened, the oyster will look mostly like a mass of flesh. Closer inspection, however, will reveal the gill, mouth, stomach, intestine, gonad, and anus.

Skeletal system. The shell is created by the mantle, a double fold of the body wall which lines the surfaces of the valves. While using elements from the water to create the shell, the mantle also protects the inner organs.

The shell consists of three layers: the periostracum, the prismatic layer, and the nacreous layer. The periostracum is made of an organic material called *conchiolin*. The periostracum protects the rest of the shell from the erosion caused by sand and water. Other organisms are often found attached to this outer layer. The middle layer, or prismatic layer, is made of alternating layers of calcium carbonate and conchiolin.

The inner layer, or nacreous layer, is commonly known as mother-of-pearl. It is composed of plates of calcium carbonate. A pearl is formed when the oyster secretes several layers of nacre around a foreign object, such as a grain of sand. The oyster's purpose in creating a pearl is to prevent the soft inner parts from becoming irritated. The *Crassostrea virginica* produces poor-quality pearls and thus is not usually cultivated for this purpose. Asian oysters produce most of the pearls that have value as jewelry.

Nervous system. Oysters do not have a true brain. Instead they have two masses of nerves called ganglia. Small nerves radiate from the ganglia to all parts of the oysters.

Sensory system. Oysters have no need for eyes or ears since they do not move around. They can detect changes in light however and will close their shells when a shadow falls over them. The most notable evidence of senses in oysters is the formation of a pearl when a foreign substance enters the shell.

Circulatory system. Oysters have a three-chambered heart located just above the adductor muscle. Blood is forced out the ventricle to the gills for oxygenation. Some blood returns to the heart from the mantle, while the rest is purified in the very small kidney before returning to the heart. Oyster blood has a bluish tint because of the copper content.

Muscular system. Oysters do not have much of a muscular system. Because oysters are sessile animals (they do not move around in search of food), they have no need for a well-developed muscular system. The primary muscle of the oyster is the adductor muscle, which, as mentioned earlier, keeps the shell closed when necessary.

Respiratory system. The gills accomplish respiration in oysters just as in fish. Oxygen (O_2) diffuses from the water flowing across the gills into the bloodstream, to be carried to the rest of the body. At the same time, carbon dioxide (CO_2) is released into the environment. Some oxygen - carbon dioxide transfer is accomplished in the mantle.

Digestive system. Oysters are filter feeders. The mantle filters out large particles and allows microorganisms to pass through. The organisms, usually single-celled, are often collectively referred to as plankton. The plankton particles are trapped in mucus from the gills and moved to the mouth.

The mouth accepts the food unless there is too much or the particles are too large. Then the excess is rejected to a cavity which is emptied when the shell closes.

The food particles move from the mouth through a short esophagus to the stomach, where nutrients are absorbed. The fecal material moves through the intestine and out the anus. The anus empties into the cavity where excess food is rejected by the mouth.

GENERAL HABITAT CONSIDERATIONS

Fish

The habitat requirements for fish vary greatly, depending on the species. The general requirements can be grouped into the categories of dissolved oxygen (DO), temperature, movement, pH, and salinity.

Although catfish are usually found in streams and rivers in the wild, they adapt well to pond culture. They are freshwater fish primarily, but can withstand some salinity, maybe even up to 20 ppt. Catfish will live for a short time in water with a dissolved oxygen content as low as one part per million (ppm) but need about 4 ppm to grow. Although they can withstand the cold temperatures of the northern U.S., they will grow much faster in warmer water. They often stop feeding when the water temperature gets too cold.

Trout are cold water fish and perform best in running water, although they can adapt to ponds. They must have running water for spawning. Trout also need a high level of DO to survive, at least 5 ppm.

Tilapia require warm water — most species will die if the water temperature gets below 50°F. The tilapia can survive a DO of 1 ppm. They need water with a fairly high pH.

Bream require shallow, weedy waters and prefer to gather around stumps, logs, and brush. The water should be freshwater, not saltwater. Bream can withstand very cold to very hot temperatures, but they grow better in warmer water.

Hybrid striped bass grow well in a wide range of water quality variables. They survive in both warm and cold water and can withstand a pH range of 5.5 - 10.0. Like catfish, they can survive DO levels of 1 ppm for a short period, but they need water with a DO of 5 ppm or higher to grow well.

Baitfish usually are found in small, weedy ponds but adapt well to pond and tank culture. Fathead minnows can withstand salinity of up to 10 ppt and wide ranges of pH.

Most saltwater fish have very few habitat requirements. They adapt well to the various conditions found at sea.

Figure 5–14. One fish that requires warm water is the *Tilapia aurea*. It will die if the water gets too cold—around 50°F.

Striped mullets require warm water and are usually found in tropical and sub-tropical areas. Striped mullets are usually found in bays, marshes, rivers, or the open sea, but they can be grown in ponds. They are very tolerant of changes in water salinity, surviving in water anywhere from 0 to 38 ppt.

Decapods

Crawfish are found naturally in shallow, weedy swamps, ponds, and ditches. They also adapt well to polyculture in rice fields. Crawfish are very sensitive to chemicals and prefer hard water.

Blue crabs require saltwater or brackish water, but the percentage of salinity varies. Males tend to stay in low salinity water, while females migrate to higher salinity areas to hatch their young. The larval crabs, called *zoeae*, cannot survive in water with a salinity less that 2 ppt. Freshwater prawns also need some salinity for the larval stages to survive.

Lobsters and shrimp have few habitat considerations. Shrimp do come into brackish water or nearly freshwater of a relatively low salinity to spawn.

Oysters

The primary habitat consideration for oysters is that they have at least a small water current flowing through them to provide oxygen and food and to carry away wastes. Water that is silty will clog the food transport system and slow the growth and respiration processes.

REPRODUCTIVE PROCESSES

Fish

All of the important aquaculture fish reproduce by spawning. The female lays her eggs and the male deposits sperm over the eggs. After fertilization, one of them usually guards the eggs until they hatch. This is not always the female. In catfish, for example, the male guards the eggs. In some species of tilapia, the female holds the fertilized eggs in her mouth until they are hatched.

Fish vary widely in their preferred place of spawning. Some spawn on the bottom of a pond. Some, such as trout, require running water to spawn. Catfish need a cover and in the wild usually spawn in a submerged hollow log. In ponds, the catfish used for reproduction usually spawn in artificial barrels or boxes. If the preferred method of spawning is not available, many fish simply will not spawn.

The tilapia is an example of a tremendously reproductive species. This causes some problems for the aquafarmer in that tilapia can reproduce before they reach harvesting size. As a result, much of the energy that the aquafarmer wants used

Figure 5–15. Discarded ammunition boxes are commonly used to spawn catfish. In farm ponds, the ammunition boxes are the closest thing to their natural spawning habitat, a hollow log.

Figure 5–16. These workers are collecting egg masses from where catfish have spawned. The egg masses will be taken to a hatchery. (Courtesy, Gary Fornshell, Mississippi State University)

in growth is used in reproduction. One way aquafarmers have found to reduce this problem is to place a net that keeps tilapia from the bottom so they cannot spawn.

Some fish such as salmon live most of their lives in saltwater but return to freshwater to spawn. These fish are called *anadromous.* Fish that spend most of their lives in freshwater but spawn in saltwater are called *catadromous.*

Many cultured fish are reproduced in hatcheries. With catfish, the fertilized egg mass is removed from the tank or barrel and placed in the hatchery. In some species, the eggs are removed from the female and the sperm are removed from the male, with the fertilization being done artificially. Trout used in aquaculture are usually spawned this way.

Decapods

The female secretes a substance (sex pheromone) that alerts the males that she is receptive to breeding. Immediately after a molt, the breeding occurs. Most decapods breed by copulation. The male places the sperm in or near the female's seminal receptacle. Depending on the species, the male sometimes remains a few days to protect the female from predators and other males. The female holds the fertilized eggs until they hatch.

Most decapods must reach certain sizes before they become sexually mature. A male lobster, for example, must have a carapace of about 40 – 45 millimeters long before he can produce sperm, but he may not be able to mate with a mature female until he is about 70 millimeters in carapace length.

Oysters

Crassostrea virginica are protandrous hermaphrodites. This means that the same oyster may be male or female at different times. Although not always, the oyster is usually male, producing sperm its first spawning season or two. After this period, the oyster will usually become female and produce eggs for the rest of its life. Although very uncommon, some oysters have been known to produce both sperm and egg cells at the same time.

Sperm produced by males and eggs produced by females are released into the mantle cavity, where the combination of water movement and cilia action moves them out into the water. Because fertilization and development of the offspring take place in the water, *Crassostrea virginica* are called nonincubatory oysters. In incubatory oysters, the females hold the eggs in their mantle cavity until sperm enter to fertilize them.

Each oyster produces millions of sperm or eggs, but the survival rate is low. During the spawn, the gonad becomes so enlarged that the rest of the inner parts of the oyster are extremely small. After the spawn, the oyster increases the energy spent in digestion to build back up its energy reserves.

NUTRIENT REQUIREMENTS

Fish

Numerous companies manufacture feeds suitable for feeding the various species of fish that have been discussed in this chapter. For the most important freshwater species, catfish and trout, several companies produce feeds mixed specifically for each. Some of the requirements for fish diets are discussed as follows:

Proteins. Proteins are formed by simple organic compounds called amino acids. Proteins are broken down by the digestion process so the fish can use them to form new tissue and to repair damaged tissue. Ten amino acids have been identified as essential amino acids; because fish cannot produce these amino acids, they must be included in their diet. Table 5-1 shows these essential amino acids.

Figure 5–17. When an aquacrop, such as catfish or trout, becomes established, companies will produce feed designed specifically for that aquacrop.

Protein is by far the most critical ingredient in a fish feed. Common prepared feeds range from 25 to 40% protein.

Proteins in prepared rations usually come from fish meal, but they may also come from other animal waste products such as blood meal or meat scraps. Some of the protein may come from plant sources such as soybean meal. Fish that are natural carnivores require a feed that derives at least 50% of its protein from animal products. Fish that are naturally herbivores or omnivores can get by on less, but still may need around 30% to come from animal products.

Table 5–1

Amino Acids Essential to Fish and the Minimum Percentage Required for Rainbow Trout

Amino Acid	Percentage Required
Arginine	2.5
Lysine	2.1
Isoleucine	1.5
Valine	1.5
Cystine	1.0
Leucine	1.0
Threonine	0.8
Histidine	0.7
Methionine	0.5
Tryptophan	0.2

Source: W. O. McLarney, *The Freshwater Aquaculture Book*. Point Roberts, Washington: Hartley & Marks, Inc., 1987, p. 156.

Fats. Fats are composed of fatty acids. In fish rations, fats usually come from animal wastes. Fish require fats to maintain proper health and growth. Any artificial feed ration for fish should contain supplemental fats.

The amount of fat in a diet usually depends on the temperature of the water in which the fish are cultured. This is because the melting point of the fat must be below the water temperature if the fat is to be digested completely. For this reason, feeds prepared for cold water fish such as trout usually contain unsaturated fats from vegetable oils. Feeds for warm water fish such as catfish and tilapia usually contain saturated animal fats.

The total amount of fat in the diet is usually between 4 and 15%.

Carbohydrates. Carbohydrates come from sugars, starches, and celluloses. Carbohydrates provide the fish with energy. Fish vary greatly in their ability to digest carbohydrates. Fish that eat mostly plants, such as carp, can digest a lot of carbohydrates because their bodies produce the enzyme amylase, which breaks down the carbohydrates. Fish that eat mostly animals, such as catfish and trout, usually need less than 10% carbohydrates in their diets.

Minerals. Several minerals are important to the health and growth of fish. Because many of the minerals are necessary in very small amounts but might be fatal in large amounts, they are often referred to as trace elements.

The following minerals have been identified as necessary for fish: calcium,

Figure 5–18. This young worker inspects feed he is about to distribute over several catfish ponds. The truck is designed specifically for feeding catfish in ponds. (Courtesy, Delta Pride Catfish, Inc.)

Figure 5–19. Several different types of feed may be used to feed fish, depending on the size and type of fish. The smaller granules are for fry, with feed sizes increasing as the fish get larger. (Courtesy, Louisiana Agricultural Experiment Station–Louisiana Cooperative Extension Service, Louisiana State University Agricultural Center)

iron, silicon, manganese, magnesium, boron, cobalt, copper, iodine, molybdenum, selenium, and sodium.

A knowledge of the characteristics of the water in which the fish are being grown is necessary to determine which minerals are deficient. A choice must usually be made between adding the minerals to the water as part of a fertilization program and adding the feed as a supplement. Adding the minerals is usually easier, but in raceways and other moving water systems, it may be better to add minerals as a feed supplement. Some commercially prepared feeds already contain sufficient quantities of some minerals.

Vitamins. Vitamin deficiencies can be very serious. Deficiencies of vitamins may cause poor growth, anemia, skin lesions, clubbed gills, and many other problems. Fortunately, most commercial feeds contain vitamin supplements.

The two types of vitamins important to fish are water-soluble vitamins and fat-soluble vitamins. Water-soluble vitamins are taken into the body, used, and excreted. Fat-soluble vitamins are stored in the body; therefore, an excess in the diet may cause problems just as a deficiency might. Table 5-2 lists the vitamins important to fish.

Table 5–2

Vitamins Essential to Fish and the Minimum Daily Requirement (MDR) for Rainbow Trout

Vitamin	MDR
Water-soluble vitamins	
Thiamine (B_1)	0.15–0.2 mg/kg of fish
Riboflavin (B_2)	0.5–1.0
Pyridoxine (B_6)	0.25–0.5
Pantothenic acid	1.0–2.0
Niacin or nicotinic acid	4.0–7.0
Biotin (H)	0.04–0.08
Folacin or folic acid	0.10–0.15
Cyanocobalamin (B_{12})	0.0002–0.0003
Ascorbic acid (C)	450–500
Inositol	18–20
Choline	50–60

(Continued)

Table 5-2 (Continued)

Vitamin	MDR
Fat-soluble vitamins	
K_3	15–20 mg/kg of fish
E	125 IU/kg of feed
A	8,000–10,000
D	1,000

Source: W. O. McLarney, *The Freshwater Aquaculture Book*. Point Roberts, Washington: Hartley & Marks, Inc., 1987, p. 158.

Decapod Crustaceans

The nutrient requirements of decapods vary greatly. All species require a source of phytoplankton or zooplankton as larvae. Adult crawfish are omnivorous but usually require some type of forage supplement such as wastes from a rice field or a commercially prepared feed when they are cultured. Shrimp are usually fed by fertilization of the pond, but commercially prepared chicken feed or shrimp feed is sometimes used as a supplement.

Oysters

Because they are filter feeders, oysters are not fed when cultured. Very little is known about their nutritional requirements. The primary concern with oysters is that they have plenty of phytoplankton and that the waters in which they are cultured be free from pollutants. Oysters are extremely susceptible to pollution; their bodies absorb the pollutants just as they absorb nutrients. Thus, pollutants are a primary concern to aquafarmers who produce oysters.

GENETIC POTENTIALS IN AQUACULTURE

Because most of the species used in aquaculture are no different genetically from their counterparts in the wild, the potential for genetic improvement seems almost unlimited. The primary source of this potential is breeding programs that select for traits important to the culture of the different species.

The cultured rainbow trout is the best example of the genetic potential of aquacrops. The rainbow trout that is commercially hatched and cultured in raceways has been genetically improved through breeding programs to accept commercially prepared feed better, to respond to handling better, to grow well in intensive production systems, and to reproduce well in captivity.

Other species will probably improve much like the rainbow trout, as breeding programs result in genetically superior strains that react better to common production techniques. Also, improved genetics through selective breeding programs may produce species that grow well in polyculture systems. One of the greatest potentials is to genetically produce species that require more or less salinity in the water or that can adjust better to changes in salinity so that the potential areas for production can be increased.

Another example of using genetics to improve aquaculture is the hybrid striped bass, a strain of fish produced by breeding two fish of different species.

Figure 5–20. A hybrid striped bass, a successful genetic improvement obtained by crossing a female *Morone saxatilis* and a male *M. chrysops*.

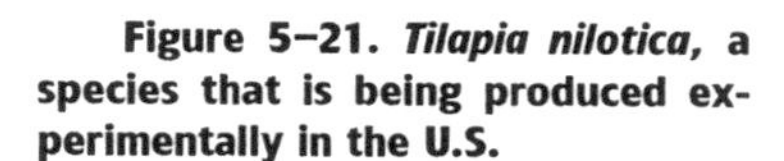

Figure 5–21. *Tilapia nilotica*, a species that is being produced experimentally in the U.S.

There may possibly be other crosses that will turn out just as well waiting to be accomplished.

Genetic breakthroughs may also help aquafarmers overcome some of the problems that often cost them profits. For example, off-flavor in catfish can cost producers a lot of money. If this problem could be eliminated genetically, the catfish producers' biggest problem would be overcome.

Geneticists have already developed a system of breeding tilapia so that a single sex of offspring can be produced. All the members would be male, for example. This would eliminate the problem of tilapia reproducing before they reach harvest size, as was mentioned earlier.

Genetic engineering that reduces the likelihood of disease is also an area where great potential exists. Finding individuals that are resistant to disease and then determining why they are resistant may make many diseases problems of the past.

Figure 5–22. An example of a genetic mutation, although not the result of human efforts, is the albino strain of *Ictalurus punctatus*, shown here with a normal catfish.

SUMMARY

Knowledge of the life processes is important to producers of aquacrops. Knowing the biological systems of the different species of aquacrops often allows the aquafarmer to make decisions concerning their habitat and nutrition.

When an aquafarmer selects a type of aquacrop to produce, the biology of the particular species often plays an important role. An aquafarmer without access to saltwater or brackish water cannot produce oysters. Northern climates are not acceptable for raising tilapia.

In order for an aquafarm to be productive, it must be adapted to the different biological requirements of the species being cultured.

QUESTIONS AND PROBLEMS FOR DISCUSSION

1. Why is a system of scientific classification important?
2. Using Figures 5-1 through 5-7, construct a list of the most important aquaculture species in your area according to their scientific classification. Why are some in different phyla, classes, orders, etc.? How are they alike? How are they different?
3. To which class do all of the important aquaculture fish belong? How many different orders were represented in the chapter? List the orders.
4. Find out if any important aquaculture plants are produced in your area (you may have to talk to your teacher or your local county agent). What is the scientific classification of each of these plants?
5. Explain what is meant by the following terms: dorsal, ventral, anterior, posterior, lateral.
6. What are the three major body parts of fish?
7. What are the three major body parts of a decapod?
8. Explain how an oyster makes a pearl.
9. In fish, what does the muscular system do?
10. In most aquatic animals, what purpose do the gills serve? How do they accomplish this purpose?
11. What does filter feeder mean?
12. What often determines the shape of a fish's mouth?
13. What is the most important habitat consideration of oysters? Why is this so important?
14. Explain the difference between *anadromous* and *catadromous*.
15. What happens to catfish and many other warm water fish when the water temperature gets too cold? What happens to tilapia?
16. Define *protandrous hermaphrodite*. Give an example of this type of animal.

17. What are the five major nutrients required in fish diets? Briefly discuss why each of these is important.
18. How are most genetic improvements accomplished? What are some other ways of accomplishing genetic improvements?

Chapter 6

THE IMPORTANCE OF WATER

Aquacrops live and grow in water. How well they grow depends on the quality of the water. An aquafarmer must realize the importance of an abundance of quality water. Decisions about aquaculture should involve information about the amount of water that is both needed and available. Aquacrops have preferences for certain "kinds" of water. Those that grow well in warm water will perform poorly in cold water and vice versa. Likewise, saltwater crops will do poorly in freshwater and vice versa. Even if the "right" water is available, the aquafarmer must know how to manage it. Getting, using, and disposing of water requires knowing what is involved.

OBJECTIVES

This chapter is intended to help students develop a general understanding of the role of water in aquaculture. The emphasis is on freshwater, with some discussion of saltwater. Upon completion, the student will be able to:

- Explain how water is important in aquaculture.
- Describe kinds of water facilities.
- Explain considerations in selecting a growing facility.

- Explain water quality.
- Identify sources of water.
- Assess suitability of water from various sources.
- Describe important water management practices.
- Calculate water volume.
- Explain how to control weeds and algae.
- Explain pollution in aquaculture.
- Explain how to add oxygen to water.
- Explain approved ways of disposing of water.

WATER IN AQUACULTURE

Water is the environment in which aquacrops grow. Crops have different water requirements. The "right" water must be available for the crops to thrive. To the beginner, it often appears that the earth has an abundance of water. A little investigation shows that this is not necessarily so!

Nearly three-fourths of the earth's surface is covered with water. Of this, some 97% is salty—and suitable only for those aquacrops that thrive in the saltwater of the oceans and lakes. Only about 3% of the water is considered fresh, with two-thirds of it being frozen in glaciers and ice caps. This leaves only about 1% of the earth's water available for use in many ways that support human and other life. Water for aquaculture, homes, industries, irrigation of crops, operation of power plants, and other uses comes from this 1%! With freshwater aquaculture, there is considerable competition for the available water. In areas where aquaculture has been underway for several years, scientists are beginning to notice a decline in the water available from the natural reservoirs. Underground supplies which were thought to be endless are beginning to dry up. Conservation efforts are underway to try to maintain desired levels of freshwater. Of course, aquacrops are not the only crops that use water. Rice and other crops that are irrigated use large amounts of water.

The challenge to the aquafarmer is to match the available water supplies with the proposed aquacrop and system used to produce it. Attention in recent years has focused on crops that require freshwater. Since there is a huge supply of saltwater, it appears that crops that thrive in saltwater could be grown in larger quantities, with less stress on the supply of essential freshwater.

WATER FACILITIES

Water facilities are the structures that contain the water in which aquacrops grow. Sometimes these are referred to as growing facilities. The enclosures which hold the water and / or aquacrop are impoundments. These allow the farmer to manipulate the environment of the aquacrop so that efficient growth occurs. A knowledge of water facilities helps the farmer to make good decisions about the kind of system to be used and the aquacrop to be produced.

Common kinds of impoundments are ponds, raceways, tanks, vats, and aquaria. In each of these, which are often constructed specifically for aquaculture, the farmer is able to control the volume of water as well as other features of the water environment. Pens and cages confine the aquacrop so that the farmer can exercise some control over the growing environment, such as the use of feed. In addition, pens and cages may be used in large lakes, oceans, and streams where it is impossible to control the water. Abandoned quarries where stone or minerals have been mined may also be suitable for cage aquaculture.

Figure 6–1. Alligator farming requires adequate water of the right kind. This adult brood "mama" alligator thrives in water that would be unsuited for other species. (Courtesy, MFC Services)

Figure 6–2. Pens are used in lakes to confine fish to an area of the pond, such as these pens used for broodfish production. (Pens and cages can be distinguished very simply: cages float on the surface of the water and pens are attached to the bottom of the lake.)

Ponds

Ponds are artificial impoundments, usually made of earth. Heavy bulldozers and other equipment construct earthen dams or levees that hold the water. The size may range from less than an acre to 50 or more acres. Research has found that ponds in the 20-acre size range may be best. This size is large enough to minimize construction costs and small enough to be easily harvested.

Shape is important. Rectangular ponds are often easier to manage than square- or irregular-shaped ponds. The bottoms of ponds should be smooth and free of holes. Ponds with uneven bottoms are difficult to harvest and get all of the fish in the seine.

Ponds built on level land are easier to harvest than those built on hilly land. It is easy to build a pond by placing a dam across a hollow between two hills, but harvesting is difficult.

Pond sites should have soil that will hold water. Soils high in clay are best. Those high in sand are poorest because they allow water to seep out of ponds.

The amount of water required depends on the pond size (both depth and surface acreage), soil type, and production system used (whether or not the water is exchanged).

Food Fish Production	200.0 Acres
Rearing Pond	4.4 Acres
Spawning & Holding Ponds	1.5 Acres
Storage & Handling Area	13.3 Acres
Total Land	219.2 Acres

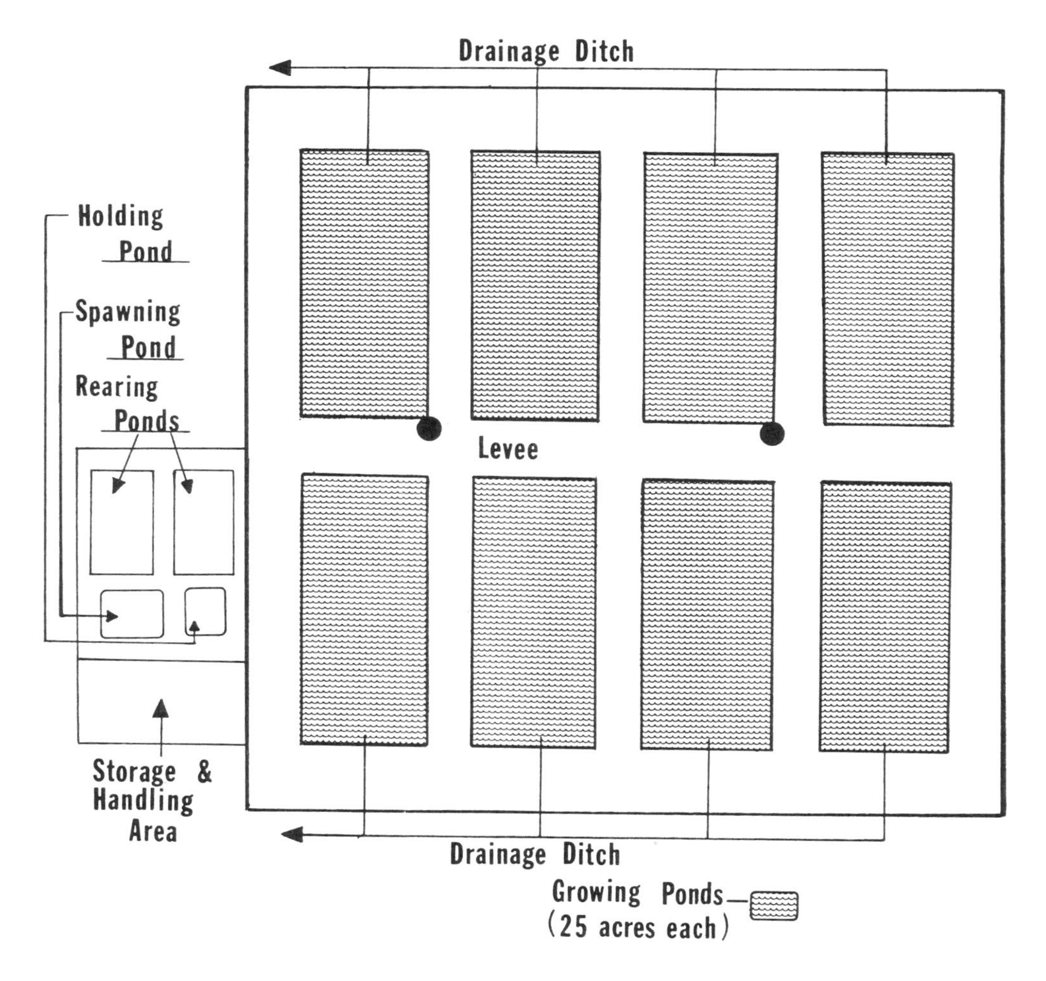

Figure 6–3. General pond layout of a food fish farm. This design provides for holding, spawning, rearing, and growing ponds as typically used with catfish.

Raceways

Raceways are typically long, narrow structures built in a series so that water flows from one to another. Raceways built on hillsides allow water to flow naturally. Those built on land that is almost flat require pumps to move the water.

Raceways vary in the rate of water flow and the way the water is handled at the end. Greater rates of flow require more water and allow more intensive (higher concentration) production of aquacrops. Raceways with very little water flow are sometimes known as semi-raceways.

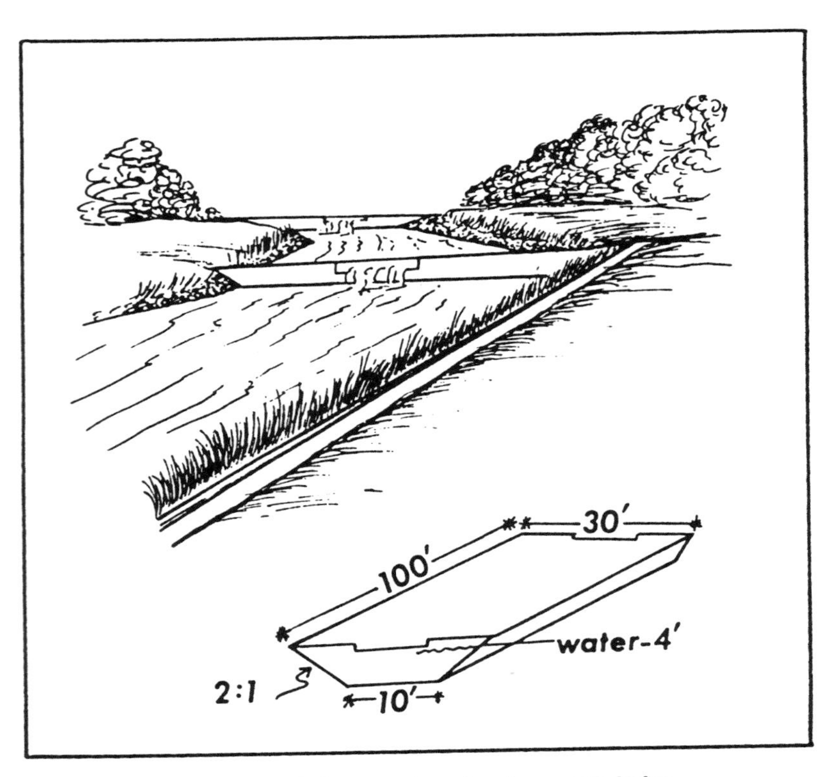

Figure 6–4. Raceway layout and segment design.

Raceways may be closed or open. A closed raceway is one in which the water is recycled. Filtering and other treatment of the water may be required. With open raceways, the water is disposed of after it is used. Disposal may require treatment of the water before it is released into streams or lakes.

Overall, raceways require more water than ponds. An advantage is that aquacrops may be grown at a higher stocking rate than in ponds. Certainly, hilly land may be better suited to raceways than to ponds.

Tanks

Tanks are water impoundments made of concrete, metal, fiberglass, or other materials. They are usually round or rectangular. Tanks constructed of light materials may be portable. Aeration of the water is often essential, or the aquacrop will be lost.

Round tanks range considerably in size, often from 6 to 30 feet in diameter and 2 to 6 feet deep. The water is run into the round tank at an angle on one side so that it flows in a circular pattern. The water is often removed in the middle where the flow is lessened and the sludge (solid material consisting mostly of uneaten feed and feces) settles. Aquacrops are usually stocked at a heavy density

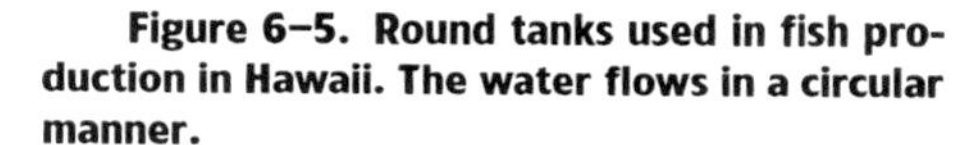

Figure 6–5. Round tanks used in fish production in Hawaii. The water flows in a circular manner.

Figure 6–6. Round tank covered with dome.

in tanks. Some way of properly disposing of the used water is needed. Environmental regulations do not allow dumping used water into streams if the water would pollute the streams.

Rectangular tanks range from a few feet to 25 or so feet long. They are often 3 feet wide and 30 inches deep. Water enters at one end and is removed at the other. Aquacrops are stocked at a high density. Used water must be properly handled.

Figure 6–7. Rectangular tank used in fish production. The water flows from one end to the other.

Vats

Vats are much like tanks. They are usually constructed of concrete or concrete block and are not portable. They are often long, rectangular-shaped structures. Water normally flows continuously into a vat at one location and is removed at another. This used water must be disposed of properly. Some aeration of the water is needed. More often, vats are used as temporary holding facilities for aquacrops.

Aquaria

Aquaria have limited use in most aspects of aquafarming. Certain pet fish may be grown in aquaria. Researchers may use aquaria in the production and

Figure 6–8. Lined concrete vats used in fish production.

observation of small populations of aquacrops. Aquaria are typically made of glass and have a capacity of 5 – 50 or more gallons of water. The water may be flowing or aerated, depending upon the system used.

SELECTING A WATER FACILITY

The aquafarmer must consider a number of factors in selecting the kind of facility to use. Several of these factors are (1) the supply of available water, (2) the disposal of water, (3) the characteristics of the water, and (4) the species to be grown.

Supply of Available Water

If a large supply of good water is economically available, facilities that require large quantities, such as ponds and open raceways, can be used. When water is in short supply, smaller systems will likely be preferred. Recirculated water in tanks and vats may be used. Aquacrops are more intensively managed when the water supply is limited.

Figure 6–9. Part of a treatment system used to recirculate water.

Disposal of Water

Some aquafarms use water and dispose of it without recycling. Some approved way of disposing of the used water must be available. Since water from aquaculture contains uneaten feed, feces, and other materials, it cannot usually be run into a natural stream or lake. The selection of a water facility requires attention to the treating of water before its disposal. Some aquafarmers use culture systems that produce very little or no water for disposal. For example, fish ponds usually have no water disposal unless the pond is drained. On the other hand, raceway and tank systems may produce large amounts of water that require disposal.

Characteristics of the Water

The aquafarmer must determine the characteristics of the water that is available. Some sites have cold water, while others have warm water. Thermal well water (water that is naturally heated by the earth) can be used to produce fish that require warm water. If it is not available, the aquafarmer must consider another alternative. The kind of crop that is possible has considerable bearing on the kind of water facility to be constructed.

Species to Be Grown

Species of aquacrops have varying water requirements. Farmers must study the species carefully to be sure that the one they choose will be well suited. A good procedure is to find out what other farmers in the area are growing. Of course, many aquacrops are new, and there are very few farms growing them. Also, farm sites within a geographic area have different features. Neighboring farms may have different possibilities.

WATER QUALITY

Water quality consists of all of the chemical, physical, and biological characteristics of water that influence how an aquacrop grows — the suitability of water for aquaculture. Certainly, water can be unfit for aquaculture. This may be due to naturally existing qualities, pollution, or other factors. Any characteristic that influences how water is used in the production of an aquacrop is a part of water quality.

Water Chemistry

Water chemistry refers to the composition of the water. Each drop of water contains millions of tiny molecules. These molecules are made up of atoms of hydrogen and oxygen. Normally hydrogen and oxygen are gases. When they combine, they form a chemical compound known as water. The compound is written as H_2O. Scientific processes are needed to determine variations in water chemistry.

Various minerals and compounds may be present in small amounts in water. Dissolved oxygen, nitrogen compounds, hydrogen sulfide, carbon dioxide, and iron are often present in water. Dissolved oxygen (DO) is particularly important for many aquacrops. For example, fish must be able to get oxygen from the water, for without adequate oxygen, the fish will die. Some chemical variations in water make it unsuitable for aquaculture. These to some extent depend on the environment needed by the crop being grown. An example of a hazardous component of water is a nitrite compound, which can kill a crop if it is present in a large amount.

With some aquacrops, the pH of the water is important. Numbers are used to describe pH. A pH of 7.0 is neutral. Numbers below 7.0 indicate that the water is acidic, while numbers above 7.0 indicate that the water is alkaline. For example, a pH of 5.2 indicates that the water is acidic. Traces of minerals in water give it the acidic or alkaline quality. Most aquacrops prefer a pH of 6.3 - 7.5, but some will thrive in water with a wider pH of 5.0 - 8.5. Easy-to-use test kits are available for determining pH.

Water may be classified on the basis of its salt content. Freshwater has less than 3 ppt (parts per thousand) salt. Water from most wells and streams is freshwater. Saltwater is the water found in oceans and seas, which typically has a salt content of 33 - 37 ppt. Water where the two flow together has a lower salt content and is known as brackish water.

Physical Characteristics

The physical features of water are largely a matter of temperature. Water may be a solid, liquid, or gas. It becomes a solid when the temperature reaches 32°F. and it changes to ice. It is a liquid when the temperature is between the freezing and boiling points of 32° and 212°F., respectively. Water becomes a gas (vapor) through vaporization. This process can be speeded by heating water to the boiling point (212°F.).

It is obvious that aquacrops cannot grow in extreme water temperatures. Most aquacrops grow in water between 50° and 90°F. However, in deciding how the physical features of the available water will influence the selection of an aquacrop, the aquafarmer must first know the temperature requirements of each aquacrop.

Water Biology

Water may contain both living and non-living plants and animals. This includes the aquacrops as well as other small and large plants and animals that may be present. In a pond or other growing facility, these are collectively referred to as a "biomass."

Tiny plants and animals present in water are usually not visible without magnification, such as with a microscope. Bacteria, fungi, and algae are common examples. Some of these are beneficial in the production of certain aquacrops; others are harmful and cause diseased conditions in the crops. The aquacrops themselves are also a part of the biology of the water. The presence of and the processes carried out by the organisms and aquacrops make up the biology of water.

Not all microscopic organisms in water are harmful. Plankton are the tiny plants and animals which float in pond water. Algae plankton (phytoplankton) produces oxygen through photosynthesis and helps keep oxygen available for aquacrops. Phytoplanktonic plants also shade the bottoms of ponds and help reduce the growth of aquatic weeds. Animal plankton (zooplankton) is a natural food of some aquacrops; however, it is usually not present in sufficient numbers to feed heavy stocks of fish.

SOURCES OF WATER

Sources of water include wells, streams, lakes and oceans, springs, industrial effluent, municipal water systems, and surface runoff. Selecting a source of supply is an important decision. Not only must the water be of good quality, but it must also be available at a reasonable cost. Suitable water is likely available from each source, though its use may be impractical in some situations.

Wells

Wells are often considered to provide the best water for freshwater aquaculture. The water is usually free of pollution from chemicals, parasites, disease organisms, and trash fish. Shallow wells, however, may have traces of pesticide residues or other hazardous compounds. Well water may sometimes contain undesirable gases and minerals. It may also be low in oxygen; thus, some way of adding oxygen will be needed.

Water from wells cannot always be pumped directly into the aquaculture facility. If it is cold, it may need to be pumped into a holding pond for solar warming before being used in aquaculture. This is also a good time to add oxygen. Some wells produce warm (thermal) water. This is particularly advantageous for warm water crops. For example, tilapia thrive in water that is about 85°F. The farmer who has a year-round supply of 85°F. well water is fortunate.

The capacity of a well to produce water is measured as gallons per minute (gpm). How rapidly a well produces depends on the size of the pipe in the earth and the size, speed, and power of the pump. Pumping rates may range from less than 100 gpm to 2,000 gpm. The amount of time required to fill a vat, a pond, or other facility depends on gpm. For example, it will take a 1,000-gpm pump 16.2 hours to put 3 feet of water in a 1-acre pond.

Well depths may range from relatively shallow to deep. Shallow wells are 30 feet or so deep. Other wells may be hundreds of feet deep. Depth depends on the distance to the aquifer (the layer or stratum of rock, sand, or gravel that holds water beneath the earth's crust). Aquafarmers must consider the cost of drilling wells and the cost of the pumps and energy to operate them as important costs in setting up aquafarms. Most farmers contract with commercial well drillers to dig their wells. Well cost is normally based on a per foot rate plus other items such as the pump and the motor. Wells occasionally need to be replaced.

Sometimes the water level may drop below the pipe that goes into the earth, or the well casing (pipe used to enclose a well) may rust out or break. In either case, it is no longer possible to get water from the well.

To drill wells, aquafarmers usually need permits, which can be obtained from local government agencies. A few years ago permits were not required. As more and more wells were drilled, the ground water level began to decline. Consequently, regulations were developed to help conserve valuable water. Some abuse of ground water has also occurred. Unfortunately, chemicals have sometimes entered the water through wells when individuals have pumped hazardous chemicals into wells to dispose of them. How unfortunate!

In a few sites, artesian wells provide good aquaculture water. An artesian well is drilled like any other well. Due to natural pressures in the earth, a pump is not needed. Internal pressure causes the water to flow like a fountain. These wells are normally fairly deep. If the water is of good quality, the farmer with an artesian well can save money on pumps and electricity.

Streams

Some aquafarms have access to rivers and creeks, which sometimes provide good water for freshwater aquaculture. Careful assessment of the water is needed to determine if it is suitable for aquafarming.

Streams are made up of water from several sources: springs, runoff from rain or melted snow, industrial effluent, municipal sewage systems, and overflow from farm irrigation such as runoff from rice farms. Water that comes from springs and the runoff from land that does not have chemical residues are more likely to be suitable. Water from municipal sewage plants, livestock feeding operations, and certain industrial uses is unsuitable for aquaculture.

Water from streams may be seasonal. During the summer and fall, the streams may dry up, leaving the aquafarmer without a water supply unless there is a backup well or another source. The quantity of water that can be removed from a stream is regulated. The general rule is that the removal of water from a stream cannot noticeably reduce its flow. Regulations must be followed in getting water from streams.

Stream water may need to be treated in some way. Water taken from streams may introduce diseases and trash fish to the aquafarm. The water can be filtered to remove objectionable items. A meshed screen may be put over the end of intake pipes, or a filtration system can be installed. Filtering costs money and reduces the profit of an aquafarm.

Pumps, canals, piping systems, and other facilities may need to be constructed to get the water from streams to where it will be used. Energy is required to operate the pumps. Regular inspection of the facilities is needed to insure that they are working properly.

Overall, water from streams is considered less desirable than water from wells. In some cases, it is unfit for aquafarming.

Lakes and Oceans

Water may be pumped from natural lakes and oceans for aquafarming. Some lakes contain freshwater; others have saltwater. All oceans and seas are salty. Hydroelectric or flood control reservoirs that have been constructed across streams are similar to freshwater lakes.

Just as streams do, the water from lakes and oceans may contain chemical residues, parasites, diseases, and trash fish or weeds. A filtering and treatment system may be needed.

An alternative to pumping water from lakes and oceans is to grow aquacrops in cages or other structures in these bodies of water. Cages typically float in the water. They are normally anchored in some way to prevent drifting away. Stationary water structures may be attached to the bottom of the lake or ocean if the water is shallow. Cages also can be used in deep water.

Using water from lakes and oceans requires careful analysis of the water and knowledge of the water environment in which a proposed aquacrop can thrive. Regulations must be followed in removing any water from a lake or an ocean. Obtaining the needed permits can be a lengthy process.

Springs

Springs are natural openings in the earth that produce water. The water is often similar to well water. It may be cold or thermal, depending on the spring. Spring water is frequently low in oxygen. It may be high in certain minerals. Of course, the use of spring water in aquaculture is limited to areas where springs are found.

Careful assessment of the volume of water produced by a spring is important. Some springs tend to dry up in drier weather. Aquafarming usually requires a year-round, dependable supply of water.

Springs often flow into streams or reservoirs. The water must be caught in some way for use in aquafarming. It is possible that trash fish, pollution, and other foreign matter can enter spring water shortly after it flows from the earth.

Industrial Effluent

Industrial effluent is the water released by manufacturing plants. Some of this water is excellent for aquaculture. For example, power plants use water for cooling and do not alter it otherwise. This water may be warmed in the cooling process. As warm water, it can be very beneficial to certain crops. On the other hand, water released from certain chemical or other industries may be unfit for aquaculture.

Only locations near industries that utilize and release large amounts of water should be considered for aquaculture that uses industrial effluent. The quality of the water must be determined. Often, this requires laboratory analysis. The quantity of water released is also a consideration. There must be sufficient water for the culture of the crop.

All regulations pertaining to the use of the waste water must be followed. The owner of the industry must agree to allow the use of the water. In a few cases, industries have initiated aquaculture.

Municipal Water Systems

Municipal and rural community water systems provide water for household and business use. The water may be taken from streams, lakes, and reservoirs, or wells. It is treated with various chemicals to make it suitable for human consumption. Chlorine and fluorine may be added. Both of these may be harmful to aquacrops!

As a source of water for aquaculture, municipal and rural systems are usually too expensive for large-scale operations. Small aquafarms that use only a few gallons may find it suitable. However, the water must be de-chlorinated. This is done by letting it stand in an open container for at least 24 hours, adding sodium thiosulfate, or filtering it through charcoal. Some aquafarms that primarily use other sources for water rely on municipal systems in case the other sources fail.

Individuals with small volumes of ornamental fish often use municipal water. Homeowners with aquaria of pet fish almost exclusively use municipal water. They run it a day or so before using and allow it to stand in large tubs. While standing, the water loses its chlorine and warms up. Of course, a small amount of water is lost through evaporation while it is standing.

Surface Runoff

Rain, melted snow, and other types of precipitation form surface runoff, which may be collected in reservoirs for use in aquaculture. Certainly, much of the water in streams is from surface runoff.

The typical use of surface runoff is for the small watershed pond. This kind of pond is built to be filled by surface runoff from hillsides surrounding the pond. Several acres of watershed may be needed to fill a pond. This approach to water supply is not well suited to commercial aquaculture operations. It is best for the hobby farmer or recreational lake operator.

Figure 6–10. Rip-rap on levees reduces erosion and protects water from muddy runoff.

Runoff tends to reflect the qualities of the area from which it comes. Runoff from pasture land may be relatively free of residues except for animal feces, vegetative matter, and any chemicals applied to the land. Runoff from cropland may contain eroded soil, pesticide residues, and other undesirable material. Sometimes runoff is collected from roads, residential areas, and business parking lots. However, such runoff is largely unfit for aquaculture.

Before runoff can be used in aquaculture, it must be carefully tested for residues that could be harmful to the aquacrop or the consumer of the crop. Laboratory analysis of a runoff sample is usually required.

Reservoirs holding runoff are sometimes used as sources of water for livestock. This practice is acceptable if the water is pumped to the livestock. It is not recommended if the livestock have access to the reservoir. If cattle or hogs wade out into the water, they may make it muddy or otherwise lessen its quality.

SUITABILITY OF WATER

Some water is of good quality for aquaculture; other water is not. The aquafarmer must assess the water. This involves comparing the qualities of the water to the preferences of the proposed aquacrop. Of course, some water is unfit for any type of aquaculture!

Water quality is determined in various ways. Laboratory testing may be required. Farmers may purchase small, portable kits for testing the water. Certainly, observing how aquatic life responds in the water is a good clue to its quality.

Several things to look for in determining the suitability of water are amount of dissolved oxygen; temperature; acidity, alkalinity, and hardness; pollutants; nitrogen compounds; and carbon dioxide.

Dissolved Oxygen

Oxygen is required for animal aquacrops to carry on body functions and to remain alive. Water must contain dissolved oxygen (DO). For example, fish draw water into their mouths and force it out over their gills. The gills remove the oxygen by diffusion and pass it into the bloodstream. Without enough dissolved oxygen, the fish become sick or die. They may be seen gasping at the surface of the water or dead with their bellies turned up.

Dissolved oxygen in water is stated as parts per million (ppm). For example, 1 pound in a million pounds would be 1 ppm. Tests can be run to determine the amount that is in water. Cold water holds more oxygen than warmer water. This means that the highest amount of oxygen is at 32°F. At 60°F. water holds up to 9 ppm dissolved oxygen, while at 90°F. it holds up to 7 ppm.

For animal aquacrops, a 5 or above ppm oxygen level is preferable. Most will survive in the 1 – 5 ppm level, but growth will be slow and disease outbreaks are more likely. Low oxygen causes the fish crop to be stressed. Below 1 ppm the aquacrops are likely to die if exposure is for more than a short time. Required oxygen levels vary among different species. Small fish tend to be more tolerant of low oxygen. For example, the fathead minnow will die if the level goes below 1 ppm. Channel catfish may die if it goes below 2 ppm, except that the small fish

Figure 6–11. Dead fish are shown floating on the water in this pond. Oxygen depletion is the cause of death. (Courtesy, Fish Farming Experimental Station, Stuttgart, Arkansas)

will survive at a lower level. Rainbow trout will die when the level reaches the 3 ppm or below level.

Sometimes water can be supersaturated. This means that it contains more oxygen than it will normally hold. Water is not likely to be supersaturated in ponds. If it is, running normal aeration equipment will cause a reduction in oxygen rather than an increase in it in the water.

Fortunately, dissolved oxygen can be added to water if the water is low in it. This is a management tool available to aquafarmers. Water from wells and springs should usually be aerated before it is used. Close observation must be made of oxygen levels to prevent the loss of aquacrops.

Temperature

Water temperature is important with many aquacrops. Fish are ectothermic animals. This means that their body temperature adjusts to that of the water. Sudden changes in water temperature can cause the death of fish. Different aquacrops require different temperatures for growth.

Temperature influences feeding, metabolism, growth, and reproduction of most aquacrops. When water is below the best temperature for a particular aquacrop, the fish will stop eating. This is a result of a slowdown in the metabolism in the animal. Metabolism refers to all of the growth and maintenance processes that occur in the body of an animal. Fluctuations in water temperature cause certain aquacrops to reproduce. For example, in the spring most fish spawn naturally as the water begins to warm.

On a large scale, attempting to control the temperature of water tends to be impractical. To warm or cool large amounts of water is expensive. Except in unusual cases, it is best to select an aquacrop suited to the water that is available. Water from deep wells, springs, and melted snow is typically cold. It will warm if allowed to stand in ponds for a period of time in warm weather.

Acidity, Alkalinity, and Hardness

The acidity, alkalinity, and hardness of water are determined through chemical analysis of a sample of the water. As was previously stated in this chapter, pH is a scale used to depict acidity or alkalinity. Most aquacrops will not thrive in water with a pH range to which they are not suited. The mid-point on the 14-point pH scale is 7. Most aquacrops will survive 2 points above or below 7. Since the pH of water in ponds and other facilities may vary during a day, most aquacrops are tolerant of short-term exposure to very high or very low pHs.

Some chemicals used to treat aquacrops for diseases or other purposes are more toxic in water that is not in the pH range of 5 - 9. If these will be used, it is important to have some understanding of water acidity, alkalinity, and hardness. The aquafarmer should read the label on any chemical to see if it is approved and how it will react with the water.

Hardness refers to the amount of calcium and magnesium in water. It is associated with a high pH. Hardness is measured in ppm, with a range of 50 - 300 ppm hardness being best. A water softener can be used if the water is too hard. In most cases, surface water is within the desired acidity, alkalinity, and hardness range.

Pollutants

Many substances can pollute water used in aquaculture. When this occurs, the water is said to be contaminated. Some substances are harmful to aquacrops; others are not. In some cases, aquacrops can ingest (take in) harmful substances and pass them on to human consumers.

Pollutants may come from industries, natural sources, ordinary life activities of humans, and agriculture. Some sources of pollution do not appear to be significant, but they can become significant with repeated exposure. For example, automobiles and trucks release gases into the air, produce used oil, and emit other substances. Used oil should be properly handled; otherwise, it may find its way into water used for aquacrops. Fuel storage tanks may develop leaks that can enter sources of water.

Farming activities sometimes release toxic substances. Farmers who grow row crops near ponds or other aquaculture facilities should be very careful in using pesticides (primarily weed and insect killers). Harvested aquacrops found to contain excess amounts of toxic chemicals are unfit for human consumption.

Determining the possibility of pollution involves studying the source of the water. Water that might have come from areas where contamination could occur should undergo laboratory analysis. It is a good idea to have samples of water from any source tested before an aquafarm is started.

Nitrogen Compounds

The air contains 78% nitrogen, as compared to 21% oxygen. There will likely be more nitrogen gas in water than dissolved oxygen. Nitrogen compounds causing problems include ammonia, nitrite, and nitrate. These usually become problems in water used in intensive aquaculture, such as tanks, which have high organic waste. Spring and well water can become supersaturated with nitrogen and cause a gas bubble disease in some aquacrops. If water from an aquacrop is reused, there is a strong possibility of nitrogen problems. Various tests and treatments are available. These are discussed later in this chapter under "Management of Water."

Carbon Dioxide

Some carbon dioxide is usually found in water. The level is greatly increased in water being used in aquafarming. Carbon dioxide is produced by plants and animals living in water during the process of respiration. Safe levels of carbon dioxide range from 5 to 10 ppm. Toxicity to aquacrops depends on the amount of dissolved oxygen in the water. Most all crops will tolerate high levels of carbon dioxide if there is a high level of dissolved oxygen. Spring and well water as they are pumped do not usually have a carbon dioxide problem.

MANAGEMENT OF WATER

Once quality water has been obtained, aquafarmers must follow practices that will keep it in good condition for fish and other aquacrops. Proper water management can prevent problems that restrict production. Improper water management may result in poor growth and, possibly, the loss of an aquacrop.

Dead fish floating on the surface of the water may be the result of water problems that could have been prevented.

Major areas of concern in water management are oxygen depletion, build-up of nitrogen compounds, turbidity and color, growth of weeds and algae, and plankton bloom.

Oxygen Depletion

Water must have adequate dissolved oxygen (DO) to support the growth of an aquacrop. In any type of water facility, oxygen depletion will cause the death of fish in a few hours. Larger fish typically die first. Most of the aquacrops require 4 – 5 ppm DO. An aquafarmer must follow good practices to keep DO at an appropriate level.

Common signs. Some of the most common signs of oxygen deficiency with fish are:

- Fish pipe (gasp) at the water surface for air.
- Fish group around incoming water source.
- Fish go off-feed (not eating).
- Non-fish animals, such as crawfish and snails, crawl out of the water.
- Growth is slowed as a product of long-term oxygen deficiency but not low enough to kill the crop.
- Repeated outbreaks of health problems, such as those related to stress of fish, occur.
- Water color changes. (This indicates a plankton die-off.)
- Fish-eating birds are present. (The fish are near the surface and easy to catch.)

Causes of oxygen depletion. The term "biological oxygen demand (BOD)" describes the uses of oxygen in the water for many natural purposes: decay of uneaten feed and vegetation, decay of plankton, use by the aquacrop, and use by trash fish.

Several causes of oxygen depletion are as follows:

1. *Overstocking*—If water is stocked beyond its normal ability to provide oxygen, the fish or other crop will use up the oxygen. Most aquafarming facilities are stocked far beyond the normal capacity of the water. Careful monitoring is needed to make sure that water problems do not develop.

2. *Weather*—In production systems located outside, such as ponds, plant plankton produce oxygen by photosynthesis when the sun is shining. On cloudy days and after dark nights, the water is lowest in DO because photosynthesis has not been taking place. Limited light may result in the death and decay of certain organisms, and this ties up oxygen. Warmer water holds less DO; therefore, close monitoring is needed in warm, cloudy weather.
3. *Time of day*—Natural supplies of DO vary greatly, depending on the time of the day. DO is highest in a pond in the afternoon of a bright, sunny day. It is lowest in the early morning just before sunrise. Just as with weather, plant plankton produce oxygen through photosynthesis during the day; they do not at night. Many aquafarmers monitor DO levels all night in the warm months of the year.
4. *Decay of feed*—When an aquacrop is given more feed than it can eat, the feed decays. This process uses oxygen. Aquacrops should not be fed more feed than they will consume in a matter of a few minutes.
5. *Feeding activity*—The process of digesting food uses oxygen. As aquacrops feed, they require more oxygen.
6. *Competition*—Some water facilities have trash fish and plants that use oxygen. These compete with the aquacrops for DO.
7. *Decay of vegetation*—Dead weeds, leaves, grass, and other organic matter decay in water. This process ties up oxygen just as the decay of excess feed.
8. *Water temperature*—Cold water holds more DO than warm water. Those aquacrops that prefer warm water are more likely to exist in water where a problem occurs.
9. *Salt content*—Saltwater holds less DO than freshwater. If freshwater is polluted by saltwater, the ability to hold DO is reduced. This is seldom a problem with freshwater aquafarms. Farmers using saltwater will need to monitor the water.
10. *Chemical reactions*—Certain chemical reactions in water may use oxygen. Although this is seldom a problem in aquaculture, it can happen with the addition of chemicals for various purposes.
11. *Equipment failure*—DO problems can occur when the equipment used to aerate water fails. For example, when power outages for electric motors or breakdowns of motors or pumps happen, a pond can develop oxygen depletion very quickly.
12. *Corrosion*—Equipment used in water aeration may become corroded and

less efficient. This is particularly a problem in saltwater because barnacles attach themselves to equipment.

DO can be fairly easily measured. Several companies make kits that are available for aquafarm use. The kits may involve collecting water samples and using chemicals to test them or using a battery-operated meter with a probe than can be extended into the water. Normally, DO readings are made about 6 inches below the surface of the water. Measurements should not be made near incoming water, close to aeration equipment, or adjacent to the edge of the water impoundment. Required DO levels tend to vary for different crops; thus, it is important for the aquafarmer to know the desired level for the crop grown.

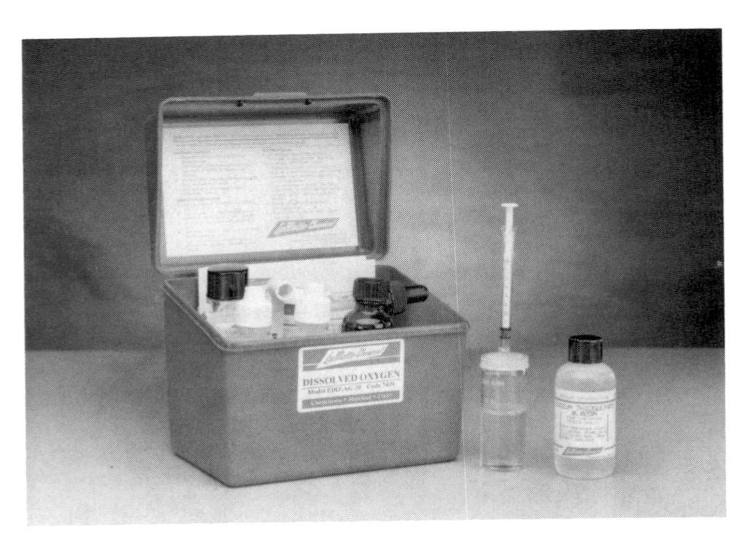

Figure 6–12. Commercial dissolved oxygen test kit. (Courtesy, LaMotte Chemical Products Company)

Handling deficiency. When an oxygen deficiency develops, steps should be taken to handle it immediately. Several ways are as follows (sometimes combinations are needed):

- Reduce the number of fish or other aquacrop in the water.
- Remove vegetation that competes with the aquacrop for oxygen.
- Inject oxygen into the water.
- Use mechanical aerators.
- Pump out bottom water and add new, aerated water to the surface.

- If the water level is below capacity, add aerated water to the impoundment.

Oxygenation of water. DO can be added to water with a variety of devices and methods. Some of the methods are more reliable than others; some are not very efficient. Several ways of adding oxygen are presented later in this chapter.

Nitrogen Compounds

Nitrogen compounds are more likely to be problems in intensive aquaculture systems. This is because of the heavy build-up of organic wastes, particularly fish excrement. As the wastes decompose, the ammonia is converted to nitrites. Two forms of ammonia are found: ammonia (NH_3) and ionized ammonia (NH_4+), also known as ammonium. Certain bacteria (*Nitrosomonas*) change the ammonia to nitrite (NO_2). Another form of bacteria (*Nitrobacter*) changes the nitrite to nitrate (NO_3).

The nitrate form is relatively harmless to most aquacrops.

Ammonia and nitrite are the most harmful forms of nitrogen. Concentrations of 0.1 ppm may kill certain species of aquacrops. Concentrations of 0.0125 ppm may reduce the rate of growth and damage the ability of the aquacrop to function. Brain damage is among the first signs of ammonia toxicity. The gills of trout may be damaged at very low levels of un-ionized ammonia. Ammonia at 0.06 ppm can damage the gills of catfish, increase susceptibility to disease, and reduce growth rate. High levels of nitrite may cause the hemoglobin in fish blood to oxidize and take on a brown color. In catfish, this condition is known as brown blood disease.

Many variables are involved in ammonia toxicity. Ammonia is more toxic in water when the temperature increases, the pH becomes more alkaline, the dissolved oxygen decreases, the carbon dioxide increases, and the salinity decreases.

Tests can be conducted to determine if nitrogen compounds exist in water. Since nitrogen forms may leave a water sample quickly, reliable methods of testing should be used. Samples should be collected in clean pint or half pint jars that can be sealed to prevent loss of nitrogen. The water samples should be taken at a depth of 1 foot.

Carefully monitoring feed consumption so as not to overfeed is an important management tool in reducing nitrogen problems. Removing feces, uneaten feed, dead fish, and other organic materials that may decay and produce the nitrogen compounds is important. Intensive systems must have considerable flow of water to remove the wastes. Tanks should be designed so that these materials can be moved to an area where removal is relatively easy, such as in the center of a round tank. Intensive systems that recycle water must have elaborate filtration and treatment systems to recondition the water for use.

Turbidity and Color

Turbidity refers to the presence of suspended particles of soil or plankton in the water. The water may appear muddy or cloudy. Some turbidity causes no problems. However, excessive turbidity reduces the amount of light that passes through the water. Thus, oxygen-producing plankton will not be able to carry out photosynthesis, thereby resulting in an oxygen deficiency.

Ponds that receive large amounts of runoff from rain are more likely to have turbidity problems because soil and vegetative matter are washed into the pond. Sometimes the vegetative matter will discolor the water, but this will not likely cause any problems. However, settling silt or soil may coat eggs, smothering them. The silt may also impair the growth of desired organisms.

Protecting aquaculture facilities from runoff by establishing grass covers on the soil around ponds is a practical way to reduce turbidity problems.

Various other treatments are also available. First, the aquafarmer should determine if the turbidity is due to silt or organic matter. With silt, some

Figure 6–13. Algae problem developing in water near pond levee.

aquafarmers add gypsum at the rate of 100 – 1,000 pounds per acre; others use alum at the rate of 15 – 25 pounds per acre. With organic turbidity, some aquafarmers broadcast 400 – 450 pounds of agricultural lime per acre. Note: The lime may increase the pH. It is a good practice to add the lime in increments, testing for pH between applications. The application of lime should stop when the pH reaches 9.0 – 9.5.

Weeds and Algae

Weeds and algae cause problems in aquaculture in several ways. Both compete with the aquacrop for oxygen, light, and nutrients. They also cause problems in harvesting. For example, weeds in ponds make it very difficult to pull a seine through the water; algae can clog a seine. Of course, weeds are not problems in tanks, vats, and concrete raceways.

Some plants growing in aquaculture facilities are desirable, particularly in the less intensive systems. Their role in producing oxygen is important. Desirable plants can create poor conditions if there are too many of them. (Another section of this chapter contains details on weed and algae control.)

Plankton Bloom

The beneficial tiny plants and animals that live in water can sometimes cause problems. When water appears to be soupy green or brown, it may have too much plankton. Applying a fertilizer to a pond could cause too much plankton to grow. Plankton can be killed off with a chemical, but this may result in DO depletion. An approach used by some producers is to remove some of the water from a pond and then add freshwater. This dilutes the concentration of plankton.

CALCULATING WATER VOLUME

Operating aquafarms requires skill in making calculations about water, in determining how long it takes to fill a pond or how much chemical to use in a tank.

The amount of water needed depends on the production system being used. Open systems where the water continuously flows require more water than closed systems. Raceways require more water than ponds.

Water requirements are typically stated as gallons, cubic feet, and acre-feet.

- 1 cubic foot of water contains 7.481 gallons.

- 1 gallon contains 231 cubic inches of water.
- 1 acre-foot contains 325,851 gallons. (An acre-foot is the amount of water required to cover 1 acre with 1 foot of water.)

Volume of Tanks and Vats

Three measurements are needed to determine the amount of water in a rectangular tank, aquarium, or vat: length, width, and depth. Using a yardstick or tape measure, the aquafarmer can determine the length and width of the inside of the tank or vat and the depth of the water. (Some allowance is needed for tanks, etc., with sloping bottoms or sides. These measurements may need to be an average of the deepest and shallowest or widest and narrowist parts of the tank.) The formula to use is:

$$\text{Volume} = \text{length} \times \text{width} \times \text{depth}$$

For example, a tank containing water that is 10 feet long, 3 feet wide, and 2 feet deep would have a volume of 60 cubic feet. The number of gallons is found by multiplying 60 times 7.481. This tank has 448.86 gallons of water. The amount of time required to fill the tank is determined by dividing the number of gallons in the tank by the gpm of the water source. Slightly less than 10 minutes would be required to fill the tank at a flow rate of 50 gallons per minute.

The volume of round tanks and vats is determined somewhat differently. Two measurements—depth of the water and diameter of the inside of the tank—are needed. The formula to use is:

$$\text{Volume} = 3.1416 \left(\frac{\text{diameter of tank}}{2}\right)^2 \times \text{depth of water}$$

For example, a round tank containing water that is 4 feet deep and 20 feet wide would have a volume of 1,256.64 cubic feet, or 9,403.92 gallons. At a pumping rate of 50 gpm (same as 3,000 gallons per hour), 3.14 hours would be required to fill the tank.

Volume of Ponds and Raceways

The amount of water in ponds and raceways is determined much like the volume of rectangular tanks. Distances are often much larger and more difficult to get. Averages of widths, depths, and lengths are often used. Slopes of levees and bottoms must be considered. In making measurements, the aquafarmer should always use water dimensions and not locations without water.

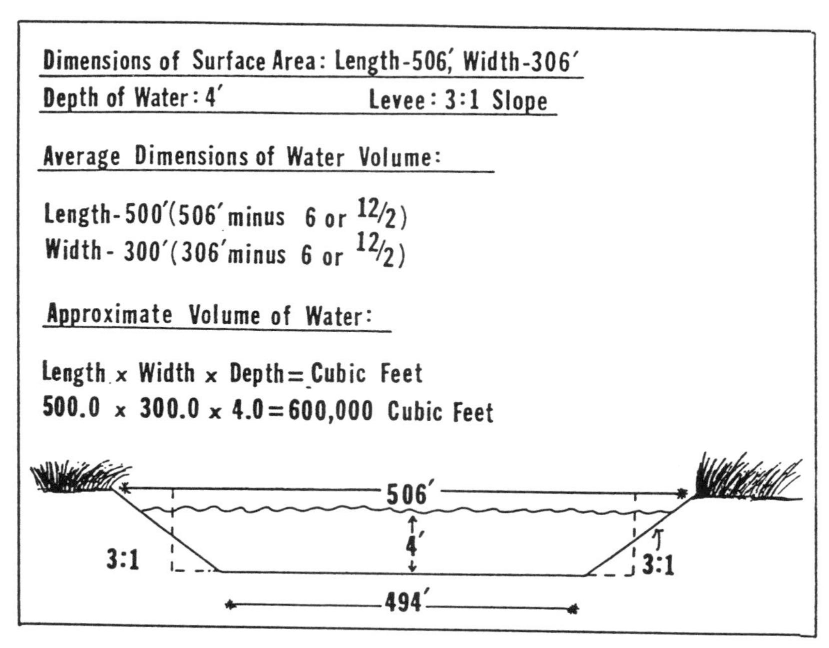

Figure 6–14. Calculating water volume of a pond.

Measuring at regular intervals along straight lines across a pond and then averaging the measurements will determine water depth. The design of ponds may provide for a certain depth of water. Information about depth, measurements, and slope of levee used in pond construction is very helpful. Volumes of rectangular ponds are more easily determined. The following is an example.

> At the water level, a pond measures 506 feet in length and 306 feet in width. The levee has a 3-to-1 slope. The pond bottom is close to level with water 4 feet deep. This means that the foot of the levee extends 12 feet into the water at the bottom all the way around the pond. The average length of the water is at the depth of 2 feet. Since the levee extends out 12 feet, the amount it extends at 2 feet deep is 6 feet (one-half of 12). This means that the average length of the water is 506 minus 6, or 500 feet. Width is 306 minus 6, or 300 feet. The volume of water in the pond is 500 × 300 × 4, or 600,000 cubic feet. The number of gallons is 4,488,600 (600,000 × 7.481). Pumping time at the rate of 1,000 gpm would be 4,488.6 minutes, or 74.81 hours. Once a pond is full, water will need to be added occasionally to keep it up to the right level.

The volume of a raceway can be determined in much the same way. Since raceways involve flowing water, continual pumping is required. Water in raceways is often replaced two times each hour. In a 30,000-gallon raceway, a 1,000-gpm supply would be needed.

Figure 6–15. Small, circular raceway using paddlewheel to provide water flow.

Amounts of Chemicals

Many chemicals are calculated as parts per million (ppm). Remember that 1.0 ppm is equal to 2.72 pounds per acre-foot, or 0.0283 gram per cubic foot. If a pond is treated, the acre-foot measurement is used. If vats or tanks are used, the cubic foot measurement is used. In a vat containing 300 cubic feet and at the rate of application of 3 ppm, 25.47 grams would be needed for the entire vat (3 × 0.0283 = 0.0849 gram per cubic foot, or 300 × 0.0849 = 25.47 grams for the vat). (Several useful conversions are presented in Appendix A.)

CONTROLLING WEEDS AND ALGAE

Weeds and algae often resemble each other. Both are plants. Algae are tiny, one-celled plants; weeds are larger, many-celled plants. The free-floating phytoplankton algae are very useful in pond aquaculture because they produce oxygen. The filamentous algae are big pests because they compete with the desired plankton for light and nutrients. The long, stringy strands that are formed are sometimes known as "pond scum."

There are three kinds of aquatic weeds: submersed, emersed, and floating. Submersed weeds, such as hydrilla and fanwort, are rooted to the bottom of the pond and grow to the surface of the water. Emersed weeds, such as cattail and water lily, grow in shallow water. Their roots are attached to the bottom, and their tops go above the water. Floating weeds are not attached to the bottom; thus, they float on the water. They sometimes give the appearance of a green blanket on the water. Examples of floating weeds are water hyacinth and duckweed.

Weeds and undesirable algae should be eliminated before a pond is stocked. It is much more difficult to control them after the aquacrop is underway. Four methods of aquatic weed control that may be used are chemical, biological, mechanical, and environmental manipulation.

Chemical methods. Chemical control involves using herbicides for weeds and algicides for algae. Several herbicides and algicides are on the market, but only a few are approved for use in aquaculture. Examples include the herbicide diquat and the algicide copper sulfate.

Before a chemical is applied, it is essential that the weed or algae to be controlled be correctly identified. **The directions on the labels on containers should always be read and followed.** Applications should be made according to

Figure 6–16. Dried algae that has been removed from a tank.

the regulations of the U.S. Environmental Protection Agency. Herbicides and algicides should not be used around water with food fish. Most should not be used within 60 days of harvest. The chemicals are usually sprayed on the weeds or applied to the surface of the water. Thus, water chemistry plays an important role in the use of chemicals.

Many aquafarmers feel that the use of chemicals is the safest, most economical, and most practical way to control aquatic weeds and algae. A note of caution: As the dead weeds decay, oxygen is used; as a result, DO levels may go below the level required for the aquacrop to live.

Biological methods. Biological control involves using plant-eating fish or other animals to control weeds and algae. This method does not pose the hazards of chemicals. However, it has not proven to be practical on large-scale aquafarms. The plant-eating fish must be sorted from the desired aquacrop at the time of harvest. They also compete with the crop for food and oxygen. Examples of fish that have been used are grass carp, white amur, and Israeli carp.

Mechanical methods. Mechanical control involves removing the weeds and algae from the water. In small ponds, the weeds and filamentous algae can be raked out and hauled away. Willow trees can be cut off and the tops removed. In most cases, any roots or parts that remain in the water are likely to sprout and grow.

Environmental methods. Environmental control is a practical approach to weeds and algae. It involves creating conditions unfavorable to weed and algae growth.

Designing ponds with a minimum of shallow water discourages the growth of certain weeds and algae. The edges of ponds should have water that is a minimum of 18 inches deep. A water depth of 2½ feet or more discourages the growth of several kinds of weeds and algae.

Other environmental methods are draining all of the water from a pond and allowing the bottom to dry and lowering the water level during cold weather to kill the weeds and algae by freezing. These methods never result in pollution but do interrupt the growth of the aquacrop.

CONTROLLING POLLUTION

Pollution is a part of aquaculture in two ways:

1. The water used in aquafarming may be contaminated.

2. The water released from aquafarming may contain materials that pollute streams and lakes.

The water used in aquaculture should be free of harmful pollution. Contamination occurs when chemicals and other substances get into the water. All water used for aquafarming purposes should be free of contamination that would impair the growth of the crop or make it unfit for human consumption.

The aquafarmer can reduce the chance of water being contaminated by using only water from known sources that have tested free of harmful substances. Many farmers rely on water from deep wells, but even this can be contaminated. Good water also can be contaminated after it is pumped. Careless handling of chemicals, failure to control erosion on pond levees, and drift from nearby row crop application of pesticides can pollute water. **Pesticide containers should be disposed of properly.** Allowing them to be thrown into ponds, streams, or lakes causes contamination that could have very easily been avoided.

The water that is discharged by an aquafarm may contain uneaten feed, live fish, fish excrement, residues of chemicals used to treat the aquacrop, and dead fish. If run directly into a stream or lake, it can contaminate that water. As a general rule, the water released by an aquafarm should not appreciably alter the natural stream or lake. Holding reservoirs or other treatment methods may be needed to prepare the water before it is dumped. Just as the water used to grow an aquacrop should be tested, discharged water should also be tested.

ADDING OXYGEN TO WATER

There are several ways of adding dissolved oxygen (DO). Some are more practical than others, depending on the situation. Water in tanks is oxygenated differently from that in ponds. Common ways of getting DO into water include the following:

1. *Splashing the water*—New water being added to a facility can be splashed against a concrete slab or another structure. The purpose is to expose as much surface area (molecules) as possible to the air to acquire oxygen. Some splashing occurs when water is being pumped into a pond or through a raceway.
2. *Pumping air or oxygen gas into the water*—This method is most often used with tanks or vats. Cylinders of oxygen can be placed on trucks with haul tanks. Sometimes aquacrops are shipped in large, heavy plastic bags. Pure oxygen can be injected into the bags after the fish and water are in place. Two ways of injecting air into larger bodies of water, such as ponds,

Figure 6–17. Small, round tanks are aerated by bubbling oxygen into the water.

are (a) bubbling compressed air through aeration tubes at the pond bottom and (b) using pumps made with rotors or blades (impellers) to force air down into the water.

3. *Spraying the water into the air*—This is a common method of aeration. The water is broken into as many droplets are possible to expose the maximum surface to the air. Various types of equipment including electric-powered paddlewheel aerators and floating pump sprayers are used to throw the water. Some farmers have portable paddlewheel aerators powered from the power shaft of a farm tractors. These are on wheels and can be easily moved from one location to another. Portable units are particularly good to handle emergencies.
4. *Using chemicals*—Using chemicals to restore DO can cause about as many problems as it solves. Potassium permanganate has been used at the rate of 1 - 3 pounds per acre in freshwater ponds. Fertilizers are sometimes used to increase plankton bloom, as this produces more oxygen. The rate is typically 40 - 50 pounds of superphosphate per acre. Fertilizers containing nitrogen should not be used with low oxygen. Chemicals tend to take longer than mechanical aerators for results.

DISPOSING OF WATER

Properly disposing of used water is essential. Such water, also known as

Figure 6–18. Floating electric-powered aerator. (Courtesy, Master Systems)

Figure 6–19. Floating paddlewheel aerator.

Figure 6–20. Farm tractor with a portable paddlewheel aerator backed into the water. (Courtesy, Delta Pride Catfish, Inc.)

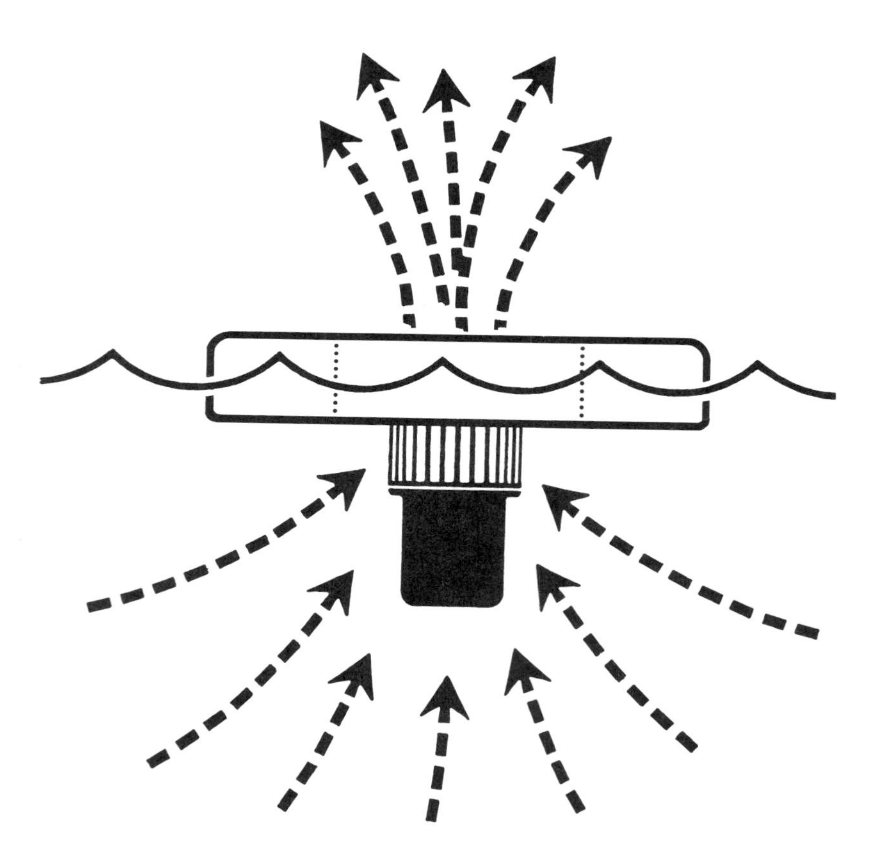

Figure 6–21. Floating aerator design showing movement of water from the pond, through the aerator, and into the air. (Courtesy, Kasco Marine, Inc.)

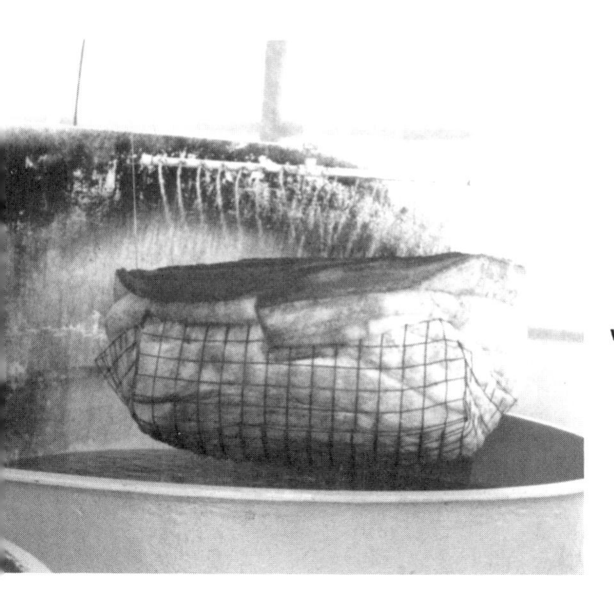

Figure 6–22. Filtration system in a small water recirculation system.

Figure 6–23. Water treatment system in a recirculating fish farm operation.

effluent, may come from aquafarms, processing plants, or other sources. On farms, it has often been used to produce aquacrops. It may contain various residues from the production process. Processing plants use water in washing the products, cleaning the facilities, and in other ways. This water contains blood, fish tissue, excrement, and other materials that cannot be released into a stream or a lake.

The water must be prepared for release. Solids must be removed from it. This is often done by allowing the water to stand in a lagoon so that the solids settle out or are otherwise removed. After removal, the solids may be disposed of in a land fill. Some treatment may also be needed to remove gases and chemicals from the water. Other filtration systems are sometimes used.

Water from an aquaculture facility should never be dumped into a stream or lake without being tested. All legal regulations must be strictly observed.

SUMMARY

As the environment in which aquacrops grow, water is very important.

Water quality refers to the suitability of water for use with an aquacrop. Potential aquafarmers must carefully study the water that is available from wells, streams, lakes and oceans, springs, industrial effluent, municipal systems, and runoff. The farmers must determine if it is suitable for their proposed aquacrops.

How water is managed is a key to successful aquafarming. Important areas to be managed are dissolved oxygen content, nitrogen compounds, turbidity and color, weeds and algae, and plankton bloom.

Skills with simple mathematical calculations are needed to determine water volumes and amounts of chemicals to use. Excess and used water should be properly disposed of by the aquafarmer.

QUESTIONS AND PROBLEMS FOR DISCUSSION

1. Why is quality water important?
2. Why is it important for the aquacrop to be carefully matched with the characteristics of the water?
3. What is a water impoundment? What kinds are used?
4. What factors should be considered in the selection of a water facility?
5. What three characteristics of water influence its use in how an aquacrop will grow?
6. What is water chemistry? Explain pH and salinity.
7. What are the physical characteristics of water?
8. What is water biology?

9. What are the sources of water for aquaculture? Briefly describe each.
10. How is the capacity of a water pump measured? Why is this important?
11. What is dissolved oxygen? Why is it important? What levels are needed in water for aquaculture?
12. Why is water temperature important with aquaculture crops?
13. What is water hardness?
14. What are the sources of water pollution in aquaculture?
15. What is the origin of nitrogen compounds in water?
16. What are the symptoms of possible oxygen problems in water?
17. What are the causes of oxygen depletion?
18. How is oxygen added to water?
19. What can be done to prevent problems with nitrogen compounds?
20. What is turbidity? How can it be a problem in aquaculture?
21. What is plankton? How is it important in aquaculture?
22. How many cubic feet of water are in a tank that is 15 feet long and 3 feet wide with water 3 feet deep? How many gallons are in the tank? How long will it take to fill the tank at the rate of 100 gpm?
23. What methods are used to control weeds and algae? Briefly describe each.
24. What three kinds of weeds are found in aquaculture? How are they different?
25. What are the two ways water pollution is a concern in aquaculture?
26. How is used water released from an aquaculture facility?

Chapter 7

PEST CONTROL

Aquacrops have pests just as do the livestock and row crops traditionally grown on farms. Knowing what is involved can help reduce losses. Good managers are always on the alert for problems. Of course, the key is to prevent problems. It is much easier to prevent pest problems than it is to eradicate them after an outbreak.

Many factors are involved in pest control. These include getting only problem-free stock, using quality water, and following good cultural practices. Some-

Figure 7–1. Healthy hybrid striped bass. (Courtesy, Gary Fornshell, Mississippi State University)

times visitors carry pests from one aquafarm to another on their shoes. For this reason, some farms limit visitors. Quarantine and sanitation are also an important part of pest control.

OBJECTIVES

This chapter focuses on controlling pests in aquaculture. Upon completion, the student will be able to:

- Explain the meaning of aquaculture pests.
- Explain how pests cause problems.
- Define "disease" and explain how fish show disease.
- Describe the kinds of fish diseases and control measures.
- Define "parasite" and explain how parasites affect fish.
- Explain how parasites may be controlled.
- Define "predator," explain how predators cause losses, and describe ways of control.
- Define "trash fish," explain how they cause losses, and describe ways of control.
- Explain what to do if fish appear sick.
- Explain how to select and ship fish samples.
- Describe the role of regulations in pest problems.

(Note: Weed control is covered in Chapter 6, "The Importance of Water.")

COMMON AQUACULTURE PESTS

An aquaculture pest is usually a plant or an animal that is detrimental to the aquacrop. This means that a pest in some way interferes with production. Pests may be large organisms or one-celled plants or animals. They may compete with the aquacrop for nutrients, consume the aquacrop, or attack the crop so that normal life functions are changed. In some cases, pests attack the water structure in which the crop is being grown.

All aquacrops are susceptible to damage by pests. Plant aquacrops may be damaged by insects, large animals, plant diseases, and other pests. Animal aquacrops may be damaged by diseases, parasites, water birds and other water animals, and similar pests.

The environment in which aquacrops grow tends to be a good place for pests to grow also. Water with high crop populations has many potential problems. Several reasons for this are:

1. High concentrations of aquacrops allow for more rapid transfer of pests from one crop to another.
2. Handling aquacrops (such as in hauling) causes stress, and stress in turn reduces resistance to some pests.
3. High populations in ponds, open tanks, and raceways are easy for predators to catch.
4. Uneaten feed and fish excrement and excess fertilizer provide nutrients for the growth of weeds.
5. Soft earth near water and uneaten feed attract animals that burrow into levees and dams.

PESTS CAUSE LOSSES

Pests are of concern in aquaculture because of the losses they can cause. In some cases, the losses are small. In other cases, the losses may involve the entire crop. Any losses cut back on the profit that can be made from an aquacrop.

Pest damage may occur in two major ways: direct and indirect. Sometimes the aquacrop is attacked directly, as when a disease causes the death of fish. At other times, pests compete for nutrients, oxygen, and other essentials for life. This indirect damage causes the crop to grow more slowly. Farmers' investments may be lost to the pests, such as when trash fish eat feed intended for a crop of fish.

Examples of pests and the losses they can cause are:

Direct Losses

1. Water birds and other predators — catch and eat or injure fish.
2. Diseases — impair the body functions of fish, resulting in reduced growth rate or death.
3. Parasites — take nourishment from the fish, reducing growth; make fish (the host) more susceptible to diseases.
4. Poaching — theft of aquacrops by humans. (Yes, humans are pests when they steal fish!)

Figure 7–2. All of the steelhead trout shown here are the same age. Variations in diet and other health conditions have resulted in different rates of growth. (Courtesy, Maine Agricultural Experiment Station)

Indirect Losses

1. Weeds—grow in water, tying up nutrients and making harvest difficult.
2. Trash fish—undesirable stray fish that consume feed intended for the fish crop; also compete for oxygen and can introduce diseases and parasites.
3. Rodents and other burrowing animals—burrow into earthen levees, possibly causing water leaks or weakening them so that they collapse.

DISEASES

A disease is a condition that develops in an aquacrop that damages it in some way. The fish or other crop may be unable to carry out normal body functions. Diseases may cause fish to fail to grow or to grow at a reduced rate. In some cases, the fish may die. Diseased fish are unfit for human consumption. Processing plants will reject them. In any case, the owner of the fish loses money when disease outbreaks occur.

How Fish Show Disease

Fish that are healthy exhibit a certain behavior and appearance. They appear alert and active. When they become diseased, they change. Aquafarmers must regularly observe aquacrops for these changes, or symptoms, which are evidence that a disease exists. Symptoms are not causes of disease. Causes result in or bring about the symptoms. Causes may require laboratory analysis for identification. For example, many diseases are caused by germs. These germs are tiny bacteria, viruses, or fungi. A microscope is needed to see them.

The most common ways fish show disease are by not eating, developing sores (lesions), lacking vigor, developing abnormal body shapes, changing their behavior, and dying.

Going off feed. Fish that are not healthy will not eat when they are fed. The alert fish farmer pays careful attention at feeding time. Healthy fish normally eat a few minutes after feed is put in the water. If not, the cause should be quickly determined. Sometimes it may not be due to a disease but to a water problem. For example, fish may stop eating when the oxygen is low. Feed that floats on the water is particularly good in that the fish must come to the surface to eat. Of course, some fish are bottom feeders and do not respond well to floating feed.

Lesions and discoloration. Diseases may show up as ulcers, lesions, abscesses, or cysts on the skin or gills or as discoloration. Ulcers are open sores on the skin that fester and contain pus. They are different from lesions, which are the result of injuries or wounds. Abscesses are swollen areas in the tissue of the body. Pus usually gathers at abscesses. Cysts are abnormal pockets (sac-like structures) filled with fluid or diseased flesh.

The experienced fish farmer knows the normal color of fish. For example, gills that are pale show that the fish has a problem. When color changes, a disease may be present. Certainly, sores on a fish indicate that something is wrong.

Lack of vigor. Most fish can move vigorously through the water. When they stop doing so, this may be the sign of a disease. This is known as reduced vigor. Fish quickly swim away from apparent danger. To them a splash in the water may indicate danger. Fish are not always moving vigorously, as they are sometimes in a resting state. Fish that appear sluggish, lose their balance, and have drooping or folded fins lack vigor. Knowing the normal vigor of a fish is very beneficial in sensing changes that are due to disease.

Abnormal body shape. Abnormal body shape may indicate that a fish is diseased. Growths on the body, particularly around the mouth and fins, may result from disease. Swollen bellies and protruding eyes are symptoms of disease. Sometimes water pollution can result in abnormalities. At other times, abnormalities may be the result of genetic variations.

Figure 7–3. Eroded ulcer on body of fish caused by *Flexibacter columnaris* disease. (Courtesy, Eddie Harris and Pete Taylor, Mississippi Cooperative Extension Service)

Figure 7–4. Tiny pinpoint hemorrhages on the skin of this fish are caused by bacteria. (Courtesy, Eddie Harris and Pete Taylor, Mississippi Cooperative Extension Service)

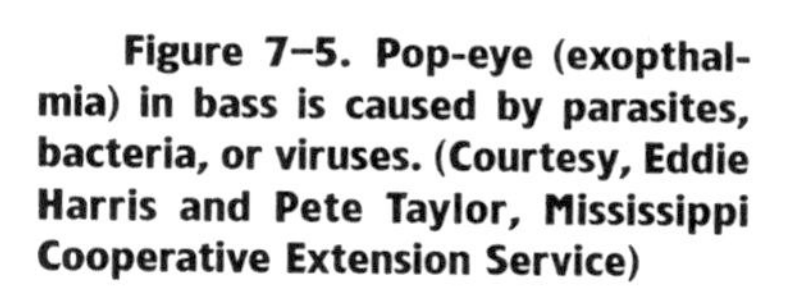

Figure 7–5. Pop-eye (exopthalmia) in bass is caused by parasites, bacteria, or viruses. (Courtesy, Eddie Harris and Pete Taylor, Mississippi Cooperative Extension Service)

Change in behavior. Any time fish change behavior, a problem should be suspected. It is important to know normal behavior. This can be learned through watching fish. When fish gather near incoming water, scratch their bodies on vegetation or rocks, go into shallow water, or "gulp" at the surface of the water for air (sometimes called piping), they are showing that a problem may exist. The alert fish farmer will check out the situation.

Dead fish. When fish die, they usually float on the top of the water for a while. Any time dead fish appear, some concern is justified. The cause of death should be determined as quickly as possible. This may require a laboratory analysis. The aquafarmer should determine if there is plenty of oxygen in the water and then look at the fish to see if they appear normal. All dead fish should be immediately removed from the water. Occasionally a fish will die when there is no disease problem, and this should not be cause for alarm.

Kinds of Disease

Fish diseases can be put in two broad categories: infectious and noninfectious.

Infectious diseases. Infectious diseases are caused by germs or pathogens (disease-causing organisms) and are capable of being transferred from one fish to another. They are living organisms that attack the tissues of fish. The germs produce poisons which may kill the fish. There are four kinds of infectious diseases: bacterial, fungal, viral, and parasitic.

1. *Bacterial diseases*—These diseases are caused by bacteria. Most are in-

ternal diseases, though a few may cause problems on the skin and gills. Laboratory analysis is needed to confirm a bacterial disease. Bacteria are microscopic and are about 3⁄25,000 of an inch long.

2. *Fungal diseases*—These diseases are caused by tiny plants known as fungi. Fungal diseases usually indicate that fish have other problems, as the fungi grow on dead flesh. A fish that has been injured in some way is particularly vulnerable to these diseases.
3. *Viral diseases*—These diseases are caused by viruses that are often smaller than the power of an ordinary microscope can detect. They are difficult to treat with drugs and chemicals.
4. *Parasitic diseases*—These diseases are caused by parasites. (More detail is presented in another section of this chapter.)

Noninfectious diseases. Noninfectious diseases are not caused by pathogens and are not transferred from one fish to another. These diseases may be due to diet, environment, chemicals, and / or physiological changes.

1. *Nutritional diseases*—These diseases are the result of an inadequate diet. Usually, the fish do not receive enough food with the proper nutrients. Sometimes too much feed can cause problems.
2. *Environmental diseases*—The water environment in which fish grow may

Figure 7–6. These fish are suffering from a nutritional disease caused by a deficiency of vitamin C. (Courtesy, Marguerite G. Gravois and Fish Farming Experimental Station, Stuttgart, Arkansas)

bring about diseases. Gases in the water may cause gas bubble disease. Other substances can result in fish that lack vigor and develop secondary problems.

3. *Chemical diseases*—A range of toxic substances may get into water. Pesticides from agricultural crops may drift into fish ponds or water supplies. Fish may become stunted or die from chemical toxicity.
4. *Physiological diseases*—Malfunctions of organs or other life processes result in physiological diseases. Sometimes the blood may have a sudden change in pH, or feeding too near the time of harvest can cause problems.

Disease Control

Prevention is the best way to control fish diseases. Good management is the key. Poor management may allow disease problems to occur that could have been avoided. After a disease occurs, some kind of treatment is necessary. Very few treatments are approved for use on fish for human consumption. Those that are available tend to be expensive and sometimes difficult to administer.

Prevention. Knowing good management practices is important in prevention. Proper nutrition can help fish to be more resistant to diseases. Stress reduces the resistance of fish to disease. Stress often occurs when fish are hauled and handled. Being careful not to inflict injury can go a long way in controlling stress. Fish become stressed when their environment is suddenly changed. High water temperature, low oxygen level, crowding, pollution getting into the water, and excessive accumulation of wastes can also trigger stress.

Two procedures routinely used in prevention are sanitation and quarantine.

Sanitation involves keeping water and facilities clean. Properly washing tanks after use can remove any disease agents that may remain in the tanks. Water from tanks should not be dumped into growing facilities. People can transfer disease from one pond or farm to another on their feet. Farms with disease problems should inform people who visit there or who may buy fish. Visitors should disinfect footwear upon arrival or departure. Farms that are disease-free may require visitors to dip their footwear into a disinfectant solution, such as iodine. Some fish farms do not allow visitors.

Quarantine involves isolating fish from each other. Isolation is used when new fish are brought onto a farm. It involves keeping them in separate facilities for a while (normally two weeks) to see if any diseases develop. Fish that develop disease should be kept separately. Whole farms may be quarantined when a disease outbreak occurs. Quarantine is needed even if the fish are being treated for the disease.

Figure 7–7. Visitors can transfer diseases on their footwear. This farm has protective fencing and a gate with a "keep-out" sign. The dip pan inside the gate is for people who enter to treat their shoes or boots in a chemical solution that disinfects the footwear.

Several important management considerations in preventing disease are:

1. Use plenty of disease-free water.
2. Stock only healthy fish.
3. Quarantine new fish two weeks before adding to other fish.
4. Use prophylactic treatments when fish are being hauled. (These are treatments applied to prevent disease. Small haul tanks make it easy and economical to apply treatments.)
5. Feed fish properly.
6. Control aquatic weeds.
7. Disinfect equipment when it is moved to prevent the spread of disease from one site to another.
8. Properly treat empty tanks and ponds. Allowing tanks and ponds to dry thoroughly between uses destroys some organisms that may cause disease.
9. Restrict visitors and provide for sanitation of footwear (as with dipping shoe soles in an iodine solution).

Treatments. The kind of treatment to use depends on the disease. This means that it is very important to have an accurate diagnosis of the disease before beginning treatment. Very few treatments have been approved by the U.S. Food and Drug Administration for use on food fish. Specific drugs to use should be determined as the need arises. Individuals should contact their local agriculture teacher, county extension agent, or land-grant university for these recommendations.

Several methods of treating fish are available. Some are practical only under certain situations. Several treatment methods are described here.

1. *Dipping* — This involves dipping fish into concentrated solutions for a few seconds (usually 15 - 45 seconds). Dipping can be used with a small number of fish, such as with broodfish or fingerlings. It is impractical to try to dip large numbers of fish, as may be found in a pond.
2. *Feeding* — This involves adding medications to feed. The medication should be thoroughly mixed in the feed so that all fish get the treatment. Large quantities of fish can be treated in this way. When a few large fish are involved (such as broodstock), a capsule with medication can be placed in the stomach of fish with a balling gun.
3. *Bathing* — Chemicals may be added to the water in tanks or vats for a period of time (an hour or so). It is important to correctly calculate the amount to use and the length of time the bath should last. A large quantity

Figure 7–8. Fish can be bathed in vats containing water with chemicals added for a period of time. (Courtesy, Master Mix Feeds)

of freshwater is added to flush out the chemicals after the treatment. Disposal of water containing the chemicals may be a problem.

4. *Injecting*—Some drugs can be administered to small numbers of fish with a hypodermic needle and syringe. This method is appropriate only with fish large enough to hold. Medications act more rapidly when given by injection than when put in feed or baths.
5. *Indefinite (or prolonged)*—Low concentrations of chemicals can sometimes be added to water for long periods of time. This method can be used with ponds and tanks. It is important for the chemical to be evenly mixed in the water when it is applied. Such treatments are useful only with chemicals that break up and disappear with the passage of time.

PARASITES

A parasite is a plant or an animal that lives in or on another plant or animal. The plant or animal on which a parasite lives is known as the host. Normally parasites cause harm by chewing or sucking fluids from the host or by taking nutrients from the food in the digestive tract. Parasites are often considered a form of infectious disease.

External parasites attach themselves to the outside bodies of fish. They may be found on the skin, gills, and / or fins. They get food by eating or sucking on the fish. Fish with external parasites frequently try to scratch themselves on vegetation or rocks to remove the parasites.

Internal parasites live in the organs or digestive tract of fish. They cause damage to the organs and make the fish susceptible to other diseases by weakening them. In the digestive tract, they live on the nutrients in the feed that the fish has eaten.

Kinds of Parasites

Fish may have several kinds of parasites. Some are very small and visible only with a microscope; others are large and visible to the eye. The most common parasites are briefly described here.

- ***Worm parasites***—Fish may be attacked by several kinds of worm parasites. Tapeworms, roundworms, leeches, and flukes are common examples. All are usually internal except for leeches, which attach themselves to the outside of fish and suck blood. Worm parasites can cause serious damage to fish. Some forms of flukes may live as external parasites, particularly on the gills of fish.

- ***Crustacean parasites*** — These parasites are sometimes known as fish lice. They resemble insects and have a hard outer shell. One kind attaches itself to the gills of fish. These parasites burrow into the skin or gills and can usually be seen with the naked eye. Crustacean parasites can transmit infectious diseases among fish. These parasites can also be transferred by birds and other animals from one pond to another.
- ***Protozoan parasites*** — These are one-celled parasites that live in water. They may become a problem when fish are stressed, as when fish have inadequate nutrition or when the water quality is poor. Symptoms of protozoan parasites include small, bloody spots on the fins, ragged fins, and loss of appetite. Some kinds of protozoa (known as sporozoa) form cysts in the organs or skin of fish. These cysts may break open, releasing many spores. An example of a protozoan parasite is "Ich," or "white spot."

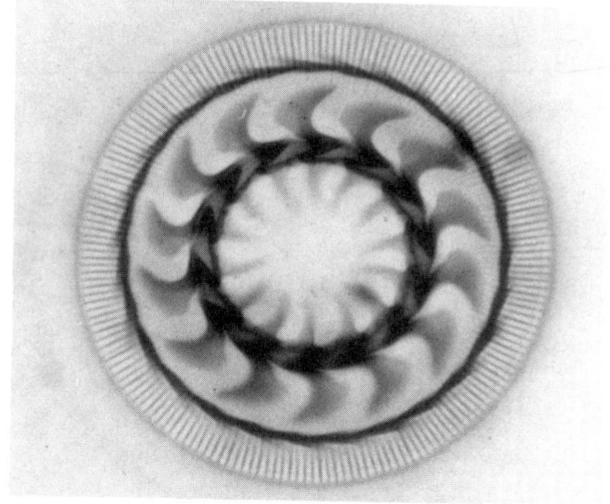

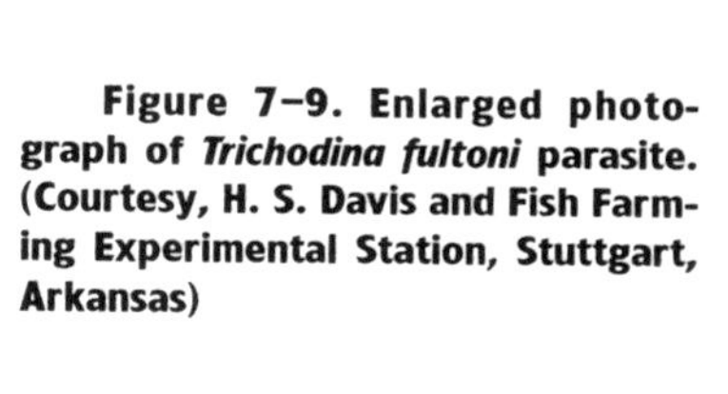

Figure 7–9. Enlarged photograph of *Trichodina fultoni* parasite. (Courtesy, H. S. Davis and Fish Farming Experimental Station, Stuttgart, Arkansas)

Parasite Control

As with diseases, the control of parasites is best accomplished by prevention. Several ways of controlling parasites are included here.

- ***Quarantine*** — This involves isolating any new fish from existing fish. If parasites are observed, some form of chemical treatment might be appropriate to use. A quarantine period of two weeks is usually considered adequate.

- ***Treatment of facilities***—Some parasites can be controlled by drying and disking pond bottoms between fish crops. Calcium hypochlorite can be applied to pond bottoms to sterilize them. Tanks and vats can be thoroughly dried and treated with a disinfectant.
- ***Dipping***—External parasites are treated by dipping the fish into a chemical solution. For example, fish with leeches can be dipped for a short while in a 3% solution of salt (NaCl). The fish should be removed from the solution when they begin to show signs of stress. (Dipping in saltwater causes the leeches to turn loose from the fish and drop to the bottom of the container.) Other parasites may require that fish be dipped for one hour in a 200 ppm solution of formalin.
- ***Elimination of birds***—Birds often transfer parasites from one pond to another. In some cases, a part of the life cycle of a parasite is spent inside a bird. Controlling the birds and removing roosts where the birds light is helpful. Sometimes screens are put over fish facilities to keep birds away.

Figure 7–10. Emptying ponds, allowing them to dry, and disking the bottoms will control some parasites.

Some species of birds are protected by law and, therefore, cannot be killed without a permit.

- ***Indefinite*** — As with some diseases, large quantities of fish are treated in ponds at a low level of concentration for a longer period of time. For example, fish with Ich may be treated with formalin at the rate of 25 ppm on alternate days for two weeks. Care should be used with such treatment. Chemicals react differently, depending on the water chemistry. Also, food fish should be treated only according to the regulations of the Food and Drug Administration. (Calculating treatment amounts was presented in Chapter 6.)

- ***Feed additives*** — Feed additives may be used to treat fish infected with certain internal parasites, particularly tapeworms, roundworms, and flukes. These additives should be carefully used and only as approved.

PREDATORS

Fish are subject to attack by other animals, which prey on fish. Farmers can experience considerable losses from predators. Common predators are other fish, birds, snakes, insects, turtles, and alligators.

- ***Other fish*** — Some species of fish prey on other fish; others prey on the young of their own species. Knowing something about the nature of fish will help in understanding how to respond in managing them. Various techniques are available to separate fish. Sometimes undesirable trash fish get into a pond and become predators. Using water from a good source is the best way to control this problem. After harvest, the water should be cleaned to remove predatory trash fish before the next crop is stocked.

- ***Birds*** — Several species of birds are known to prey on fish. These can cause large losses over a growing season. Cormorants, kingfishers, herons, grebes, and mergansers are some examples. The double-crested cormorant has been found to be a particular pest. This bird can eat 1 or 2 pounds of fish a day. Over a growing season, the cormorant could consume 200 or more pounds of fish. If several were present, the losses could quickly add up to thousands of pounds of lost fish. Some species of birds are protected by law and cannot be killed without a permit. Control measures include placing a screen or net over the fish facilities, using loud, exploding cannons, and frightening away with scarecrow-type devices. Constructing ponds with a minimum of shallow water so that there are no places for

Figure 7-11. This gas cannon produces a loud noise to frighten away birds. It does not harm the birds. (Courtesy, Aquacenter, Inc.)

birds to stand can be helpful. However, controlling water depth will not deter birds that swim and dive to catch fish.

- ***Snakes*** — In addition to preying on fish, snakes pose certain hazards to people who work around ponds. Many fish farmers do not consider snakes to be consumers of large quantities of fish. Mowing levees closely to prevent the growth of tall weeds and grass and removing places where snakes can hide will help control them. Killing them when they are accessible is also another means of control.

- ***Insects*** — Water insects are a minor problem in most fish farm operations. Some insects may consume eggs or fry. Spraying with insecticides is sometimes done but is usually not recommended. Insecticides are poisonous and can make fish unfit for human consumption.

- ***Turtles*** — Turtles cause problems in several ways. Some species prey on fish. All species compete with fish for feed and space in the growing facility. Turtles are particularly objectionable at harvest time, as they must be removed from the fish crop. Hand removing can be hazardous because some turtles bite. Trapping or shooting with a rifle is one means of control. However, some species of turtles are protected by law; thus, a game warden or conservation officer should be consulted about the regulations.

- ***Alligators*** — Rarely do alligators cause problems as predators, but they have been observed with fish in their mouths. The greatest danger may be to people who work around fish growing facilities in areas where alligators grow. The alligators should be trapped and then moved away from the fish farm.

- ***Bullfrogs*** — Bullfrogs cause losses by killing fry and fingerlings. They also compete with the aquacrop for space and food, particularly while in the tadpole stage. Control is limited to placing a poison (copper sulfate) on the egg masses and trapping or catching.

TRASH FISH

Undesirable fish that are present in a crop of fish are known as trash fish. An example is the presence of tilapia in a catfish pond. Sometimes trash fish are referred to as wild fish. Trash fish can cause losses in several ways:

1. They eat the feed provided for the desired fish crop.
2. They compete for oxygen and other nutrients.
3. The fish crop grows more slowly because of the competition.
4. Labor requirements at harvest are greater because the trash fish have to be separated from the fish crop.
5. They may bring diseases and parasites to a fish farm.
6. Some may also be predators and consume the fish crop.

Trash fish reduce the profit to the fish farmer. More feed, labor, and energy to oxygenate the water are required. The costs for controlling diseases and parasites are also greater.

Sources of Trash Fish

Trash fish get into a fish facility in several ways. Through good management, they can usually be kept out.

Only good-quality water should be used. Water from wells is free of trash fish. Water from streams, lakes, and other surface runoff may be contaminated with trash fish.

The construction site for a pond may contain wild trash fish. A pond built across a creek may get trash fish from the water in the creek. Puddles of water present in a construction site may be a source of trash fish. Rivers may overflow their banks and take trash fish into fish farms during the rainy season.

Fish stock may contain trash fish. Fingerlings may have undesirable fish growing with them. Buying stocker fish only from reputable dealers will help minimize the problem.

Occasionally, a fish-eating bird will transport a trash fish to a fish farm. The bird may catch the fish in a nearby creek or lake and fly to the fish farm where the fish escapes uneaten from the bird's beak.

Control of Trash Fish

Control involves keeping trash fish from entering the fish farming facility. Water from streams or lakes needs to be filtered before it is pumped into the aquaculture facility. Sock-type filters may be placed over the ends of pipes. In some cases, box filters constructed of wooden frames with screens are used. Both types of filters should be cleaned regularly to remove the screened out trash fish. If not, the screen could get overloaded and break, thus releasing the trash fish into the growing facility.

After harvesting, fish ponds can be emptied to remove the trash fish. Re-filling the pond can be a costly, time-consuming operation.

After the fish have been harvested, some farmers treat the water in a pond with a chemical to kill unwanted fish. Rotenone has been used for this purpose. When used at the rate of 0.5 - 2.0 ppm, it will kill most trash fish. Rotenone and other chemicals should be used only as approved by regulatory agencies.

Biological control is sometimes used. It has not always been practical. For example, when some farmers stock their ponds, they stock a species of fish that preys on the trash fish but not on the fish crop. However, this requires sorting at harvest time and can result in some of the other problems caused by the trash fish themselves.

WHAT TO DO IF FISH GET SICK

If fish begin to show signs of being sick, quick action is needed. Accurate diagnosis of the situation is essential. Delaying can result in the loss of an entire fish crop. Here is a list of things the fish farmer should do:

1. Stop selling fish to farmers, processors, and others until the problem is solved.
2. Follow quarantine and sanitation procedures to reduce the chance of the problem spreading.
3. Get the advice of a local aquaculture specialist. (The local agriculture teacher or extension agent may assist in this.)
4. Contact another fish farmer in the area about the problem.
5. Contact a local supplier of feed or chemicals or a local processing plant.
6. Contact fishery specialists at the state land-grant university.
7. Contact the nearest fish disease diagnostic laboratory to make the necessary arrangements to have a sample of the fish analyzed.

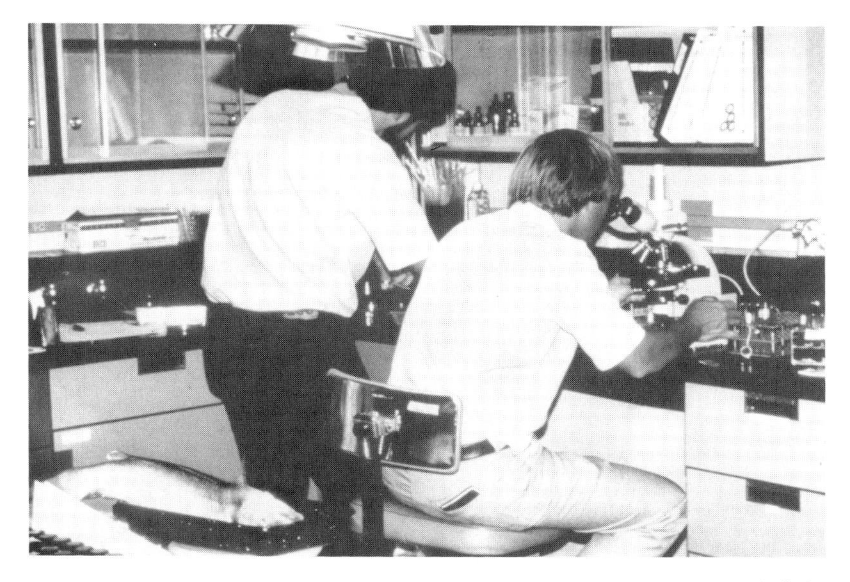

Figure 7–12. Accurate diagnosis usually requires careful laboratory analysis by trained scientists. (Courtesy, Louisiana Agricultural Experiment Station – Louisiana Cooperative Extension Service, Louisiana State University Agricultural Center)

8. Follow the necessary procedures to solve the problem. (Always read and follow the instructions on any chemicals that may be used.)
9. If the crop is insured, contact the insurance agent.

Some farmers set up small diagnostic laboratories on their farms. A few analyses can be made with microscopes and rather simple procedures. Of course, training in how to make analyses is essential.

Once a diagnosis has been made, treatments require consideration of a number of factors. The effectiveness of all treatments depends on water chemistry, temperature, accurate and uniform application, and dosage level.

SELECTING AND SHIPPING FISH SAMPLES

Samples of diseased fish must often be carefully examined in a laboratory. These fish are identified on the basis of the symptoms they exhibit. Diagnostic laboratories are usually equipped to examine both live and dead fish. It is extremely important that the sample examined be representative of the problem among the fish.

Selecting a Sample

Live fish provide the best sample. Even though they may show symptoms, they may be hard to catch. Only fish that have the symptoms should be sent to a laboratory. Such fish can be caught with a net or seine, but a little patience will be needed—catching them will not be easy. Fish taken at random from a seine may not have the disease. Fish caught by hook and line from different areas of a water facility are poor samples.

Figure 7–13. This fish with *Pseudomonas* lesions has almost passed the stage of usefulness in laboratory analysis. Samples of diseased fish should be good enough to allow accurate diagnosis. (Courtesy, Fish Farming Experimental Station, Stuttgart, Arkansas)

Dead fish are not as good for laboratory analysis. The probability of an accurate diagnosis is reduced. Dead fish that have red gills and somewhat normal color and mucus might be okay. Dead fish that have lost body color, have no mucus, and have white gills are of no value and should not be sent. Of course, it is much easier to collect a sample of dead fish!

Although water samples are unusable with most diseases, sometimes when fish show signs of problems, water samples are analyzed to determine if there is a water problem. Also, because treatment procedures often are based on the chemistry of the water, an analysis of the water can help provide needed information for administering treatments.

Shipping Fish Samples

Fish samples must arrive at the diagnostic laboratory in a suitable form for analysis.

Live fish can be placed in a plastic bag that is sealed up. The bag can be put in an ice chest with crushed ice. If the fish will be hauled only a short distance, they can be placed in a container with well-oxygenated water and a little ice to keep the water cool.

Sometimes fish are frozen for shipping. Frozen samples are very difficult to work with in a laboratory, but they are satisfactory for pesticide analysis.

Dead fish samples should be iced down immediately to retard further tissue breakdown. It is impossible to analyze dead fish that have deteriorated. Thus, only dead fish that are still in good condition should be sent. Dead fish should be shipped in an ice chest with plenty of crushed ice.

OTHER PESTS

Aquafarms are sometimes subjected to damage by other pests, such as muskrats and beavers, crawfish, and livestock. Muskrats and beavers primarily cause problems by burrowing into levees, thus causing water leaks and weakening water structures so that they might break.

Closely mowing areas around ponds will control burrowing animals. Using rock rip-rap on levees at the water line may be helpful; however, these rocks may be a problem at the time of fish harvest. Traps can be used to catch burrowing animals. Sometimes other methods that are legal and approved by the local game authorities can be used to eradicate the animals.

Livestock should be kept out of aquaculture facilities. Not only do they muddy the water, but they also may damage levees and create deep paths that make the area around a pond rough. Fences will keep livestock out.

It is interesting to note that some pests are valuable for their own worth. Crawfish are grown in some areas as a valuable aquacrop. Beavers produce quality fur. Livestock are important sources of food. Some pests are "animals out of place," much as weeds are plants growing where they are not wanted.

REGULATIONS IN PEST CONTROL

Farmers cannot control pests in just any way. Established regulations must be followed. These regulations are intended to protect the environment, consumer, and species of animals that may be endangered.

Several areas of regulations in pest control are briefly described here.

- ***Food safety***—Fish crops that are to be used for human food must be protected from contamination with hazardous chemicals. Only a few chemicals have been approved by the Food and Drug Administration for use on fish crops. And those that are approved must be used according to the instructions! Before processing, processing plants will condemn contaminated fish. All regulations of the Food and Drug Administration, the U.S. Department of Agriculture, and other regulatory agencies must be followed.

- ***Feed safety***—Just as human foods should be free of harmful contamination, feed for animals should be also. Many times fish by-products are used in manufacturing feed for fish. If the by-products contain hazardous substances, the feed will be unfit for use with fish. If it is eaten, the fish crop will be contaminated.

- ***Water pollution***—Water that has been treated with certain chemicals is polluted and cannot be released into streams and lakes. A system for holding and cleaning up the water must be in place. Normally, the water cannot be run into a municipal sewage system for disposal.

- ***Protected animals***—Some of the pests in aquaculture are protected animals. Individuals are restricted to certain practices in attempting to control them. Protected animals cannot be killed without a permit, and some permits are difficult to get. The alternative is to use loud noises, sound waves, bright flashes, or other means to frighten the animals away. Another alternative is to catch and move them to another location.

- ***Worker safety***—Some chemicals are hazardous to those who use them. They must be handled in a safe manner. Protective clothing must be worn. Exposure to these materials might result in harm to the user. If exposed, the individual should immediately take the necessary precautions. Exposed skin should be promptly washed.

- ***Chemical storage***—Specific storage instructions apply to some chemicals. These instructions should be carefully followed.

- ***Container disposal***—Containers in which chemicals have been held must be properly disposed of. Empty cans should not be thrown into creeks, ponds, or behind buildings to be ignored. The approved procedures for disposing of pesticide containers should be followed. **It is important to read the labels for details!**

- ***Cleaning tanks***—Vats and tanks in which fish have been treated must be cleaned. The water should be properly handled and not run into streams or lakes. Individuals should be careful to avoid getting the water on their skin. The runoff water from washing tanks should not be used as drinking water for livestock or pets. The water should not be allowed to run into fish growing facilities. Haul tanks can be sources of disease, parasites, and trash fish. Thus, water from these tanks should never be emptied into aquaculture facilities.

- ***Release of illegal trash fish*** — States may have laws banning certain fish. Sometimes these may be trash or exotic pet-type fish. They cannot be dumped into streams as a means of disposal. For example, striped bass farmers view tilapia mixed in their crop as trash fish. Some states may have laws against the release of tilapia; therefore, a farmer cannot dispose of them in a nearby creek.

Information on legal regulations is available from several sources. The local agriculture teacher or county extension agent is one source. All states have land-grant universities with staff qualified to help. Most states have state regulatory agencies that can provide assistance. Some areas have private consulting laboratories that can help with problems. Regardless, there are many sources that can provide the needed assistance with regulations. Ignorance of the law is not an excuse!

SUMMARY

Most aquacrops are subject to damage by pests. The water environment is a good place for pests; conditions good for the growth of aquacrops are also good for some of the pests. Other pests thrive on stressed or weakened aquacrops. Pests cause both direct and indirect losses. Diseases, parasites, and predators directly attack aquacrops. Weeds, trash fish, rodents, and other animals cause indirect losses by competing with the aquacrops.

Fish show pest problems by going off-feed, developing sores and abnormal body shapes, losing vigor, becoming discolored, changing behavior, and / or dying. The fish farmer must know enough about fish to know their normal behavior (observing their behavior at feeding time is very important). Prevention of pest problems is best. Quarantine of new fish and sanitation are important in prevention. Treating problems is more difficult. Treatments may be administered by dipping or bathing the fish in a chemical solution, adding medications to feed, injecting with medications, or exposing fish to lesser concentrations on a long-term basis.

Predators, trash fish, and other pests may also cause problems. Good management can help to prevent losses from them.

Diseases and parasites must be properly diagnosed for treatments to be given. Samples of fish should be carefully selected and sent to a diagnostic laboratory. Regulations pertaining to using chemicals on aquacrops should always be carefully followed.

QUESTIONS AND PROBLEMS FOR DISCUSSION

1. What is a pest? Why are aquacrops subject to pests?
2. How do pests cause losses?
3. What is a disease?
4. How do fish show that they are diseased? What are the symptoms?
5. What kinds of diseases may attack fish? Briefly describe each.
6. How are diseases prevented?
7. What treatments are used with fish? Briefly explain each.
8. What is a parasite? What are the differences between internal and external parasites?
9. What are the common kinds of parasites?
10. What are the methods of parasite control? Briefly explain each.
11. What are predators? What kinds cause problems in fish?
12. What are trash fish? How are they controlled?
13. What steps should be taken if fish are observed to be sick?
14. How are samples of diseased fish selected for laboratory analysis?
15. How are samples of fish shipped for analysis?
16. How do burrowing animals cause problems in aquaculture?
17. What regulations should be observed in controlling pests in aquaculture?

Chapter 8

AQUACULTURE FACILITIES

Once an aquafarmer has determined the type of aquacrop to produce, the next step is to decide on what kind of facility in which to produce the aquacrop. Many variables influence this decision. Certain aquacrops require certain types of facilities. The amount of capital available to the aquafarmer determines whether certain types of facilities are affordable. The available land, sources of water, and other resources must be considered. Also, the amount and intensity of production will determine what type of facility is necessary. The soil type and local climate will also be important factors.

This chapter describes common facilities used for aquaculture. Included in the discussion are the site selection for and the construction and maintenance of the various facilities; suitable species for different types of facilities; and an overview of the various types of equipment needed with the different types of facilities.

OBJECTIVES

The following objectives should help the student develop an understanding of the role of facility type in aquaculture. Upon completion of this chapter, the student will be able to:

- Describe the common types of facilities used in aquaculture production.

- Discuss factors that influence the site selection process for aquaculture facilities.
- Determine the various methods of constructing aquaculture facilities.
- Associate species with facilities that are suitable for their production.
- Describe the equipment necessary for aquaculture production in the various types of facilities.

COMMON FACILITIES USED IN AQUACULTURE

The five types of facilities used in aquaculture are levee-type ponds, watershed ponds, raceways, cages, and tanks. These facilities all have similarities, but some factor or factors make each type unique. A brief description of the different types of facilities, which includes some of the important species grown in those facilities, is given as follows.

Levee-Type Ponds

Levee-type ponds are different from watershed ponds in that they are not usually built to dam up a natural stream or a watershed. Levee-type ponds are usually built to hold water that is pumped from ground wells or another source of surface water. Sometimes the source of surface water is a watershed pond. Levee-type ponds may also be built to take advantage of natural springs, but this is less common. These ponds are often built on relatively flat ground where the earth is moved to the outer edges of the ponds and forms a levee. These ponds generally have four raised levees and a flat bottom.

Many of the early catfish farmers in the U.S. believed that intensive catfish production was only possible in levee-type ponds located on flat, nearly level land. In the late 1980s and early 1990s, the largest increases in catfish production however took place in the hills of Mississippi and Alabama. Many aquafarmers in the hills took advantage of the natural movement of water on the ground by building watershed ponds.

The most common aquacrop produced for food in levee-type ponds in the U.S. is the channel catfish. Baitfish, ornamental fish, and freshwater shrimp also are produced. The other food fish commonly produced in levee-type ponds are tilapia, hybrid striped bass, trout (only in northern areas), and bream.

Figure 8–1. Levee-type ponds used for catfish production. (Courtesy, Ralston Purina Company)

Watershed Ponds

Watershed ponds take advantage of natural movements of water in streams or as surface runoff. The primary source of water for watershed ponds is rainfall; however, the rain water is sometimes supplemented with water from a spring or a well. Watershed ponds may be used as the primary facility for aquaculture, or they may be used as feeder ponds. When they are used as feeder ponds, the water from them is pumped or flows by gravity to one or more levee-type ponds that serve as the primary facility for aquaculture production.

Watershed ponds may have one or more dams or levees, taking advantage of the natural topography of the soil to keep water in the ponds. The bottom of watershed ponds is usually uneven, which makes harvesting of the aquacrops more difficult. Thus, the ponds are usually drained to nearly empty to facilitate harvesting. Many aquafarmers, however, use watershed ponds with success. One factor in their favor is the low cost of providing water for aquacrops.

Figure 8–2. Watershed pond. Note the gently sloping terrain, which provides the watershed.

Both levee-type ponds and watershed ponds can serve as standing or static systems. This means that the water is replaced with new water very slowly. Once the ponds are filled with water and production begins, water is only added to keep them full. Water loss due to evaporation and percolation into the soil that is not replaced by rain water is supplemented by water from ground wells or other sources. As a result, ponds do not use as much water as some of the other types of facilities.

The same species produced in levee-type ponds can be produced in watershed ponds, other factors (such as temperature and pH) being equal.

Raceways

Raceways are enclosures where the water moves through at a rapid rate, carrying wastes out at the lower end of them. The most common raceways are rectangular in shape and used with cool or cold water crops. Round tanks are sometimes used in the same manner. Production is usually much more intensive with higher stocking rates in raceways than in ponds because the wastes are removed and the water is oxygenated.

Raceways are known as flow-through or open systems, because water is replaced every hour or so. A large supply of water is needed to provide for this exchange. Raceways usually take advantage of natural springs or streams because few ground wells will have the necessary water required for an extended period of time.

Most raceways utilize the slope of the terrain to move the water out at the bottom, although round raceways use a manufactured slope. Very few rectangular raceways are built on level or nearly level terrain because pumping the water through the system increases the cost. Raceways are usually constructed from concrete, but other materials may be used.

The three species of trout (brown, brook, and rainbow) are the most commonly produced fish in raceways. Some species of salmon also do well in raceways, as do catfish (in warmer areas), but the culture of these is usually more economically feasible when other methods are used. The trout require a high level of dissolved oxygen and cold water, which makes raceways an ideal method of culture.

Cages

Cage culture is really more of a method of production than a facility. Cage culture is often used in levee-type ponds, watershed ponds, lakes, rivers, oceans, and estuaries where intensive production outside the cages is difficult due to the size of the water facility or to predators. This type of culture is used primarily with fish.

Cage culture began in Japan as a way of producing saltwater species because the ocean was too large to try to raise aquacrops without cages or some type of enclosure. In the U.S., most cage culture has been used to grow freshwater crops in ponds that are not particularly well-suited to aquaculture. Cages often allow for the productive use of at least some of the water in these types of ponds.

Cages are sometimes used to produce aquacrops in the same facility where other crops are being grown in a polyculture that would not work in a natural setting. For example, a pond with bass and bream may be suitable for cage culture of channel catfish, but if catfish fingerlings were put in the open water with the bass, the bass would eat the catfish.

Cage culture is usually more intensive than pond culture. With periodic movement of cages and maximum stocking rates, aquafarmers can produce about as many pounds of aquacrop per acre as they can in open ponds with intensive culture.

Very large cages, sometimes used in oceans or estuaries for saltwater and brackish water production and in rivers for freshwater production, are often called net pens because they are constructed with nylon or plastic netting. The stocking rates and requirements vary, based on the rate of flow of water through the cages and the species cultured. Although net pens are commonly used in Southeast

Asia, very few are used in the U.S. The most common use of cages in the U.S. is in farm ponds not particularly well-suited to levee-pond aquaculture production methods.

One concern with cages is that the fish cannot forage for food throughout the water facility. This means that the fish have to be fed more. They also have to be fed a feed that is more nutritionally complete, because by not being able to forage for food, they may be missing an important part of their diet.

Figure 8–3. Permanent cages for holding catfish as broodfish.

The type of species that will grow in cages depends mostly on the type of water in which the cages are placed. Catfish, tilapia, bream, and several other freshwater species seem to grow very well in cages. Saltwater and brackish water species such as red drum, salmon, and shrimp do well if the cages are placed in the right type of water. With shrimp, it is important to find a flow of water fast enough to compensate for the size of the mesh or netting, which must be very small to keep the shrimp from escaping.

Tanks

Tanks used for aquaculture may have many different characteristics. Some tanks are used as flow-through systems with many of the same characteristics of raceways. Round tanks are particularly well-suited for this purpose if they have a bottom that slopes to a drain in the middle. This allows for excellent removal of the wastes.

Figure 8–4. Tanks used in a fish hatchery. Note the mesh basket into which the egg mass is placed. Fiberglass tanks are commonly used in hatcheries.

Other tanks may be used in static systems much like pond culture. Tanks are the primary water facility in closed systems, those that recirculate over 90% of the water used by filtering out the wastes and then pumping the water back through the systems. Tanks may be constructed from galvanized steel, concrete, fiberglass, or several other materials. Several manufacturers offer different-size fiberglass tanks that are reasonable in price and suitable for many types of aquaculture production.

Tanks are the only facility commonly used in closed systems. In closed systems, the water from the tanks is pumped out of the tanks and through filters and then recirculated back through the systems. These systems make efficient use of water but are expensive to set up and operate. The status of closed systems for large-scale aquaculture is still considered experimental.

Baitfish and ornamental fish adapt to tank culture extremely well, usually requiring aerators and flow-through systems to enhance the intensive stocking rates. Tilapia, catfish, and hybrid striped bass adapt well to tank culture. Soft-shelled crawfish and soft-shelled blue crabs have been cultured experimentally in closed systems with fairly good results.

SITE SELECTION FOR AQUACULTURE FACILITIES

Although the site selection process may vary widely for the different types

Figure 8–5. Some aquaculture facilities make use of greenhouse technology to conserve energy. This facility contains a hatchery. Note the access to electricity, with lines running alongside the facility. (Courtesy, Gary Fornshell, Mississippi State University)

Figure 8–6. An electric control panel needed for an electric aerator. A source of electricity is often important in site selection.

of facilities used in aquaculture production, several general considerations must be made regardless of the type of facility. Of course, an adequate supply of water is needed. Water requirements will be discussed concerning each type of facility.

One factor that must be considered is the size of the facility needed and the amount of land necessary to support the facility. Determining if other land is available adjacent to the site for future expansion is also important.

Water used for aquacrops is often reused for other purposes, such as irrigating other crops. If this is the case, the facility should be located close to the secondary use if possible to reduce the cost of moving water from one use to the other.

The location of the site in relation to others is very important. For example, there should be service people close enough to the operation so that equipment can be serviced or repaired in a timely manner. The suppliers of medicines, chemicals, equipment, and materials should be close enough so that these goods can easily be transported. Also, the markets for the aquacrops to be grown should be in fairly close proximity to reduce transportation costs.

A suitable site can be accessed by roads that are in good repair. Telephone lines and electrical lines should also be available, although diesel generators may be used to supply some of the power needed. Some specific site selection criteria for the different types of aquaculture facilities are given in the following sections.

Levee-Type Ponds

The three basic site requirements for levee-type ponds are an adequate supply of clean water, soil that holds water, and suitable terrain for pond construction.

For levee-type ponds, surface runoff is not normally a source of water. These ponds use either well water or spring water, with well water being the customary source. If well water is the source, a test well is usually drilled to evaluate the quality and quantity of the water. The location may also be a factor in that the deeper down the water is in the earth, the more it costs to pump it to the surface. Well water is often low in dissolved oxygen, but splashing the water into the air as it is being pumped into the pond will correct this. Well water should also be checked for pollution, although this is seldom a problem.

If spring water is the source, it should be observed several times during the year to make sure it has an adequate flow. Some springs almost stop flowing during certain times of the year, often in the fall. Spring water may be too cold for some warm water species, such as catfish, freshwater shrimp, and tilapia. If so, the water can be held in a warming pond where the sun warms it before it is pumped into the production pond. The spring water should be checked for nitrogen gas. Sometimes spring water is supersaturated with nitrogen gas, which can lead to fish kills.

The soil type dictates whether or not a pond can be built on a particular site. Even if the topography is good and an adequate source of water is found, the soil type will determine if the pond will hold water well enough to produce an aquacrop. A quick look at a soil map for the area under consideration will tell a lot about the soil type and whether or not it is suitable for aquaculture. Soil for ponds should have a slow infiltration rate (slow permeability) and a high runoff rate. To have a slow infiltration rate, the soil will usually have a high percentage of clay. A common recommendation is soil that is at least 20% clay. After selecting

Figure 8–7. A valve used to fill a levee-type pond. A source of water is a major factor in selecting a site for an aquaculture facility.

a tentative site, the aquafarmer can consult the U.S. Soil Conservation Service to provide soil tests and analyses to help determine if the site is suitable.

The topography for a levee-type pond site should be relatively flat. Two important factors that should be considered are drainage and flooding. The pond should be able to drain by gravity during any season of the year. If it will not, drainage ditches may have to be dug, which will increase the cost of construction.

The pond should not be built in low-lying areas that are subject to periodic flooding. Flooding is hazardous to an aquafarmer. The aquacrop can leave the pond during flooding, and trash fish or other undesirable animals can enter. Care should be taken to make sure that rivers, bayous, and drainage ditches are at a

Figure 8–8. An electric paddlewheel aerator operating in a levee-type pond. Note the level terrain and the side-by-side construction of the ponds. Two ponds often share the same levee in commercial operations.

lower elevation than the drain pipe of the pond. Also, low-lying areas may be classified as wetlands, and building a pond may not be allowed.

Watershed Ponds

Site selection for watershed ponds includes the same three factors as those for levee-type ponds, but the desired characteristics are a little different. For watershed ponds, the topography of the site is the source of water. Streams or surface runoff provides most of the water for watershed ponds. As a result, the ponds must be on land with at least a gentle slope that will carry the water into the ponds by gravity. The dam for the ponds is usually built perpendicular to the stream or natural flow of water to create a reservoir. The least expensive watershed ponds to build are in gently sloping valleys. Deep valleys require larger dams, which cost more to construct, and the deeper water is less productive than the shallow water.

Figure 8–9. A watershed pond built in a gently sloping valley. Note the uneven edges, common with watershed ponds. These edges make the pond more difficult to harvest than a levee-type pond.

The amount of watershed that is required to fill the pond is also a consideration. Usually between 5 and 25 acres of watershed is needed for each acre of water surface. This varies, depending on the soil type of the watershed, the amount of rainfall, and the number of trees on the watershed. The watershed should be sufficient to fill the pond during rainy times but should not let the level of the pond fall more than 2 feet during drier times of the year.

Watershed ponds also require soils that will hold water well. If a proposed site does not have enough clay in the soil, additional clay is sometimes brought in and added to the pond bottom and dam. Care must be taken to locate a site that does not uncover limestone, sand, or gravel areas, which will cause the pond to leak.

Raceways

The primary concern for a site for a raceway is a source of water that can be consistently used to flow through the system. The two most common sources are streams that are diverted through the raceway and natural springs, although some aquafarmers use wells as their water source.

Another important site criterion is land that gently slopes so gravity will move the water through the raceway. A ground slope of 1 – 3% will accomplish this requirement. In a series of raceways, the drain pipe of an upper raceway should be at least a few inches higher than the water level in the next raceway in order to oxygenate the water between the raceways.

Cages

As mentioned before, cages are used in many types of water facilities. When selecting a site for a cage, the aquafarmer should make sure two primary considerations are met to insure water quality. First, the cage must have natural movement of new water through it since the fish cannot really move through the water. In ponds and estuaries, this movement is accomplished by placing the cages where prevailing winds will keep water moving through them. Of course, in rivers, the natural flow does this job. In the ocean, the tide and the other natural movements of water are usually sufficient.

The second important criterion in cage culture is having water that is deep enough for the waste matter to move well below the cage. Feces and waste feed

Figure 8–10. Cages built with PVC plastic. Note the walkway to provide easy access for feeding and observing the aquacrop.

Figure 8–11. Cages being stocked with catfish fingerlings. The wooden top of the cage serves as a feeding ring to keep the feed from floating away from the fish. (Courtesy, Master Mix Feeds, Inc.)

use oxygen as they decay and would take up valuable space in the cage if it was placed on the bottom. A water depth of at least 6 feet is recommended for the most common type of cage, one that is 4 feet deep. With any cage, at least 2 feet of water space should be maintained below the cage.

Several other factors also should be considered, although these may not be quite as important as water movement and depth. Cages should be placed at least 10 feet from other cages. Cages too close together may lead to low or even fatal dissolved oxygen levels. Cages should not be placed near coves and weed beds. Both may restrict the natural flow of water through the cages. If possible, cages should be kept away from high traffic areas where people and animals might disturb the aquacrop. This has to be balanced with being able to feed and check the aquacrop on a daily basis.

Tanks

As mentioned earlier, tanks come in a wide variety of sizes, shapes, and uses.

Figure 8–12. This tank was built by putting a plastic lining in a wood and wire frame. This is a relatively inexpensive way to get started in aquaculture.

The general criteria for selecting a site for a tank are access to roads, suppliers, and utilities; a source of water; and the use of the discharged water. Future expansion needs must be carefully considered, since tanks are often built in smaller areas.

The primary source of water for tanks is usually from ground wells. This availability must be considered in the selection of a site. Because they usually are smaller and require less land than ponds, tanks are often the choice when land prices are prohibitive, such as near large cities. Tanks may give proximity to markets and suppliers where ponds are not feasible.

CONSTRUCTION OF AQUACULTURE FACILITIES

Once a suitable site has been selected for an aquaculture facility, the facility must be built before aquaculture production can begin. Although the type of facility should have been decided on before the site was selected, the shape, size, and placement of the facility (direction or actual placement on the site) may not have been determined.

This section discusses the construction of levee-type ponds, watershed ponds, raceways, cages, and tanks, including some design considerations for each that would allow the facility to be better utilized in the production of an aquacrop.

Levee-Type Ponds

Levee-type ponds typically have a levee on all four sides and a nearly flat bottom. This makes harvesting the aquacrop easier. On flat ground, these features usually mean lower costs of construction than that of watershed ponds with similar production capacities.

One of the first decisions an aquafarmer must make is that of the size of the pond or ponds. The most common size of a levee-type pond used for catfish production is a 17½-acre pond built on 20 land acres. This size is somewhat of a compromise between ease of management and cost of construction. Smaller ponds are easier to manage and harvest, while larger ponds cost less per surface acre of water to construct.

The normal depth for a levee-type pond is between 4 and 4½ feet, with a freeboard of about 1½ feet. (Freeboard is the height of the levee minus the normal water level.) Of course, this means that the levee is built to a height of about 6 feet from the pond bottom. The freeboard should not exceed 2 feet, as this increases the cost of constructing the levees and makes it difficult to get equipment into and out of a pond. A levee with a freeboard of less than 1 foot is very susceptible to erosion.

The shape of levee-type ponds varies widely. A square pond gets the most acreage out of the amount of its levees, but it requires a longer seine for harvesting than a rectangular pond of the same area. The most common shape is rectangular. However, property lines often force an aquafarmer to have irregularly shaped ponds at the edge of the property. Whatever the shape and size, it is recommended that if several ponds are built, they should be somewhat uniform. Otherwise, an aquafarm might need several different seines to harvest the different ponds.

Figure 8–13. Two workers with a seine pulled to a corner of a levee-type pond. The rectangular shape of the pond allows for the use of a much smaller seine than a square pond of the same acreage. (Courtesy, Gary Fornshell, Mississippi State University)

The location of a pond is usually determined by the prevailing winds. The longer levee is usually placed parallel to the direction of the prevailing winds in order to minimize the erosion caused by wave action against the bank. Some aquafarmers argue, however, that this decreases the amount of oxygenation the pond receives due to the wind, so they recommend putting the longer levee perpendicular to the prevailing winds. At present, there is not a hard-and-fast rule for deciding the orientation. This judgement may be left up to the aquafarmer's own unique situation, depending on how easily the pond is manually aerated, how concentrated the production, and how intensive the prevailing winds are in that area.

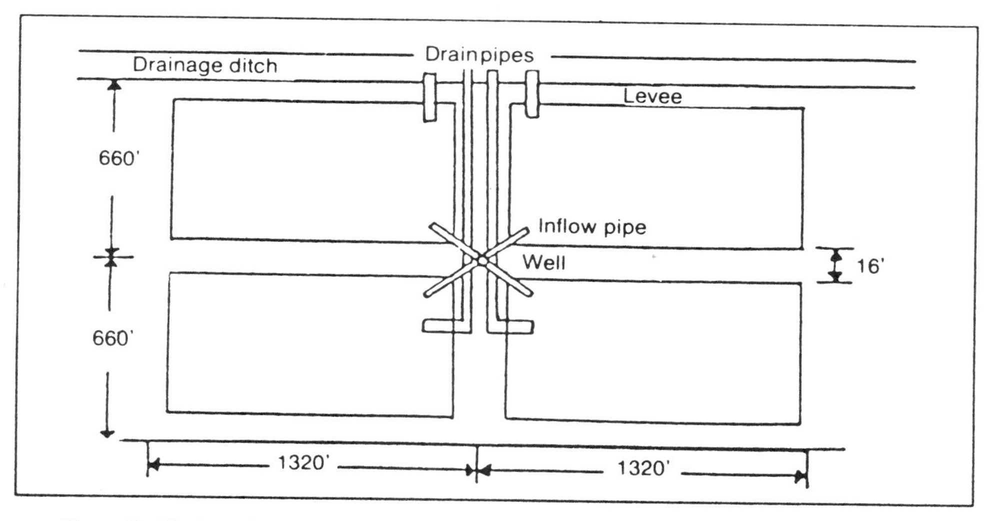

Figure 8–14. A typical layout of four levee-type catfish ponds. (Source: T. L. Wellborn, "Construction of Levee–Type Ponds for Catfish Production," Southern Regional Aquaculture Center Publication No. 101)

Other factors may be more important than prevailing winds in determining a location within a site. Putting ponds as close as possible to roads usually minimizes the chances of poaching and / or vandalism. Of course, this is not always possible. Frequent checking of the ponds by night crews, in irregular patterns, is probably the best deterrent.

The main levees (where the wells are located) should have a top width of 20 feet or more. This width will support the traffic that occurs when the aquafarmer is feeding, harvesting, checking oxygen, and taking samples. Other levees should be at least 16 feet wide, to allow for safety and to reduce the required maintenance. Often the soil type dictates that at least two levees be graveled (if the levees become impassable to two-wheel – drive vehicles when it rains, they should be graveled).

Although levee-type ponds may not look like it, most have a little slope at the bottom and contain a shallow end and a deep end. The bottom slope allows for better draining. A typical slope for the bottom would be about 0.2 feet of drop for every 100 linear feet along the long axis of the pond.

A pond with a good soil type for holding water will usually have a levee with a slope of about 3:1. This means that for every 3 feet the levee extends horizontally (out into the water), it drops 1 foot. Therefore, a 6-foot high levee will extend 18 feet horizontally from the point at which the slope reaches the bottom of the pond. This point is called the *toe*. Soils that do not hold water as well may need to have a slope of 4:1 or even 5:1. More cubic yards of fill (soil) must then be moved in the construction of the levee, thus increasing the construction costs.

If the slope of the levee is the same on both sides, the formula for determining the amount of cubic yards of fill needed per foot of levee is:

$$\frac{\text{cubic yards}}{\text{ft. length}} = \frac{[(\text{slope} \times \text{height}) + \text{top width}] \times \text{height}}{27}$$

An example of how the formula works is given below. For a levee with a 3:1 slope, a 16-foot top width, and a 6-foot height, the calculation of cubic yards of fill needed per foot of levee would look like this:

$$\frac{\text{cubic yards}}{\text{ft. length}} = \frac{[(3 \times 6) + 16] \times 6}{27}$$

$$= \frac{(18 + 16) \times 6}{27}$$

$$= \frac{204}{27}$$

$$= 7.56$$

The following example shows how much slope requirements can increase construction costs. The cubic yards of fill needed per foot for a pond levee with a 5:1 slope, a 16-foot top width, and a 6-foot height are calculated.

$$\frac{\text{cubic yards}}{\text{ft. length}} = \frac{[(5 \times 6) + 16] \times 6}{27}$$

$$= \frac{(30 + 16) \times 6}{27}$$

$$= \frac{276}{27}$$

$$= 10.2$$

Obviously, the amount of slope is important. Consider that on a typical levee of a 20-acre pond, the total length is 3,960 feet. With a 3:1 slope, the levee would require almost 30,000 cubic yards of fill. With a 5:1 slope, the levee would require just over 40,000 yards of fill. The money used for the extra fill may be the difference between making a profit and experiencing a loss!

Most levee-type ponds are constructed by self-loading pans, also called scrapers. Before they are used, all roots, stumps, and topsoil are removed. This allows a good bond between the foundation soil and the fill material. The pans compact the soil as they carry the fill to build the levee, which is very important.

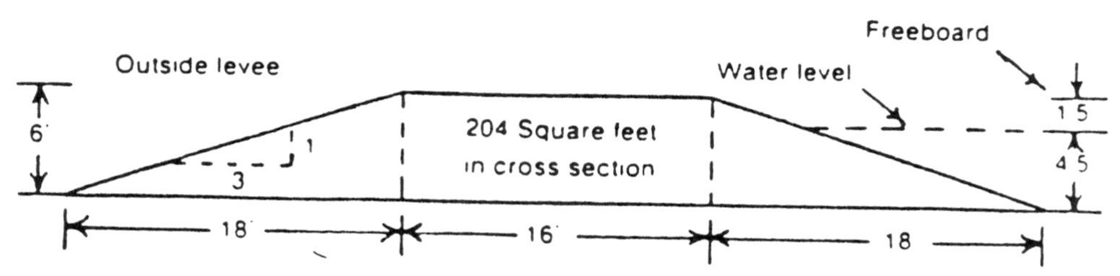

Figure 8–15. Cross section of a typical levee for a levee-type catfish pond. (Source: T. L. Wellborn, "Construction of Levee-Type Ponds for Catfish Production," Southern Regional Aquaculture Center Publication No. 101)

Water places a lot of pressure on the levee, so the soil must be compacted well. If a bulldozer is used to construct the levee, a sheep's foot roller can be used to improve compaction. For maximum compaction, the soil must be slightly moist (about 12 – 15%). When many ponds are being constructed on one aquafarm, a laser-guided pan should be used to improve the accuracy of the bottom slope and the slope of the levee.

All ponds must have some method of drainage that will regulate the water level in the pond at all times and that will be large enough to drain the pond in five to seven days when necessary. The drain will consist of a pipe that runs along the bottom of the pond through the levee and out the other side. Most drains are constructed from PVC pipe. The soil is hand compacted around the drain pipe for at least 1 foot so that the equipment used to construct the levee above it will not damage the pipe. This also creates a good seal around the drain pipe to keep the pond from leaking. The outside end of the drain pipe should extend at least 5 feet past the toe to prevent erosion of the levee during draining. The drain will have a standpipe to regulate the water level in the pond.

On inside swivel or modified Canfield outlet drains, the standpipe is located inside the pond at the desired water level. The standpipe can be turned down to drain the pond completely. On outside drains, which are the most common on new ponds, the standpipe is located outside the pond and connected to the drain pipe with a "T" joint. The outer end of the "T" has a valve to drain the pond completely. On some outside drains, the standpipe is only 24 inches high and also has a valve. The valve must be opened slightly to regulate the water level. With either outside drain, the drain pipe should extend 5 – 10 feet past the toe to the bottom of the pond to prevent clogging. The inside end of the drain pipe should have a screen to prevent losing the aquacrop through the drain.

Watershed Ponds

Before constructing a watershed pond, the aquafarmer should use a level

Figure 8–16. A modified Canfield drain, where the water level in the pond is regulated by the standpipe located inside the pond. Note the netting to keep the aquacrop from escaping through the drain.

and stakes to mark off the water line of the proposed pond. This is done to make sure that the size will suit the purpose of the pond and that the water will not encroach on the property of neighbors.

The first step in constructing the dam of a watershed pond is to remove all trees, stumps, and topsoil from the site. Then a 3-foot deep trench is dug into a good clay base on the dam site. Clay is filled and compacted into the trench, and this core is continued all the way to the top of the dam.

The dam of a watershed pond should have a slope of 3:1 or higher and a top width of 16 – 20 feet. A spillway is necessary to handle flooding problems. The spillway is usually grassed and extends from 10 to 50 feet or more, depending on the size of the pond and the expected flooding of the watershed. The spillway must have some type of barriers to keep the aquacrop from escaping the pond during flooding. These barriers must be kept from clogging, thus allowing the water to escape over the top of the dam.

Watershed ponds typically have drains in which the drain pipe has a "T" on

Figure 8–17. This spillway contains netting to keep the aquacrop from escaping during a flood, but it needs cleaning. A stopped-up spillway may cause water to run over the dam in heavy flooding, which results in wear and stress on the dam.

Figure 8–18. A drainage ditch from a levee-type pond with a modified Canfield drain. Permits must be obtained if the water from an aquaculture facility drains into a public stream or lake.

the inside with a standpipe to handle small flows of excess water. The top of the standpipe is usually about 3 feet from the top of the dam. A sleeve pipe is usually fitted around the drain pipe near the bottom and extends a few inches above the drain pipe to insure that the water being drained is the poor-quality water from the bottom of the pond and not the freshwater that is just entering the pond. Like levee-type ponds, the drain is fitted with a valve at the outside end and should be large enough to drain the pond within five to seven days.

An important step is to make sure that all areas of the pond are at least 2½ feet deep. This will prevent weeds from taking up space in the water that could be used by the aquacrop.

Because watershed ponds often have an erratic water supply, creating a series of ponds in the same valley, instead of building a very large pond that is hard to manage, works very well. This method allows for the best use of available water. The lowest level pond in a series is drained for harvest first, and water from the upper ponds is used to fill that pond so production can resume immediately. Temporarily extended drain pipes can be used to store water. When an upper level pond is drained for harvesting, the water in it is used in the lower level ponds. This keeps the aquafarmer from having to wait for a substantial rain to restock the ponds.

Raceways

The construction of raceways is normally much simpler than the construction of ponds. Once a suitable source of water is located, the site for the raceway is cleared. Because most raceways are constructed with concrete, the soil type is not as important. Some earthen raceways are used, but the rapid flow of water tends to erode them quickly. A few raceways are built of wood or other materials.

Raceways are usually fairly long and narrow. They typically range from 75 to 100 feet in length and 10 to 30 feet in width, although some may be longer and / or wider. The depth of most rectangular raceways is 3 feet or less, although some may be as much as 6 feet. The amount of water available will determine the size of the raceways.

The layout of raceways may be parallel or in a series. In a parallel layout, water moves through only one raceway, and each raceway can be drained individually. This prevents wastes from one raceway from entering the next and keeps diseases from being transmitted between raceways.

In a series, water flows through several raceways before being discharged. Although this system allows for maximum use of water, it can lead to problems with diseases. If the aquacrop in the top raceway becomes diseased, then the crops below it soon will also, because the water from the upper raceway flows

Figure 8–19. These outside drains regulate the water in a concrete raceway. The inside of the drain pipe is located on the bottom and covered with netting.

directly into them. The most common raceway layouts have a combination of parallel and series layouts. This means that three or four series of raceways with two or three raceways in each will be laid out parallel to each other. This represents a compromise between disease control and maximum use of water.

Some tanks are really raceways because they are constructed as flow-through systems. Culture is just as intensive in these tank systems as in rectangular raceways, if not more so. Often the culture is so intensive the aquafarmer has to add oxygen to the tanks even though a steady supply of freshwater is coming into them. This type of intensive culture requires a very tight control on diseases and stress, but the construction of the system allows the aquafarmer to add medicines or supplements to the water very easily.

Round tanks built as raceways slope to drains in the middle. These systems remove wastes very effectively. They sometimes have a Venturi drain with an inner standpipe to regulate the water level in the tank and an outer standpipe to draw water and wastes from the bottom. The inner standpipe can be temporarily removed so that the wastes, which get trapped between the two pipes, can be cleaned out. Some systems have a screened drain with a valve that regulates the water height by letting water escape at exactly the same rate as it enters the tank.

The effluent is the water discharged from the raceways or other water facility. Because such a large amount of water flows out of raceways, the aquafarmer may need a permit to discharge into natural bodies of water. Sometimes the water is pumped back through the raceways, but this is usually not economically feasible.

Cages

Cages come in all shapes and sizes. The most common shapes are round and rectangular, although some cages are square. Rectangular cages are usually 3 feet wide × 4 feet long × 3 feet deep or 4 × 8 × 4 feet. The most common round cages are 4 feet deep with a diameter of 4 feet. Square cages are usually 4 × 4 × 4 feet or 8 × 8 × 4 feet.

Cages may be constructed from a variety of materials, but the material must allow for adequate water movement through the cage and must keep the aquacrop from escaping. The material must also be non-toxic to the aquacrop. Materials for extended use of cage culture also should not rust or rot in water, so some types of wire and wood are not suitable.

Cages consist of a frame, mesh or netting, feeding ring, lid, and some type of flotation. The frame helps the cage hold its shape as the water moves through it. Some cages have only a top frame for the lid, but most have a bottom and side frames as well. These frames may be constructed from PVC pipe, aluminum tubing, treated wood, or fiberglass. The frame may be used to help in the flotation, as when capped PVC pipe is used.

The mesh or netting is usually made of nylon, plastic, or some type of wire that is covered or galvanized. Solid plastic netting is the most commonly used. Plastic-coated welded wire is also a popular choice for cage construction. The size of the holes should be ½ inch or larger, depending on the size of the fingerlings used to stock the cages. Netting with holes smaller than ½ inch reduces the water circulation, which is important in keeping the water oxygenated through the cage.

Lids are used to keep the fish in the cages and to keep predators out. The

Figure 8–20. These workers are harvesting catfish eggs from cages. The old ammunition boxes make good spawning areas for the catfish. (Courtesy, Gary Fornshell, Mississippi State University)

lids should be securely fastened to keep other people from stealing the fish. The lids may be constructed from the same type of material as the cages or from solid pieces of material such as treated plywood. Some studies have shown that catfish grow better in cages if the lids are made of a solid material that they cannot see through. If the lids are solid, doors are usually put in for feeding.

The feeding ring is usually a band of fine mesh 10 – 12 inches wide, extending from a few inches below the water level to a few inches above it. This keeps floating feed from moving out of the cage before the fish have a chance to eat it. Another type of feeding ring is constructed of a floating material such as styrofoam and placed within the cage. Some feeding rings are attached to the lid and extend down into the water.

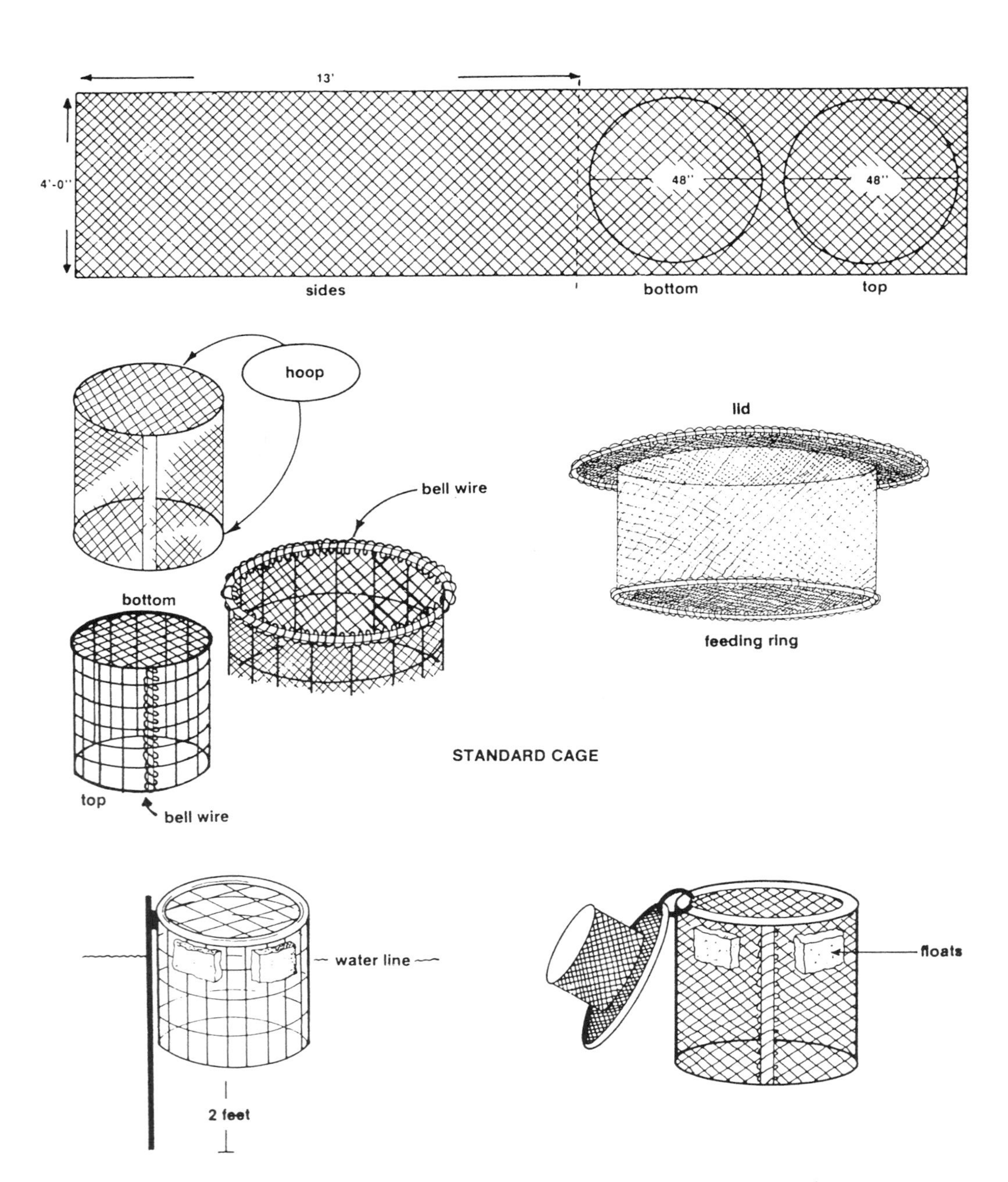

Figure 8–21. A simple cage to construct is the 4 × 4 cylindrical cage, made with ½-inch plastic mesh and bell wire. (Source: M. P. Masser, "Cage Culture — Cage Construction and Placement," Southern Regional Aquaculture Center Publication No. 162)

The flotation device is attached to the cage so that it will keep the top of the cage about 6 inches above the water in a level position. Flotation devices are very important in that they keep the aquacrop separate from the accumulated wastes at the bottom of the pond or other water facility. The most common material used is styrofoam, which should be covered with a canvas bag to keep it from breaking up and floating away from the cage. Other types of flotation devices include innertubes, capped PVC pipes, and plastic bottles. Cages are sometimes suspended from docks.

Cages are placed in the water facility at least 10 feet apart and tied to concrete blocks to anchor them or attached to docks. Docks should be avoided if the water level fluctuates greatly. Each cage should be anchored from at least two sides, four if the water flows very fast or if very windy conditions exist.

Tanks

Most tanks built by the aquafarmer are either concrete or wood. Fiberglass or plastic tanks that are produced by a manufacturer are often purchased by the aquafarmer. Tanks are usually either rectangular or round, as these two shapes are the easiest to harvest and clean.

The primary concern with constructing tanks is the drain and finding a suitable place to drain water from the tanks. Most fiberglass tanks have the drain built into the side or bottom of them. The aquafarmers then must locate the tanks

Figure 8–22. These tanks slope to a Venturi drain in the center, which keeps them clean of wastes. Note the platform built to provide easy access to the aquacrop.

Figure 8–23. A hatchery building with tanks for rearing of fry. Many hatcheries are located in metal buildings so the climate for the eggs and fingerlings can be carefully controlled.

where they will drain into suitable drainage ditches or holding ponds. Concrete tanks require a little more planning. If tanks slope to the middle for draining, a ditch must be dug beneath the bottom and the drain installed before the concrete is poured. These drains usually include a standpipe to control the water level in the tanks.

Tanks in closed systems, those where the water is filtered and recirculated back into the tanks, require more construction. Because these closed systems reuse the same water, the wastes must be removed before the water is pumped back into the tanks. These systems each need a biofilter, which will be discussed later in this chapter. The problem with closed systems is the cost as compared to the amount of production, particularly when the amount of production increases.

EQUIPMENT USED IN AQUACULTURE FACILITIES

Because some of the types of equipment discussed below are used in several different types of aquaculture facilities, they are not presented as equipment for levee-type ponds or equipment for raceways. Each type of equipment is described and the facility or facilities in which it is needed are given.

Aerators

Some type of aeration is necessary for all aquaculture systems. The simplest form of aeration is by the splash method, commonly used with raceways and some closed systems. The splash method simply means that something is used to break water into small droplets as it enters the facility so as to oxygenate the water. With raceways, this is frequently the only type of aeration needed because the water is not in the facility long enough for the aquacrop to deplete the oxygen.

In other types of production facilities, more aeration is often needed. In levee-type ponds, electric paddlewheels and floating spray-type surface aerators are commonly used. One of these will probably be found on every pond, ready

Figure 8–24. An electric paddlewheel aerator. Very energy-efficient for the amount of oxygen they provide, electric paddlewheels are commonly used in large ponds with intensive production.

to be used if oxygen levels get too low. Both of these simply throw oxygen-poor water into the air so it can be oxygenated. Contrary to the belief of some, outboard motors do not do a comparable job of aerating the water.

In addition to the electric devices, many aquafarms will have two or more portable paddlewheels that operate from the PTO shaft of a tractor. Where dissolved oxygen problems have caused danger to the aquacrop, these portable aerators are moved to assist the regular aerators in oxygenating the water quickly. If this is necessary, the portable aerator should be placed right beside the regular aerator. This is done because the fish will be gathered close to the regular aerator

Figure 8–25. A floating spray-type aerator. These aerators are common in smaller ponds and in ponds with less intensive production. Many golf courses, parks, and other recreational areas use these for aesthetic reasons. (Courtesy, Otterbine Aerators / Barebo, Inc.)

and also so the fish will not have to search the pond for oxygen if one of the aerators stops working.

Tank systems are not usually large enough to need big paddlewheels or splash-type surface aerators, although small splash-type surface aerators may be used. Tanks that are used as round raceways do not usually need additional aeration, although sometimes the stocking rates require it. If so, compressed

Figure 8–26. A portable paddlewheel aerator that operates from a tractor PTO. These portable aerators are important because they can be moved quickly to ponds with dissolved oxygen problems. They are also commonly used during harvesting because the aquacrop is usually confined to a small area.

Figure 8–27. A portable paddlewheel aerator in operation. The operator backs the aerator into the pond and engages the PTO. A good parking brake on the tractor is a must!

oxygen is often injected or diffused into the water. This method is many times used with closed systems as well.

If compressed oxygen is diffused into the water, the device should be such that small bubbles are created. Large bubbles send most of their oxygen into the air when they reach the surface. If possible, a device to keep the bubbles from rising straight to the surface will also help. The longer the bubbles are under water, the more diffusion occurs between the bubbles and the water.

Seines and Seine Haulers

Seines are large nets that are used to harvest the aquacrop. The size of the holes in the netting determines the size of the aquacrop that will be harvested. These seines are pulled either by hand or by a tractor and seine reel.

Seines are most often used in levee-type ponds or watershed ponds. In a levee-type pond, the water may be left at its regular level and the seine will be pulled through the entire pond. This requires a seine large enough to reach from the bottom to far enough above the top to keep the aquacrop from swimming or jumping over. In a watershed pond, a large portion of the water is usually drained from the pond before harvest, so a smaller seine is required.

The seine will have weights across the bottom to keep fish from swimming under the seine. A worker still must ride the seine at the toe of the levee to keep the aquacrop from swimming under where the seine starts to slope up to the bank.

The aquacrop is usually herded by the seine to a corner of the pond and held there for harvesting. Sometimes a live car, or sock, is attached to the seine

Figure 8–28. A catfish pond that has been seined and the fish gathered in a live car, or sock. Note the metal rods used to hold the sock above the water so the fish cannot swim over it.

Figure 8–29. A typical catfish harvest scene. The boom of a backhoe is used to move the fish from the pond to the haul truck.

Figure 8–30. A portable scale for weighing catfish. When catfish are loaded onto a haul truck, the live weight of the fish is recorded with an in-line scale attached to the boom. (Courtesy, Aquacenter, Inc.)

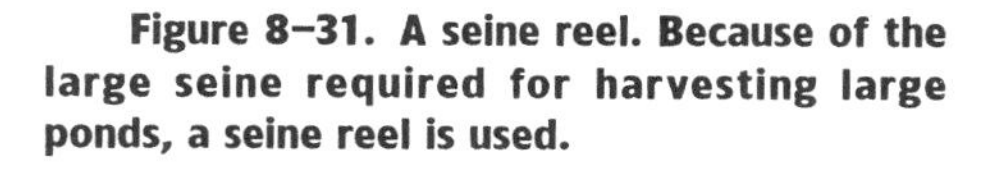

Figure 8–31. A seine reel. Because of the large seine required for harvesting large ponds, a seine reel is used.

and the aquacrop is forced into it. A live car, or sock, is a net with a bottom, which is used to temporarily hold the fish until they are harvested.

Once the aquacrop has been sufficiently crowded together for harvesting, steel rods are used to hold the sides of the seine or the sock above the water so the aquacrop cannot escape. These steel rods are usually made of reinforced steel with a "Y" welded to the end of each rod to hook the seine over. Then the aquacrop is harvested, usually with a net attached to a powered boom, although dip nets are sometimes used to harvest the aquacrop by hand.

Seines are seldom needed with raceway, tank, or cage production methods. With raceways, the aquacrop is usually herded to one end with a board and dipped out with a dip net. Dip nets are the predominant method for harvesting from tanks as well; lowering the water level makes using dip nets easier. With cages, the aquacrop is usually dumped into a boat or a hauling tank.

Seine Boats

When a seine is used, a seine boat usually travels along behind it and scares the fish as the seine crowds the fish together. It is important to keep the fish out of the seine where they might get their gills caught in the netting.

Most seine boats are converted 14 or 15 feet long aluminum fishing boats. They generally have an outboard motor and carry the sock (if used) and the rods to prop up the seine or the sock.

Figure 8–32. A typical seine boat. (Courtesy, Aquacenter, Inc.)

Figure 8–33. A feed bin located on a catfish farm. Note that the bin is built so a feed truck can drive under and load.

Feeders

Feeding time is very important in many types of aquaculture production, because this is often the only time the aquafarmer gets to check the aquacrop. The way an aquacrop feeds can reveal a lot about the health and well-being of the fish. As a result, in most small systems, such as tanks and cages, the feeding is often done by hand. Feeds that are not commercially prepared, such as scraps from meat processing plants, are also usually fed by hand, as they tend to clog up automated feeding systems.

Most larger systems however require some automation in order to get the feeding done on a daily basis. These large operations usually feed commercially prepared pellets that float on the top of the water.

The three most common types of automated feeders are feed blowers, demand feeders, and controlled auger systems. Feed blowers are often used with a large number of levee-type ponds or watershed ponds. Feed blowers are attached to trucks or tractors which drive around the tops of levees and blow

Figure 8–34. A feed blower mounted on the back of a truck. Feed blowers are very common on large farms.

Figure 8–35. Demand feeders. Note the rods running down into the water. Fish bump these rods when they get hungry, and feed is released into the water. (Courtesy, Hurley Fiberglass Company, Newport, Arkansas)

feed out into the water. This type of feeder is the most common on large-scale aquafarms that produce channel catfish. As the feed is blown on the pond, the fish respond almost immediately. If they do not, the driver knows that something is wrong with the fish.

Demand feeders store feed inside the feeders, and the fish bump a rod (pendulum) to cause feed to be released. The fish usually have to be trained to bump the rod, but most species learn quickly. Demand feeders are common in raceways where trout are produced. Some demand feeders have controls to regulate the amount of feed that is dumped during a certain period of time to reduce wasting feed. Since the aquafarmer is not always present when the fish feed, the feeders must be checked regularly to make sure they are working properly.

Automated auger systems work just like those in poultry houses or dairy barns. A control panel is used to start the feed through the auger, which drops it into the water facility through a series of openings. An electronic system is used to regulate the amount of feed that goes to each water facility. Automated auger systems are most commonly found on aquafarms that use a series of tanks or raceways. Like demand feeders, these systems must be checked regularly to make sure that all openings are clear and that the augers are working properly.

Monitoring Devices

Since the most prevalent problem in intensive aquaculture production is low levels of dissolved oxygen, a means of checking the dissolved oxygen is frequently necessary. Dissolved oxygen may be checked with a chemical kit or an electronic meter. The chemical kit is very accurate and relatively inexpensive, but the procedure is time-consuming. If many ponds or tanks must be checked, an electronic meter is essential. Good electronic meters cost anywhere from $500 to $1,500, but aquafarmers consider them a good investment. Any farm with 10 or more ponds can justify the purchase of a meter to check for dissolved oxygen.

The dissolved oxygen in ponds should be checked at least twice a night, as the photosynthesis from the algae that produces oxygen during the day halts at night. The most common time for oxygen stress is just before daybreak in ponds.

Other monitoring devices or kits are commonly used to check nitrites, nitrates, and pH of the water. These may be needed with ponds and tanks, especially if the tanks are in closed systems. A nitrite test kit will tell if the biofilter has suddenly died and new water needs to be added to the system.

If a raceway is constructed properly and an adequate amount of water moves through the system, these monitoring devices are not as important. The water

Figure 8–36. An electronic dissolved oxygen meter mounted on a four-wheeler.

should be checked periodically, however, to make sure quality water is coming from the water source.

Haul Tanks

Some aquaculture producers may find that haul tanks are never necessary. In fact, in large-scale systems, the producers usually have the fingerling supplier deliver the fingerlings to the production facility and the processing plant pick up the finished aquacrop at the production facility. For many small-scale producers,

Figure 8–37. Portable haul tank.

however, the ability to haul a live aquacrop to a specialized market or to pick up their own fingerlings may be an important way to save money.

Live-haul tanks that can be loaded into a pick-up truck or flat-bed truck often come in very handy. These tanks are usually made of aluminum or steel. Tanks that are more shallow provide for greater diffusion of oxygen during the hauling.

If the aquacrop is to be hauled very far, some means of adding oxygen to the water is necessary. Also, for even farther distances, the aquafarmer should plan on replacing the water periodically to avoid build-up of ammonia.

Care must be taken not to shock the aquacrop with drastic changes of water temperature. Some species do not react well to sudden temperature changes of 5°F. The aquacrop should be acclimated gradually by exchanging some of the water in the tank with that in the water facility in which it is to be placed.

Biofilters

In closed systems, biofilters remove the wastes from the water, which are then reused by the aquacrop. The ability of the biofilter to remove the wastes is the primary determinant of the amount of an aquacrop that can be produced.

Biofilters must remove the toxic wastes produced by the aquacrop, or as the recirculated water is reused, the aquacrop will die. The toxic organic nitrogen compounds produced in a production system are un-ionized ammonia (NH_3) and nitrite (NO_2). Biofilters utilize two types of bacteria to convert these toxic organic compounds to nitrate (NO_3), which is not toxic. Bacteria from the genus *Nitrosomonas* convert ammonia to nitrite, and bacteria from the genus *Nitrobacter* convert nitrite to nitrate.

The pH of the water used in the closed systems must be kept stable. This is necessary because most aquacrops are best suited to a particular range, usually 7 - 8, and because the bacteria in the biofilters perform best at these pH levels. Because biofilters give off carbon dioxide (CO_2) and use bicarbonate ions (HCO_3), aquaculture production in closed systems tends to lower pH. Therefore, the calcium carbonate ($CaCO_3$) levels must be increased. This is usually accomplished by using calcerous gravel or shells in the filter or by periodically adding sodium bicarbonate ($NaHCO_3$) to the system.

The three most common types of biological filters are submerged rock, upflow sand, and fluidized bed. All three of these systems work by the growth of bacteria on the rocks or sand grains in the filters.

- ***Submerged rock filter*** — In a submerged rock filter, the water passes through a bed of calcerous rock or shells. The bacteria grow on the exterior of the rocks, where they convert the nitrite and ammonia to

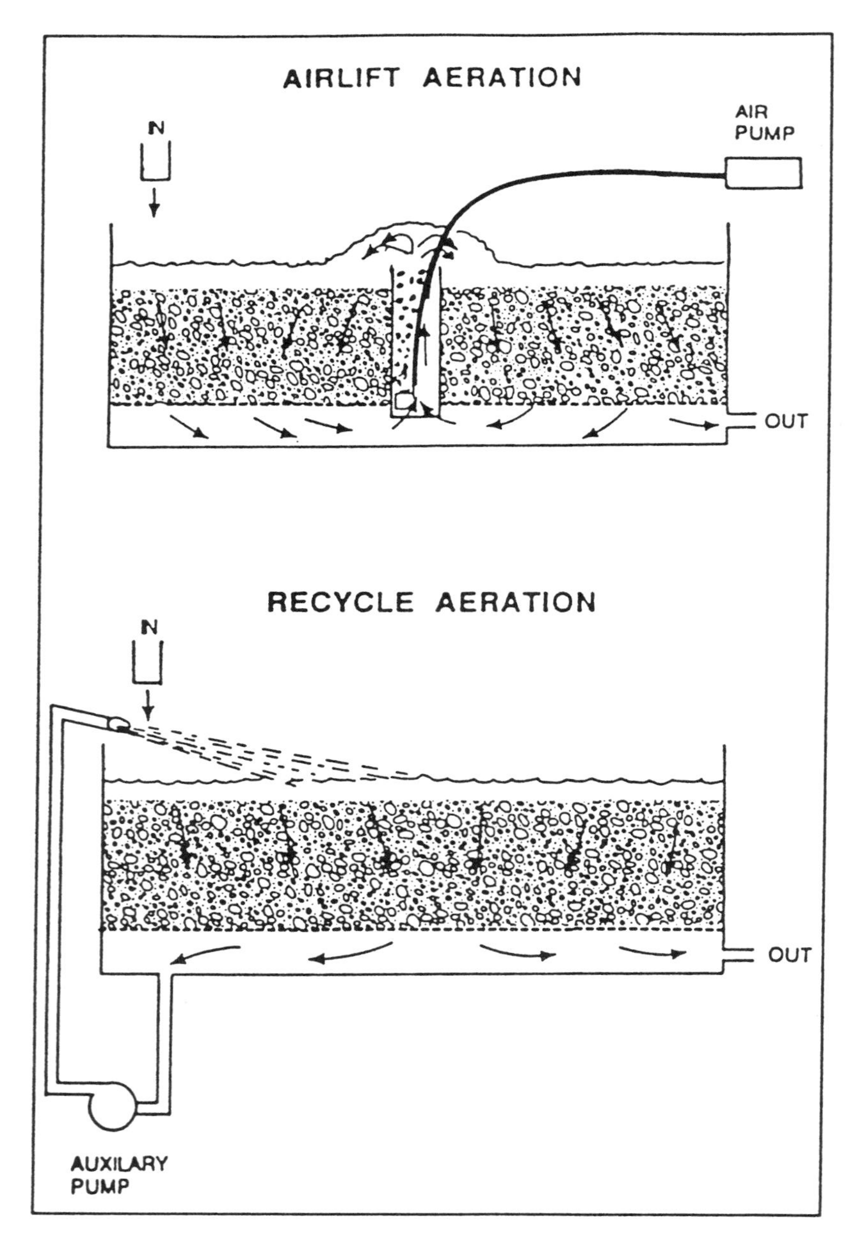

Figure 8–38. Two diagrams of a submerged rock filter. The air pump shown in the top diagram increases performance of the filter over one without the air pump, shown in the bottom diagram. (Courtesy, Ronald Malone, Louisiana Sea Grant College Program, Louisiana State University)

nitrate. Solid wastes accumulate in the spaces between the rocks and decompose. The calcerous rocks keep the pH of the water regulated.

The submerged rock filter is widely used in specialized operations such as soft-shelled crab or crawfish shedding systems. The primary disadvantage is that it is very bulky and may lower the production capacity because enough space is not available.

- ***Upflow sand filter*** — In the upflow sand filter, the water flows upward through a coarse sand bed. The bacteria grow on the surfaces of the sand, which has a greater surface area than the rock, so the filter does not have to be as large. Because there is little void space between the sand grains

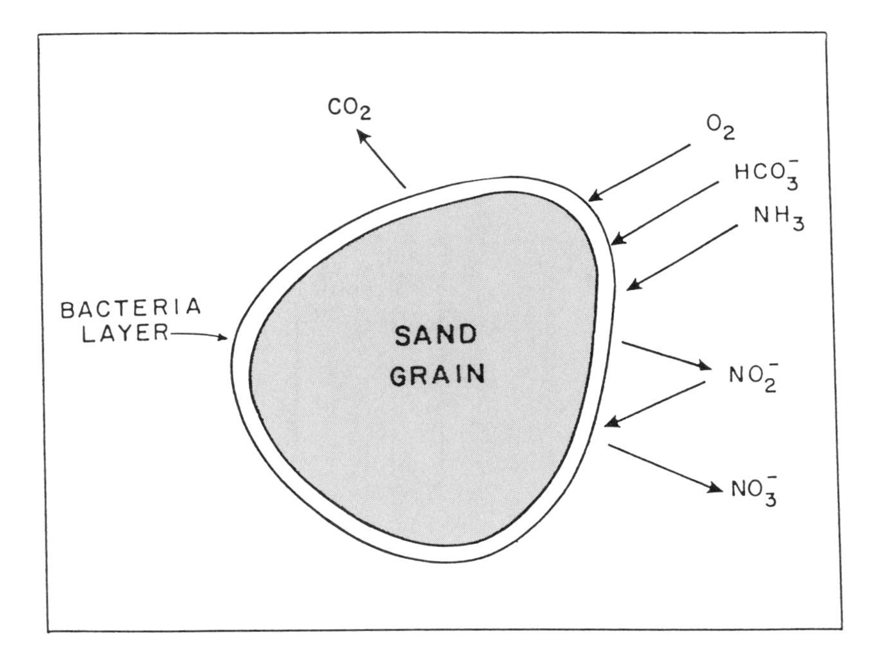

Figure 8–39. A view of how sand is utilized in the filtering of wastes. The bacterial film coating the sand removes ammonia and nitrites, making the water safe for the aquacrop. (Courtesy, Ronald Malone, Louisiana Sea Grant College Program, Louisiana State University)

for solids to accumulate, the solid wastes build up while a film of bacteria thickens, thus reducing water flow. For this reason, the bed must be expanded periodically for cleaning, usually once per day.

The upflow sand filter is expanded by increasing the water flow. When the water flow is increased, the solids, which are lighter than the sand, move on through the filter, while the sand rubs together to remove excess bacteria. When the bed is being expanded, the water is removed through

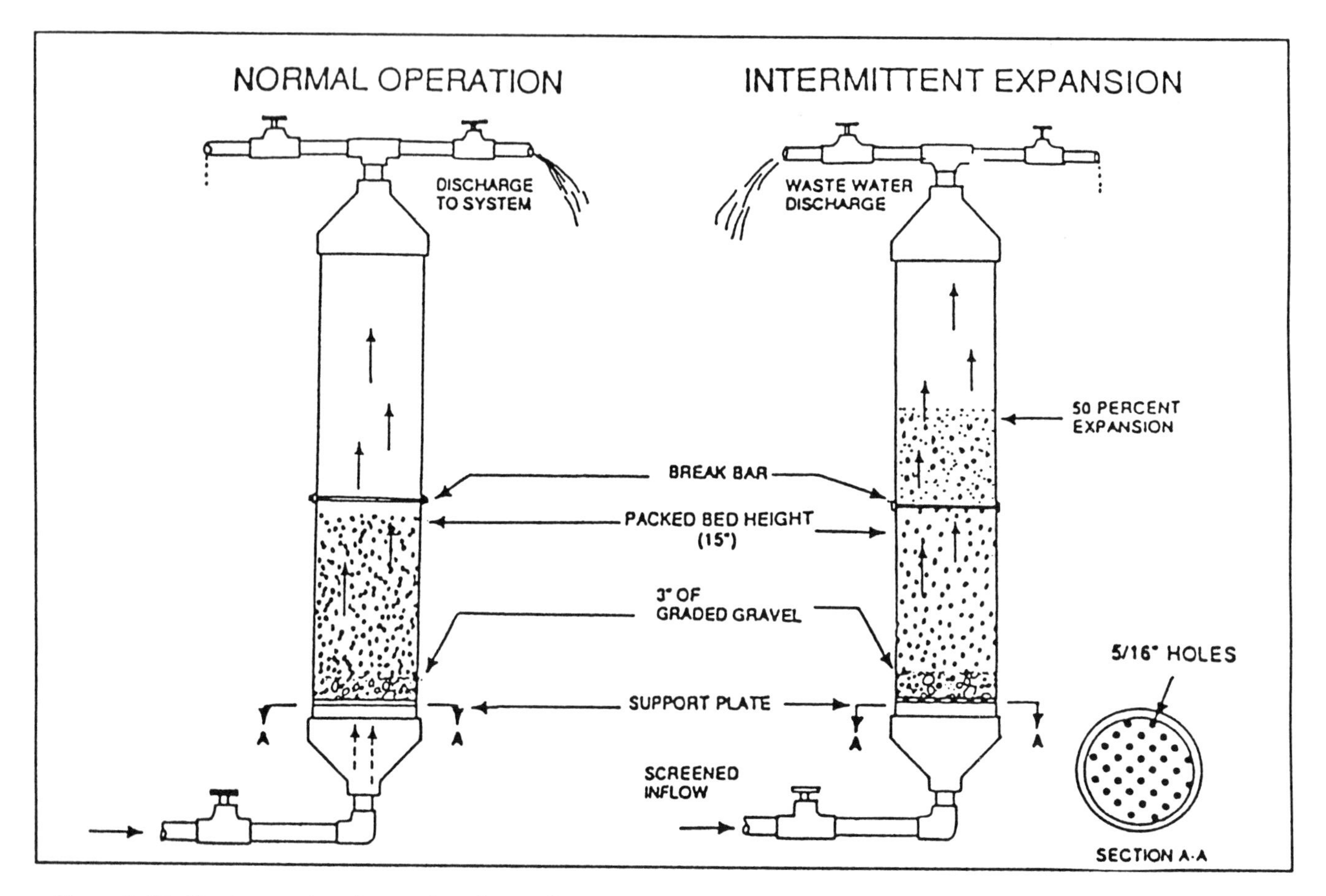

Figure 8–40. The two modes of operation of an upflow sand filter. (Courtesy, Ronald Malone, Louisiana Sea Grant College Program, Louisiana State University)

a waste valve and is not recirculated through the system. The upflow sand filter does not regulate the pH of the water, so this must periodically be checked and sodium bicarbonate added.

- ***Fluidized bed filter*** — A fluidized bed filter is an upflow sand filter that is kept in a constant state of between 25 and 100% expansion. This avoids the build-up of solid wastes and the creation of a bacteria film. Because the water is in a constant state of expansion, very little maintenance is needed with the fluidized bed filter.

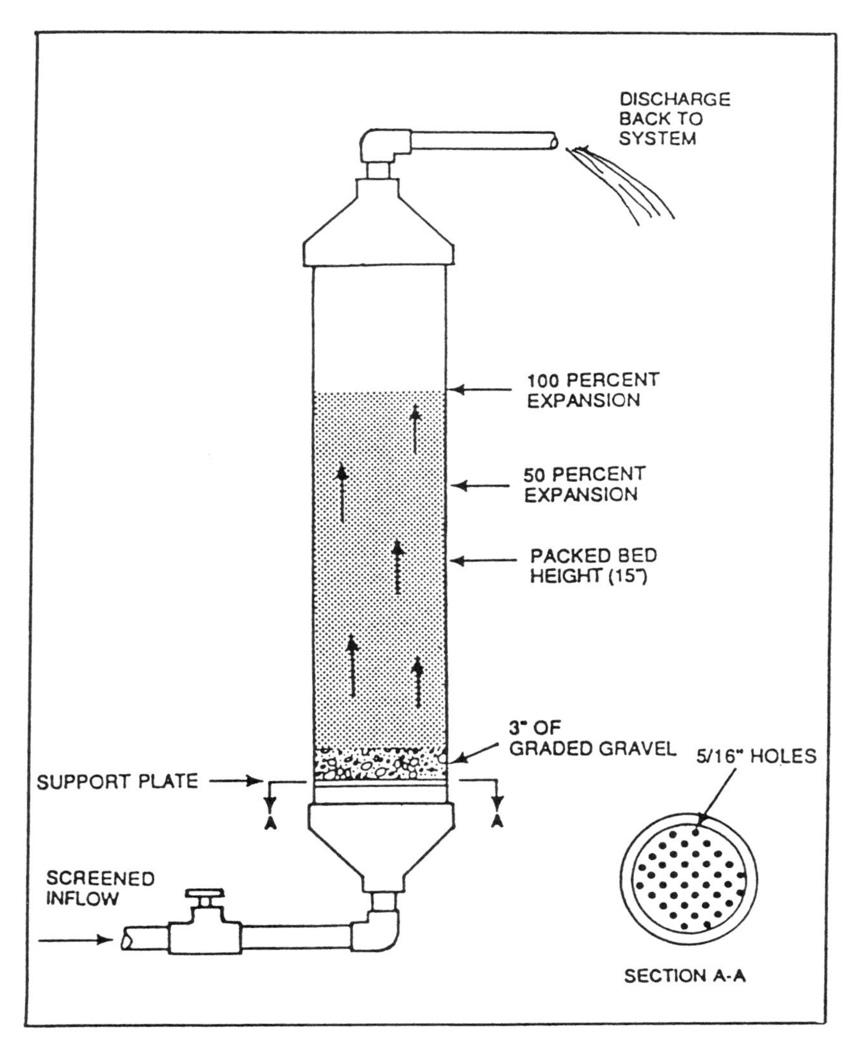

Figure 8–41. A diagram of a fluidized bed filter. (Courtesy, Ronald Malone, Louisiana Sea Grant College Program, Louisiana State University)

Because the solids are constantly being moved through the system, another filter is needed to catch the solids. While the fluidized bed filter does an excellent job of removing the nitrites and ammonia, it usually is used in conjunction with an upflow sand filter, which does a good job of removing the solids.

SUMMARY

Preparing an adequate production facility on a good site is an important part of a successful aquaculture operation. An appropriate facility depends on several factors, including the aquacrop to be produced, the level of production desired, the time an aquafarmer has available for management, and the capital he / she has to invest.

For large-scale production, ponds and raceways are the most common types of production facilities. Ponds are static systems where the water is held in the ponds and new water is added only to maintain the water level. They are primarily used for warm water aquacrops, such as catfish and tilapia, as the sunlight tends to heat up the water in ponds.

Raceways are open or flow-through systems which require a lot of water for their operation. They are generally used for a cold or cool water aquacrop such as trout.

Smaller-scale operations, which may also be very intensive, generally use cages or tanks. Cages are placed in existing water facilities such as farm ponds, rivers, oceans, and estuaries. The use of cages often allows aquaculture production in areas where production in the entire water facility is not feasible. Catfish, bream, and salmon are sometimes cultured in cages.

Tanks, although they may be used as round raceways, are also commonly used as static systems, similar to pond culture, or as closed systems. Closed systems, which recirculate the water, require the use of biofilters to remove the wastes from the water before it is pumped back into the growing facilities. Tanks may be used to produce almost any aquacrops, but ornamental fish and baitfish are two of the most common.

Site selection for a water facility involves the consideration of many factors. The most common factors are an adequate source of water, the size of the facility needed, and the proximity to local markets, suppliers, and utilities. The different types of water facilities all have unique requirements that also must be taken into account.

Ponds require the most involved construction, from determining drains to moving fill to create levees or dams. Topography is important for ponds as well as for raceways. Cages are usually built by the aquafarmer and offer a low-cost way to become involved in aquaculture. Tanks may be bought or constructed by the aquafarmer. Different materials from which tanks are built include fiberglass, plastic, wood, and concrete.

Several types of equipment are necessary for aquaculture production. In pond production, mechanical aerators, seines, and seine boats are required. For a large-scale pond operation, a feed blower and electronic dissolved oxygen monitor are almost mandatory. Some way of testing water quality is recommended for all types of water facilities. The aquafarmer must be able to test for dissolved oxygen, pH, temperature, nitrites, and ammonia. For a large system of any type, some type of automated feeder is recommended. Haul tanks are useful for many aquaculture operations. For a closed system, a biofilter is necessary for the removal of wastes from the recirculated water.

QUESTIONS AND PROBLEMS FOR DISCUSSION

1. What are the four most common types of aquaculture production facilities?
2. How are watershed ponds and levee-type ponds different?
3. What are some of the advantages and disadvantages of using watershed ponds for aquaculture production?
4. Why is soil type so important in the selection of a site for a pond?
5. What type of topography is needed for a levee-type pond? What about topography for a watershed pond?
6. What is the most important site consideration for raceways?
7. What is meant by a static system? An open system? A closed system?
8. How deep should water be for cage culture? Why?
9. How far apart should cages be placed? Why?
10. When lower-quality soil is used, how does that affect the cost of building a levee?
11. Determine the amount of fill needed for constructing a levee with a slope of 4:1 on both sides.
12. Why should the freeboard on a pond be between 1 and 2 feet?
13. What is the toe of a pond levee or dam?
14. What determines the size of a drain pipe of a pond?
15. Why should the minimum depth of a watershed pond be $2\frac{1}{2}$ feet?
16. What determines the size of watershed needed for a watershed pond?
17. How does a Venturi drain work?
18. What is the purpose of a feeding ring on a cage?
19. Why are aerators necessary in ponds?
20. What are biofilters? How do they work?
21. What type of monitoring equipment is needed for aquaculture production?

Chapter 9

MARKETING AQUACULTURE PRODUCTS

Success in aquaculture involves knowing more than how to produce a quality aquacrop efficiently. It includes marketing the crop so that a profit will be generated. An aquafarmer should never start the production of a crop until a satisfactory market for the final product has been found. *Knowing how to produce is important; knowing how to market is essential!*

Marketing is the link between aquafarmers and consumers. Since aquaculture is driven by what consumers want, it is important to spend some time looking at what consumers will buy. Early aquafarmers found that no marketing system existed. They began by developing their own markets. These were often small niche markets that wanted a steady supply of uniform product. The early aquafarmers found that careful marketing and making a profit went together. Aquafarmers are in the business to make a profit!

OBJECTIVES

This chapter emphasizes some fundamentals of aquamarketing. Upon completion, the student will be able to:

- Define "marketing" and "marketing functions."
- Explain marketing channels.

- Define "economy of scale" and relate it to aquaculture.
- Identify factors to be considered in selecting a market.
- Explain marketing planning.
- Explain how to develop a marketing plan.
- List and explain live aquaculture product forms.
- Describe the procedures used in processing.
- Describe the processed forms and cuts.
- Explain considerations in transporting aquacrops.
- Explain the role of consumer preferences in marketing.
- Describe the role of promotion.

THE CONCEPT OF MARKETING

As the link between aquafarmers and consumers, marketing involves all of the steps needed to get a desired product to the consumer. Marketing actually begins when the farmer is trying to decide what to produce. Aquafarmers cannot produce independently of what consumers will demand. Of course, through advertising and other means, demand can sometimes be created. All of the functions in moving aquacrops from the point of production to the consumer help make the concept of marketing.

Today's aquafarmers produce fish and other crops for the money they can get when the crops are sold. They produce for specific markets. This makes them commercial farmers rather than subsistence farmers. They are likely to buy their fish at the supermarket just like any other consumer.

Marketing Functions

Some individuals feel that marketing involves only the sale of an aquacrop. Though marketing includes selling, it is much more. Marketing aquacrops includes several functions:

- Planning what to produce
- Sampling the aquacrop
- Assembling and grading
- Transporting

- Advertising and promoting consumption in other ways
- Processing (also known as manufacturing)
- Selling (changing ownership)
- Storing (live and processed)
- Developing products (such as easy-to-cook forms of fish)
- Following up with consumers

Planning production. The aquafarmer needs good information about the kinds of aquacrops that the consumer will demand. This will require some serious investigation. The key to successful marketing is producing something that is in demand. The future aquafarmer should determine the aquacrops that have the highest demand. Before deciding what to produce, the future aquafarmer should consider the climate, available water, facilities, personal preferences, and other factors. Sometimes it is not possible to grow the crop with the greatest profit potential. This may be because of climate or other factors. The future aquafarmer wants to grow the aquacrops that will result in the greatest returns in the long run.

Sampling. Aquacrops may take on off-flavors. Processing plants may take samples just prior to harvesting to check on the flavor. The samples may be quickly cooked in a microwave oven and tasted by people who are trained as tasters. An aquacrop that is off-flavor should not be harvested. Steps should be taken to correct the off-flavor problem so that the crop can be harvested in a few days.

Scientists are not always able to explain what causes off-flavor. Sometimes it is attributed to feed and other times to water quality. The problem may be due to a number of different water-related factors. An aquacrop that is off-flavor should never be harvested and hauled to market.

Sampling may also be used to check on the presence of dangerous residues or organisms in the aquacrop. Industrial chemical pollution has occasionally been found in wild fish. Sometimes harmful organisms are present in other food items, such as oysters. Laboratory study is usually needed for accurate analysis. **Aquafarmers should be very careful to avoid pollution in a cultured crop.** Only wholesome food should be marketed!

Assembling and grading. Aquacrops must be gathered together in uniform lots. With fish, assembling involves loading them into tanks for hauling. It may also include bringing together the fish from several different farms. This is most often the case at a processing plant.

Grading is an important step. Buyers want fish that are of uniform size and species. Trash fish and diseased or injured fish must be removed. Fish that are

Figure 9–1. A uniform lot of fish headed for the processing plant. Note that the fish are of uniform size and species. (Courtesy, MFC Services)

too small or too large must be culled out or assembled into other lots. Processors want shipments that are of uniform size.

Transporting. As a marketing function, transporting involves getting the aquacrop from the farm to the consumer. This may involve stops along the way for other steps in the marketing process. The farmer may only be interested in transporting to the nearest processing plant. What that plant may be able to pay may be a function of the distance that the processed product must be hauled. Processing plants must haul the produce in refrigerated trucks to distributors, supermarkets, restaurants, and other outlets. Of course, recreational fee-lake operations are not as interested in transporting the product as they are in being convenient to the fishing public.

Advertising. Advertising involves calling the attention of possible consumers to a product. It is usually part of a larger promotion program. Its purpose is to convince people to try what is being sold, such as an aquaculture food item or other commodity.

Many farmers may not be directly involved in advertising aquacrops. Yet, nearly all of them have some sort of indirect involvement with it. Poor advertising can result in low consumption and lower profits to growers. Good advertising can result in increased consumption, profit, and expansion.

Farmers may help support some of the costs of advertising. For example, a

Figure 9–2. Various promotion items are used to encourage aquaculture consumption. Here a mug promotes the consumption of catfish.

fee may be added to each ton of feed for the specific purpose of product promotion.

Operators of fee-lakes are well aware of the value of good signs and other forms of promotion. Without good signs, the fishing public might not know where the fee-lakes are located.

Processing. Processing is done to get the aquacrop into a convenient form for the consumer. It may involve only a few simple steps, or it may be a complex process that results in a pre-cooked food item. What is done depends on the kind of aquacrop. Trout, for example, often have the head (with eyes) left on the whole fish. The scales and internal organs are removed. Shrimp may be whole, without heads and not peeled, or without heads and peeled (outer crustacean shell removed). Processed products may also be put into convenient packages for the

Figure 9–3. Processing procedures vary with the kind of aquacrop. Here crawfish are being peeled to be packed for shipment. (Courtesy, MFC Services)

consumer. (More on processing is presented later in the chapter in the section "Processing Aquacrops.")

Selling. Selling involves changing the ownership of an aquacrop. The seller and buyer must reach agreement on a price. To the farmer who sells to a processor, selling is simply delivering the crop to the processing plant at a fixed price. The farmer usually knows the price well in advance. The fish are weighed, the total value is determined, and a payment is made for the amount (less any harvesting or hauling costs if the processor provided these services). Certainly, the processor expects the farmer to provide a quality, uniform aquacrop. Some farmers go so far as to operate a restaurant to sell their product. In this case, selling occurs when the consumer buys the ready-to-eat dinner.

Storing. Aquacrops must often be kept for a period of time before their use. Storing may occur on the farm or at the processing plant with live aquacrops. It may occur after processing when the product is refrigerated or frozen. Storage conditions vary with different aquacrops. Individuals must know the qualities of the product they are storing so as to maintain maximum quality. For example, processed fish that are improperly stored can take on off-flavors or undergo a change in texture. If this occurs, the consumer will not buy the product.

Developing products. In order to enhance demand, food processors often develop new forms of products to appeal to consumers. These may involve new ways of packaging, easy-to-cook forms, and new kinds of products. A trip through the frozen food section of a local supermarket provides the opportunity to see many forms in which foods have been prepared for consumer convenience.

Following up with consumers. This function may be described as a kind of evaluation. It deals with trying to determine how much consumers like a product. Many times labels on food products will provide the opportunity for consumers to send in their comments. In some cases, a toll-free telephone number is given so that consumers can share concerns or suggestions. By knowing how consumers feel about a product, the producer can try to make it better.

A Marketing Essential

All individuals involved with aquacrops must be aware of the importance of allowing only quality products to reach consumers. Poor-quality products damage the reputation of those products. Fish with off-flavor may cause the customer to stop buying the product. Any foods which are improperly stored, handled, or cooked might cause food poisoning. An outbreak of food poisoning can destroy the demand for an aquacrop.

The rule to follow is: Only quality, wholesome aquaculture products should be marketed.

AQUACULTURE MARKETING CHANNELS

Depending on the aquacrop, several ways are used to link the producer with the consumer. In this chapter, the major emphasis is on marketing freshwater fish. Several channels are available to the fish farmer. They are:

- Marketing through processing plants.

Figure 9–4. Alligator skins have been removed and are boxed, ready for shipment to the tannery. (Courtesy, MFC Services)

- Marketing through local retail markets.
- Marketing by hauling to fee-lake operators.
- Marketing through restaurants and supermarkets.
- Marketing through recreational fee-lakes.

Processing Plants

The vast majority of most cultured food fish is sold by producers to processing plants. These plants are usually capable of handling thousands of pounds of fish each day. Very expensive automated machinery is used in these plants. Sanitation and safety procedures are strictly enforced to insure a wholesome food product.

The typical processing plant buys fish from the farmer, dresses them, puts them in attractive and protective packages, and sells them to wholesalers, restaurants, supermarkets, and institutional food services. Processors may have trucks that deliver the processed fish to the point of sale. Fish that are sold wholesale go to a distributor who usually sells them again to a retail outlet. Wholesalers often buy in very large quantities. Retailers may buy only a few pounds a week. Some processors may also operate retail outlets.

Some processing plants have harvesting crews that will go to the farm and harvest a fish crop. The farmer is charged for this service. Most of the time the charge is only a few cents per pound of fish. This allows farmers the freedom of investing their money in production facilities rather than in expensive harvesting equipment.

An attractive advantage of marketing through processing plants is that the plants will often sign a contract with producers, insuring that there is a sale for the crop when it is ready. The contract usually specifies the price and the amount of fish that will be bought. The price might not be as high as some of the other market outlets, but it is fairly well guaranteed. In other words, the risk of having a crop and no place to sell it is virtually eliminated. Sometimes farmers pool their resources and build processing plants. They have more control over processing and share in the profits, if any.

Local Retail Markets

Aquafarmers may set up local retail markets. These include operating roadside sales stores, operating restaurants where all or some of the crops are sold, and selling only on an order basis to individuals or groups.

Roadside stores. Some farmers operate small roadside fish markets. The fish may be sold live or dressed according to the wishes of the customer. A roadside store adds another dimension of management to an aquafarm. The store must be operated efficiently. Good workers must be employed. Equipment must be obtained. Licenses must be purchased. Tax documents and other government reports will need to be prepared. Advertising and public relations will be important.

When properly operated, roadside stores can produce increased returns to the farmers. By selling directly from the pond to the consumers, the aquafarmers can get all of the profits that would go to the processors and distributors.

One of the biggest problems with roadside markets is that considerable effort is necessary to make them successful. Regular store hours must be set and observed. The store must be in a prime location convenient to customers. Only top-quality, wholesome products should be sold.

The volume of aquacrops sold through roadside stores operated by growers is small. However, a few are able to make a good profit.

Restaurants. A few enterprising aquafarmers have attempted to operate restaurants in conjunction with their aquafarms. Most often, these have been fish, shrimp, crawfish, or crab businesses. The restaurants may be located on the farm or in a nearby town. In addition to the aquaculture items, the restaurants often have other foods on their menus. These restaurants however usually specialize in only a few choices of foods.

The big concern for an aquafarmer who has a restaurant is management. Facilities, equipment, personnel, and a good location are needed in order for the business to be successful. The farmer must also keep records and comply with regulations on food preparation, sanitation, and other areas. Skills to be a good aquafarm manager are different from those of a good restaurant manager. Regardless, a few farmers are able to be successful with this type of venture.

Selling on an order basis. A few aquafarmers sell large orders of live, dressed, or cooked products to groups for special functions. Civic clubs, church groups, and others may contract with an aquafarmer for a given amount of product either raw or cooked. In order to do this, the aquafarmer must have the facilities, equipment, and personnel to do the work. Records and other business matters must be handled.

Live Hauling

Live hauling involves the aquafarmer providing live aquacrops to others who will likely use them in recreational fee-lakes. Sometimes the producer may have a haul truck and deliver the fish to the fee-lake. More often the fish may be hauled by the owner of the fee-lake or by a custom hauler. Live haulers do not usually

Figure 9–5. Live hauling requires a truck equipped with tanks and aerators.

harvest or grade the fish, as this is done by the farmer. The fish must be in excellent condition because they must often travel long distances and be readily available for the fee-lake. Live haulers often pay slightly more per pound for fish than processing plants. The volume of fish that can be marketed this way is considerably less than by processing.

Restaurants and Supermarkets

Marketing through restaurants and supermarkets is attractive for some aquafarmers. They must harvest their aquacrop, dress or otherwise process it to the form desired by the restaurant or supermarket, and then deliver it. This may also involve the farmers making sales calls on restaurants and supermarkets to see if they will buy. The farmers must have needed harvesting, processing, and hauling equipment. Conditions must be clean and insure that a wholesome product will be produced.

The volume that can be marketed this way is often small; however, some producers develop niche markets with restaurants. For example, restaurants in large cities may cater to ethnic groups that want a specific fish. These fish may be grown hundreds of miles away and shipped in at a premium price. Good examples include tilapia and hybrid striped bass. Some restaurants will buy these fish without any processing by the farmer. Of course, grading, packing, and transporting will be needed.

Other than for special situations, the volume of aquacrop marketed directly from farm to restaurant or supermarket is relatively small.

Recreational Fee-Lakes

A few aquafarmers operate recreational fee fishing businesses. Typically, these are small ponds that have been heavily stocked with eating-size fish. Individuals pay a fee to get to the pond and a price per pound for the fish caught. For a fee, the operator may also provide dressing services for the fish caught. Usually fee-lakes are open only certain hours on certain days of the week. A few are open every day year-round.

Recreational facilities must be conveniently located. Often, major highway intersections near towns or cities are good locations. To the large aquafarmer, a recreational facility adds another management dimension. Employees must be on hand to run the fee-lake and provide whatever services the fishing public wants. Restrooms, bait sales, and refreshment vending machines should also be available. Areas around ponds must be kept well-mowed, clean, and neat. Advertising will be needed to attract customers. There must be good driveways and parking areas to handle automobiles. One further item is the necessity of insurance. Allowing the general public on a farm increases liability in case an accident were to occur.

Fee-lakes must be different from those used for growing fish. They should often be separated by some distance. Fences to restrict access may be needed. There is always the chance that a disease or trash fish may be transported by the fishing public.

ECONOMY OF SCALE

"Economy of scale" describes the size of a fish farm and the marketing alternatives in relation to the profit that is made. The volume of fish produced and the ability of the market to use them must be balanced. At a minimum, a farm must not produce more than can be marketed through the available channels. What is economical to one farm situation may not be to another. The key to looking at economy of scale is profitability.

Many factors go into making a profit. Larger farms usually have lower per pound costs for growing fish than smaller farms. Certainly, this depends on the management and costs of production. Sometimes smaller operations return a higher rate of profit because the owners are able to follow good management practices and to keep costs of production low. A few possibilities are given here.

- *Small farms might not produce enough fish to be attractive to a processor;*

Figure 9–6. Small growers must rely more on hand labor and less on the use of machinery and equipment.

therefore, smaller farmers may need to find other marketing alternatives. Smaller farmers may get into more direct marketing through roadside stores, fee-lakes, and farmer-owned restaurants. Per pound prices are often more when sold through these marketing channels. By virtue of not having to hire a lot of labor and invest in expensive equipment, smaller farms may

Figure 9–7. Large growers may use expensive machinery in the harvesting of fish. (Courtesy, Delta Pride Catfish, Inc.)

be profitable. In case of a family aquafarm, the big question to be answered is "Will the farm produce enough profit for the family to have a good level of living?"

- *Large farms can sell to wholesalers or to processors if the volume is substantial.* The prices paid are usually less than those paid by retail outlets. The larger producers should have lower per pound costs because feed and other inputs are often sold at a lower price in greater volumes. On the other hand, larger farmers may find that local sales to retail stores and restaurants are not economical. These market outlets simply do not demand enough fish to justify the work required to serve them.

In summary, economy of scale has to do with size as related to alternative markets and costs of production. Careful analysis of a situation is needed in an attempt to determine if a proposed aquafarm will be profitable.

SELECTING A MARKET

As has been described, marketing is more than merely changing ownership of an aquacrop. It is a rather detailed process whereby aquacrops get from the producer to the consumer. The marketing alternatives that are available may vary greatly, depending on a number of factors. The aquafarmer must study these in relation to marketing.

In considering how an aquacrop will be marketed, the farmer should answer the following questions.

1. What is the size of the market? How much can be sold retail and how much wholesale? How does this fit into my farming operation?
2. Where is the market located? Will some of the aquacrop be sold locally? Will the crop need to be hauled long distances? Will the returns from the crop cover the extra hauling costs?
3. What processors are available? Will the processors agree to a minimum price for the aquacrop?
4. What equipment and facilities will be needed to process on the farm? Will this be profitable?
5. What are my personal preferences? (Some people like to operate retail facilities; others prefer to sell to processors and thus avoid the hassles that go with retailing.)

PLANNING FOR MARKETING

Consumers have choices! They can buy an aquacrop or they can leave it. Marketing involves trying to determine consumer preferences and then providing a product that will be consumed. In some cases, producers can try to create a demand for their product. Various promotional efforts will be needed. Regardless, some understanding of the marketing planning process can pay big dividends to the aquafarmer.

The Plan

Aquafarmers should carefully plan how they will market their crops. The same also applies to processors and others involved with the aquaculture industry.

Marketing plans are frequently developed to guide the marketing process. Such plans may be simple or complex. A contract with a processor is part of a marketing plan. Marketing plans may range from 5 to 50 pages or more. Large processors may have elaborate marketing plans. Marketing plans must be realistic and based on sound information. Once a plan has been made, there must be a commitment to carry out the plan.

Marketing plans may be created on an annual basis for products already being produced and marketed or as needed for marketing a new product. The purpose of a plan is to help spell out how an aquaproduct is to be marketed. Plans typically have four parts: situation, objectives, strategy, and assessment.

1. *Situation*—The situation statement describes current conditions. It analyzes the market situation and answers the question "Where are we now?"
2. *Objectives*—The objectives indicate what is to be accomplished in marketing. They should be fairly specific statements. The objectives are sometimes known as the business proposition and answer the question "Where do we want to go?"
3. *Strategy*—The strategy part of the plan explains how the objectives will be achieved. Resources will need to be allocated to market the product. It will not just happen; people must work at marketing. The strategy part of the plan answers the question "How do we achieve our objectives?"
4. *Assessment*—This part of the plan spells out how the marketing effort will be evaluated. There is normally a procedure for assessing each objective. The assessment part of the plan answers the question "How well did we do?"

The Process of Planning

Marketing planning is the process used to develop a marketing plan. It provides for a focus of efforts in marketing an aquacrop. Time and other limited resources are allocated to the highest priorities. Limited production resources are utilized more effectively.

Reliable information — from reports published by government agencies, aquaculture associations, and other sources — must be used in the marketing plan development. The farmer may also wish to do a little investigating by talking to people who are already in the business; contacting processors and other possible market outlets; studying the successes and failures of others who have ventured into aquaculture, noting their strengths and weaknesses and then comparing these to his / her own strengths and weaknesses.

The objectives and strategies should be carefully developed. Deadlines should be set for reaching objectives. Ways to reach the objectives must be realistic. Resources must be allocated to marketing. At the end of the year (or other market period) some means of assessment must be specified. The most important factor to assess is how well the plan worked financially. Was a profit made? The information gained from the assessment can be used to develop a revised marketing plan for the following year.

Situations to Consider

In developing a marketing plan, evey aquafarmer has a different situation. Several factors to consider are offered here, as follows:

1. *What are the preferences of the aquafarmer?* This factor deals with the likes and dislikes of people. For example, the small aquafarmer who likes to work with the public might do well to have a retail roadside market or fee-lake.
2. *What opportunities are available in the local area?* What opportunities are available that will require working with someone a distance away? For example, the best opportunities might involve working with a distributor in a large city a long way from the farm.
3. *What aquaculture currently exists in the area?* For example, if tilapia is currently being successfully grown in the area, this crop must be adapted to the climate and other environmental factors.
4. *What are the local, national, and international trends that could impact the proposed aquafarm?* For example, a shortage of an aquacrop that could

be produced may be a good signal that the crop might be profitably grown and marketed.

5. *What resources are available to start the aquafarm and market the crops?* Can additional resources be obtained? For example, will the banks or other lending agencies provide the money needed to carry out the marketing plan?
6. *How well are the proposed aquacrops suited to the climate, water, and other available resources?* Some aquacrops may be well suited and in demand by the market.

PRODUCT FORM

"Product form" describes the stage at which an aquacrop may be marketed and how it will be processed. It varies with the kind of crop produced. Consumer demand also plays an important role. All of these product forms are live forms that may require processing to get them ready for the consumer. (Processed forms are discussed in the next section of this chapter.)

A few examples of live fish product forms are discussed here.

Eggs

The egg form of an aquacrop may be in demand by hatcheries and by

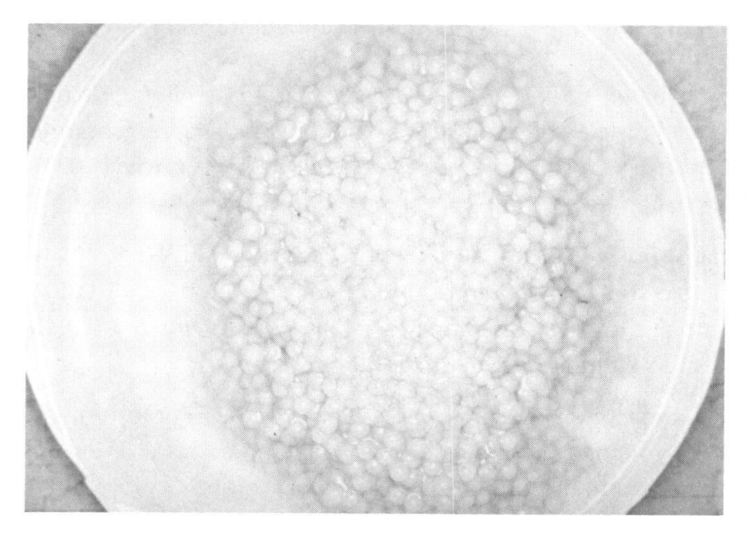

Figure 9–8. Some producers market fish eggs to hatcheries and for human food. (Courtesy, Master Mix Feeds)

humans. Hatcheries often buy "eyed eggs." These are fertile eggs that are beginning to show the development of the fish. The hatcheries continue with the incubation process and grow the fry into marketable fingerlings.

Some kinds of eggs are eaten by humans as food products. This form is often referred to as caviar, a rather expensive restaurant appetizer or entrée. It is usually the roe (egg mass) of large fish such as the sturgeon. Sometimes fish are harvested exclusively for their roe. An example is the mullet caught in the Gulf of Mexico by commercial fishers of the U.S. The roe is removed intact from the fish and exported to Japan, and the fish are used in various products or disposed of as waste.

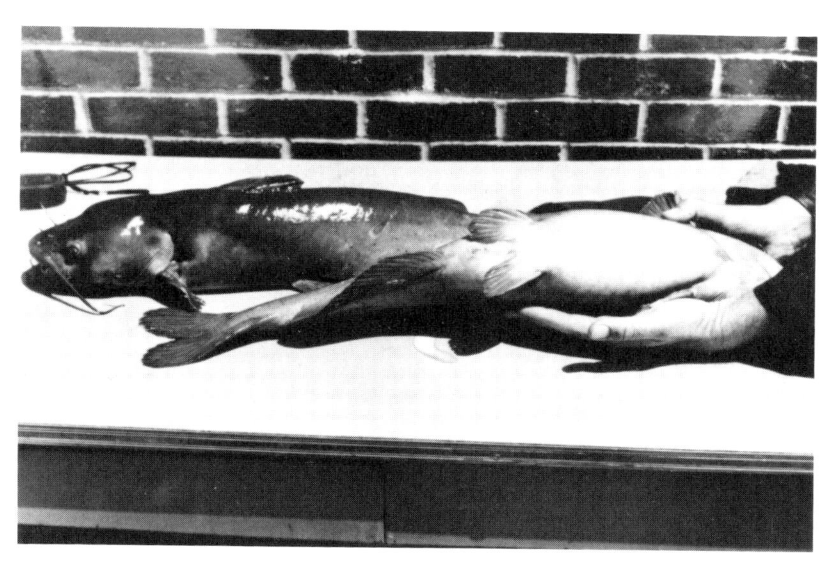

Figure 9–9. Broodfish are marketed by a few growers. The fish on the right is a female with a large abdomen, which indicates that she is near spawning. (Courtesy, U.S. Fish and Wildlife Service)

Broodfish

A few aquafarmers produce broodfish to be sold to other farmers for reproduction. This may be a good marketing alternative if the fish are of good quality, improved through selection, or of a selected strain. The volume that could be marketed with most species of broodfish is often limited. Careful study of the possibilities would be needed before a broodfish farm could be launched. Marketing often depends on having a good reputation and making personal contacts with growers.

Fry

Newly hatched fish are sometimes sold to fingerling growers. Fry are very small and difficult to see. The basis for selling is usually by the thousand. Being so small, they are difficult to count. The number is often established by collecting a small sample in a container (such as a graduated beaker or a small cup) and then counting the number in the sample. This count is then used to calculate the number of fry in a larger volume. When the fry reach an inch in length, they become fingerlings.

Fingerlings

Fish that range from 1 to 8 or 10 inches in length are usually referred to as fingerlings. They are sold as individual fish, with the price usually being an amount per fish based on the length of the fish. Where food fish are grown, there is a demand for fingerlings. Some farmers produce their own fingerlings; others buy them ready for the grow-out ponds.

Fingerling farmers must produce a quality product. A farmer who markets poor-quality fingerlings will not have much repeat business from growers. A good reputation is essential! Contacts with buyers of fingerlings may be local neighbor-to-neighbor contacts, through advertisements in aquafarming publications, or through direct calls to food fish farms to see if fingerlings are needed. The fingerling producer will likely need to provide hauling to the fish farms; however,

Figure 9–10. Harvesting fingerlings for market. (Courtesy, Master Mix Feeds)

Figure 9–11. A food-size channel catfish ready for market. (Courtesy, Louisiana Agricultural Experiment Station – Louisiana Cooperative Extension Service, Louisiana State University Agricultural Center)

some fish growers have their own hauling equipment. Hauling and treating fish to prevent disease during hauling are a part of fingerling marketing.

Food Fish

The major market for fish is for human consumption. Cultured food fish usually weigh from ¾ to 2 pounds or so. Uniform sizes are essential in processing plants. Fish weighing over 4 pounds may be penalized with a lower price. Many farmers have thousands of acres devoted to food fish production. They carefully make decisions about production on the basis of the market available. (More information is presented elsewhere on marketing food fish.)

Other Forms

Depending on the crop, several other forms of live products may be mar-

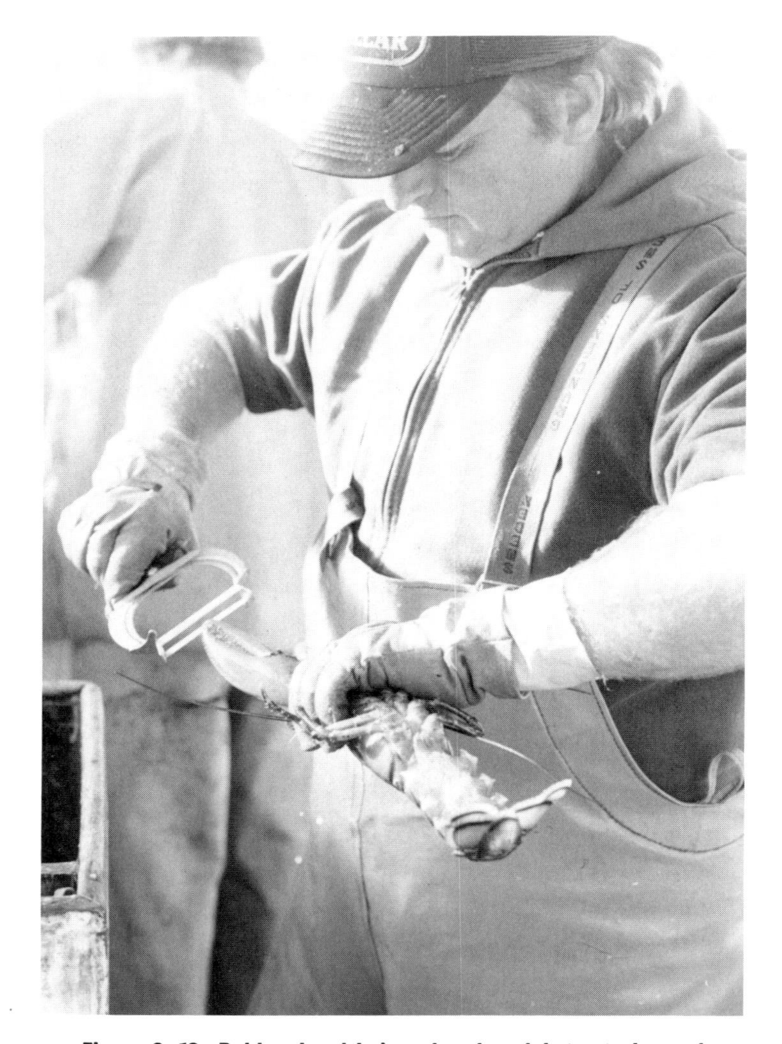

Figure 9–12. Rubber band being placed on lobster to keep claws closed. (Since lobsters are often shipped live, this prevents them from attacking each other.) (Courtesy, Robert C. Bayer, University of Maine)

keted. Some fish are marketed for use as bait by sport fishers. Other live fish, particularly exotic species of fish, are marketed as pet fish.

Bait producers may sell to retail bait shops. This most often involves hauling the baitfish to the shop. The sale price is usually on a per fish basis. This means that the baitfish are counted, usually at the point of delivery. Baitfish must be properly hauled and kept healthy.

A large number of species are used as pet fish. These are typically very small fish that are sold to pet stores, discount chains, and other outlets. They are usually sold on a per fish basis. The producer usually arranges transportation to the outlets. This may be by truck or air freight, with the fish packed in plastic bags containing water and oxygen. Sometimes a few pieces of ice are added to the bags or the bags of fish are placed on ice.

PROCESSING AQUACROPS

Processing involves getting the aquacrop ready for the consumer. It includes a wide range of activities. Specialized processing plants are needed. The trend is toward aquacrops being processed by larger plants, with less on-farm processing. These facilities are usually impractical for individual farms unless there is a large volume. Processing prepares the aquacrop and preserves it for the consumer. The focus in this chapter is primarily on processing fish.

What Happens at a Processing Plant

Large-scale processing plants are modern facilities with the latest in technology. Most specialize in one species or product. This is in contrast to the older seafood plants that handled several different products. The modern plants have various types of labor-saving equipment to perform many activities. However, even the most modern plants need considerable labor.

With fish, the term "dressing" is often used. Dressing is essentially the same

Figure 9–13. Scene in a small fish processing plant.

as basic processing. With fish it usually includes removing the head, internal organs (also known as gutting), and skin or scales. With shellfish, such as shrimp or crawfish, processing includes deheading, peeling, and deveining (removing the vein under the shell). Processing involves additional work to prepare the fish for the consumer.

Several steps in processing fish are listed here.

- ***Receiving***—This involves taking delivery of the aquacrops from the farmers. The fish are unloaded into baskets for weighing and moving to a vat or tank for holding. The fish may be held several hours or up to a day. Farmers are paid on the basis of the weight of the fish received. A receiving clerk must keep careful records of the weight received from each farm. Fish are normally kept alive until processing, which usually occurs within 24 hours of delivery.

  Flavor testing is often done before the fish are unloaded. Plants are careful to process only fish that have the right flavor. Even though the fish are checked before being hauled to the plant, they are frequently checked again before being unloaded. Samples are quickly cooked and tasted and smelled by trained individuals. Any fish that are off-flavor are not processed. A small sample of fish can provide some indication of the flavor of all of the fish from a pond, raceway, or tank. Fish with off-flavor are rejected and usually hauled back to the farm.

- ***Sorting***—In most cases, farmers are expected to deliver uniform lots of fish to the processing plant. If not, the plant must grade the fish into batches of uniform size and species for processing. Fish with diseases, with injuries, and of mixed species are usually refused by the receiving clerk.

- ***Stunning***—The first step in the processing line is often to stun the fish. An electric shock is used to paralyze the fish. Large tubs or tanks of fish are shocked at a time. Stunning allows the fish to remain alive but makes them lifeless. Dead fish deteriorate quickly. Stunning makes it easier for the beheader to handle the fish.

- ***Removing the head***—The first step in preparing most fish is to behead them. A band saw or fish head saw is used to cut the head off at the proper place. Workers usually behead the fish by pushing them into the saw blade. These workers must learn to perform this step very rapidly. Not all species are beheaded; therefore, this step may not apply.

- ***Removing the internal organs***—This step is also known as *eviscerating*. Most processing plants use a vacuum process that sucks the organs out of the body cavity through a slit cut in the belly or where the head was removed.

- ***Scaling or skinning*** — Depending on the species of fish, the scales or skin is removed. Most processing plants use mechanical methods to eliminate scales and skin. Sometimes hand or other methods are used to take off the outer covering of the fish. The machines must be carefully adjusted to avoid damage to the fish and to do a good job of getting rid of the scales or skin.

- ***Washing*** — Water is used throughout processing to keep the fish products and equipment clean. Sanitation is important in all areas of a processing plant or a food handling business.

- ***Inspecting*** — After they have been washed, the fish are inspected to see that all skin or scales and internal organs have been removed. If parts remain, they are removed. Damage is also noted. If injuries or disease evidence is found, the fish are discarded.

- ***Fabricating*** — This may include a wide range of activities and is frequently known as cutting. It involves preparing the fish into suitable pieces. Size, weight, and shape are important. Pieces must be of the size desired by the consumer. When the cuts are made, wasting valuable fish products should be avoided.

  Increasingly, fish are being prepared for easy cooking. Some fish products are seasoned, such as catfish fillets with cajun spice. Pre-cooking may be used to cook fish partially or fully. Smoked salmon and trout are marketed as gourmet, ready-to-eat foods.

- ***Packaging*** — Fish may be packaged in large boxes, individually wrapped packages for supermarket sales, cans or jars, or in other ways. The kind of packaging used depends on how the fish products are to be preserved. Packaging also includes the addition of an appropriate label on the container. The label should comply with legal regulations in describing the product, the amount of product, and the way the product is to be stored. Laser beams are now being used to count fish or fish portions to insure that each package has a uniform number of items.

- ***Preserving*** — Preservation is used to keep food from spoiling. Fish that have spoiled are not wholesome for human consumption. Spoilage is usually due to the presence of bacteria, yeast, and / or mold. An example is salmonella, a dreaded bacteria that causes food poisoning in humans, sometimes resulting in death.

  Following procedures that are unfavorable to the growth of bacteria,

Figure 9–14. Consumers are more prone to buy attractively packaged fresh fish products in the supermarket. (Courtesy, Delta Pride Catfish, Inc.)

yeast, and mold will help prevent their growth in fish products. Extreme heat or cold will kill many of these organisms. Inadequate moisture and excess saltiness or acid will also kill or prevent the growth of these organisms. Preservation methods include refrigerating, freezing, canning, smoking, drying, and pickling fish products. Fresh fish products are preserved by refrigerating or freezing. The quick freezing of individual fillets has gained wide consumer acceptance. These fillets can be kept frozen several weeks or a few months with little deterioration. Products that have been canned can be stored for quite a while before they are used.

The kind of packaging used is closely related to the method of preservation. Canned products are packaged in cans or jars. Frozen and refrigerated products are packaged in plastic or paper containers.

- ***Storing***—How fish are stored is based on how the fish have been preserved. Frozen fish must be stored at 0 - 5°F. Canned fish should be stored in relatively cool warehouses and not exposed to heat. Proper storage is essential. Fish will deteriorate rapidly when improperly stored. Improper storage causes spoilage, thus making the fish unfit for human consumption.

- ***Delivering***—Transporting fish to the consumer is an essential step in processing. Trucks must be properly refrigerated to keep the product at the appropriate temperature. Air freight may be used for high-value fresh fish products. Small batches may be shipped in insulated chests containing dry ice or some other means of keeping the products cool.

- ***Merchandising*** — Merchandising involves promoting the sale of processed aquacrops. Attractive packages in supermarkets are helpful. Posters and banners may be used by supermarkets and restaurants. Advertisements may be put in newspapers, on radio or television, or on billboards. Clean, modern delivery trucks can certainly help in promoting sales.

Processed Forms and Cuts

Fish may be processed into several different forms and cuts. The forms depend on the kind of fish and the way the product is intended to be cooked. A processor wants to get as much product from each fish as possible. About 40% of a fish is head, internal organs, fins, scales or skin, and other parts that may be cut away. This means that only 60% of a live fish is left after dressing. Some of the fancier cuts (such as fillets) result in lower dressing percentages. Several common forms and cuts of fresh fish are described here.

- ***Whole*** — Whole fish have not been dressed. The scales or skin, head, and internal organs are still in place. The fish are usually not fed within 24 hours of harvesting. They are usually washed, chilled (often 34°F. body cavity temperature), and boxed for shipment. Refrigerated trucks or insulated containers to which ice has been added are used to transport. The restaurant chef or the home preparer does the desired dressing. Tilapia and hybrid striped bass are frequently processed in the whole form.

- ***Drawn*** — The drawn form involves removing the internal organs and leaving the head. Removing the skin or scales is optional with the drawn form. The preparer of trout in this way should be careful to leave the eyes.

- ***Dressed*** — The internal organs, head, scales or skin, and fins are removed with the dressed form. This is one of the most popular ways of preparing fish. Sometimes this form is referred to as pan-dressed.

- ***Steak cut*** — The steak cut is made by slicing across a dressed fish. The slices are 3⁄4 - 1 inch thick. A section of backbone and other bones may be present in steaks. This method of cutting is not as common as a few years ago. It is more frequently used with wild fish.

- ***Fillet cut*** — The fillet cut is made by slicing parallel with the backbone. Most fish produce fillets without bones. This cut yields a low dressing

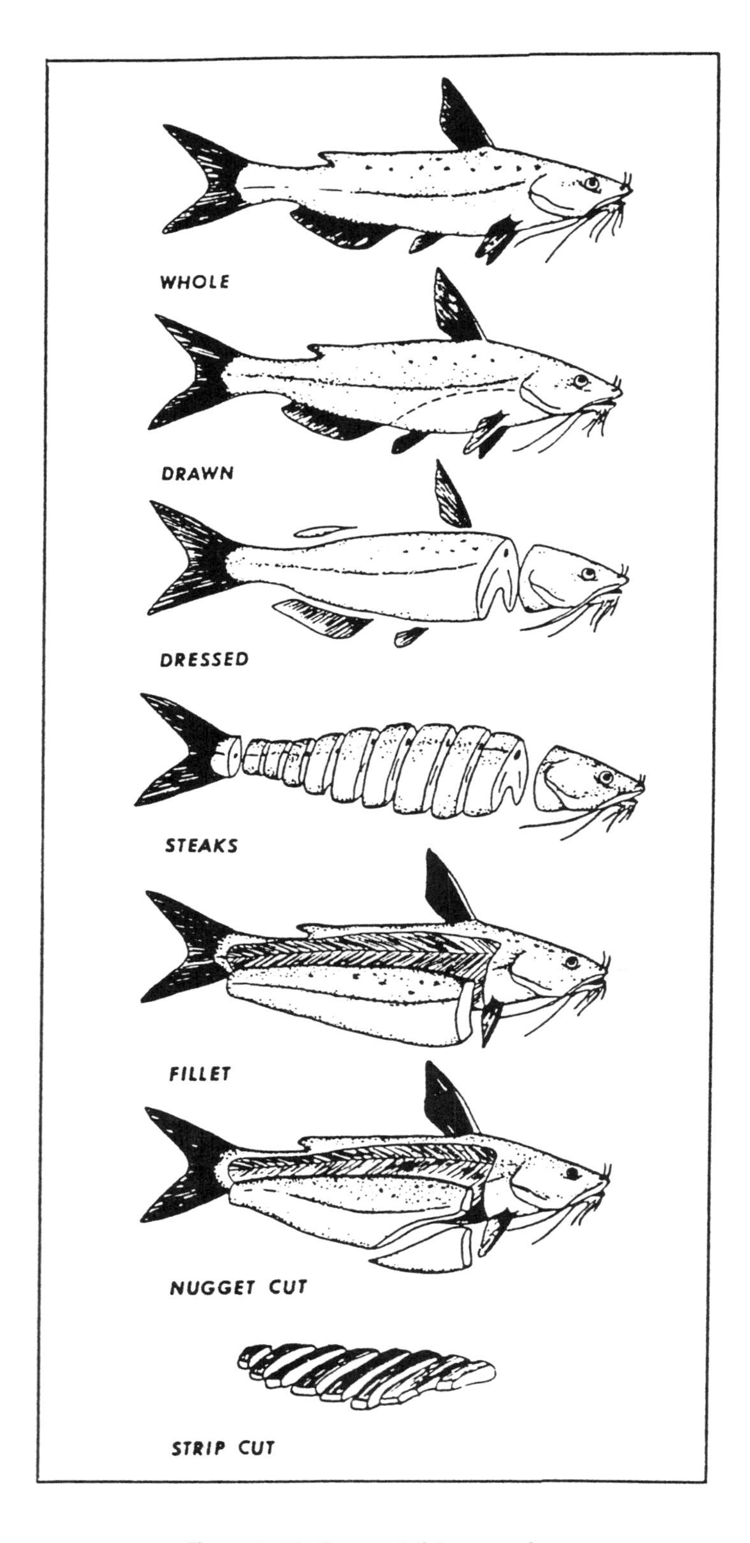

Figure 9–15. Forms of fish processing.

percentage but is a very high-quality cut. Prices for fillet cuts are usually considerably higher than for steaks or dressed fish.

- ***Nugget cut***—This cut may be made from parts of the fish that do not go into the fillet cut. The belly area of the fish is the chief source of the nugget. It is sometimes called the belly flap.
- ***Strip cut***—The strip cut is made from fillets sliced into narrow strips of fish. This is a high-quality, premium-priced cut of fish. Since strips are made from fillets, they do not have bones.

TRANSPORTING AQUACULTURE PRODUCTS

Getting aquacrops from the point of production to the consumer requires hauling. The products are hauled live to the processors, and the processed products are delivered to the market outlets. In all cases, quality products must be delivered. Live products are transported differently from processed products.

Transporting Live Aquacrops

All forms of live aquacrops may be transported. A few examples are:

1. Eggs may be shipped from a spawning operation to a hatchery.
2. Fry may be shipped from a hatchery to a fingerling grower.
3. Fingerlings may be hauled to a grow-out farm.
4. Food-size aquacrops must be hauled to the processing plant.
5. Recreational fish must be hauled from the farm to the fee-lake.
6. Pet fish must be transported from the producer to the pet store.
7. Broodfish may be shipped from one farm to another.

Considerations in Hauling

Aquacrops are hauled in different ways. How an aquacrop is hauled depends on the species, life stage of the species, distance to be hauled, climate, and medication needs. The transportation of any aquacrop should always meet legal requirements.

Water temperature and quality should be appropriate for the species. The water should be free of harmful substances. Some species can tolerate rather sudden changes in water temperature. As a rule of thumb, water temperature should vary no more than 5°F. Fish are conditioned by gradually changing the temperature of the water to that used in hauling. Haulers need to realize that species are adapted to certain temperatures. For example, trout grow in cool water, while tilapia prefer warm water. Hauling in water not suited to the species increases stress.

Life stage refers to whether or not the eggs, fry, fingerlings, or food fish are being hauled. Equipment for hauling eggs is certainly different from that needed to haul food fish. Small compartments that keep egg masses intact may be used. Large tanks are okay for food fish. The number and size of fish determines the size of the container that is needed.

Aquacrops can endure crowded, stressful conditions better for short periods of time than for longer distances, which require more time. Moving fish from one side of a farm to another is different from moving fish several hundred miles. The conditions of hauling must be good for long trips. A general rule is that no more than 3 pounds of fish should be hauled per gallon of water.

Weather has some effect on hauling. Hot weather can be particularly harmful. Hauling at night or early or late in the day reduces exposure to heat. Sometimes tanks have special equipment to provide the best temperature for the aquacrop.

Medications are sometimes put in the water in haul tanks to prevent bacterial growth and other disease problems. Only approved products should be used. Directions should be carefully read and followed. Diseased fish that are constantly subjected to medicated water are beginning to develop resistance to the medication. Treating only during hauling is probably a good way to reduce the development of resistance by the disease organisms.

Equipment Needed

Fish are typically hauled in tanks or plastic bags. Tanks are made of fiberglass, aluminum, or wood. These tanks often have partitions so that different sizes and species can be separated. The water in the tanks should be changed every 24 hours if the fish are in them that long. Aerators and / or oxygen systems must be used. Tanks may have refrigeration units attached to keep the water cool.

Plastic bags are used to ship small quantities of fish. Often they are used with exotic pet fish or eggs, fry, or fingerlings. Plastic bags also may be used to handle individual large fish, as in moving them from one pond to another. The bags have water and pure oxygen added to them. The are frequently placed on ice, or a chunk of ice is added to each bag. With some species, the bags should not be directly on ice or ice should not be put into the bags. Polyethylene bags 4 millimeters thick are usually preferred. Fish bags should be handled carefully to prevent bursting.

Transporting Processed Aquafoods

Processing plants are careful to produce only wholesome food products. The quality must be maintained as the products are moved to consumers. As they leave the plant, they are packaged for shipping.

Transportation is selected on the basis of how the products have been processed. Fresh products may be shipped in refrigerated trucks or trucks with ice. There should be no delay in moving the fresh products to consumers. These products will begin to deteriorate rapidly if the temperature rises above 40°F. Frozen products must be shipped in trucks equipped with freezing units that keep the temperature near 0°F. Canned fish products can be hauled in trucks without refrigeration; however, they should not be exposed to excessive heat or be allowed to freeze.

CONSUMER PREFERENCES

Consumption is an important part of aquaculture. Aquafarmers need to produce what consumers want. Providing for the needs and desires of consumers is important in having a demand for the product that is produced. Several considerations are presented here.

- ***Species of product*** — Consumers want certain species of aquacrops. Preferences may vary among the regions where the people live. Such differences may be based on what is available locally. For example, people in the central part of the U.S. along the Mississippi River and its tributaries often prefer catfish. People living in cooler, mountainous areas may prefer trout. Those living near oceans or lakes may care for the species that have been caught in them. Exceptions to preferences for regional products exist, however. An example is the low regard that some people who live near the Gulf of Mexico have for the mullet caught there.
- ***Form of product*** — Consumers often want aquacrops that have been made into certain forms. Whole freshwater trout are often preferred, while fillets of catfish are popular. Ease of cooking is often a consideration. Fresh, wholesome portion-cut fish are much more in demand in some markets; others may prefer fresh, whole fish that have not been processed. Certainly, consumers respond to attractive and convenient packages.
- ***Ethnicity of people*** — Some people have grown up eating certain aquacrops. They wish to continue with a product that they know about and like. Large cities often have ethnic groups with certain preferences of

Figure 9–16. Consumers often prefer attractively packaged, easy-to-cook foods.

species and form. For example, oriental markets in large cities may pay premium prices for fish that are not native to the area and that have been processed to their preferences.

- ***Socioeconomic levels***— Income also has an effect on the kinds of aquacrops consumers will buy. People with higher incomes pay more for their food. Supermarkets in inner-city areas will likely sell products that are different from those in suburban areas. The same tends to be true with restaurants. High-priced restaurants typically demand a higher-quality product and a species that their customers will buy.

PROMOTING AQUACROPS

Promotion involves trying to get consumers to buy a particular product. When consumers buy, there is a greater demand for the product. Prices should be good, and growers should find production more profitable.

Various promotion efforts are used. Retailers, wholesalers, processors, marketing associations, and government agencies often get involved in the promotion process. Several methods are described here.

- ***Advertising***— Radio, television, newspapers, magazines, and other media may be used to try to influence people to buy a product. Some advertisements target national audiences; others are intended for the local area served by a restaurant or a supermarket. Advertising costs can run into a lot of money.

- ***Chef training***—Associations of aquacrop producers or processors may hold cooking schools for restaurant chefs. The schools train the chefs in ways of preparing the product. This is a good promotion strategy. It helps chefs know how to prepare a tasty aquafood so that consumers will repeat as buyers. It also encourages chefs to try preparing new foods or other ways of preparing a particular food.

Figure 9–17. Attractive signs are helpful in promoting specialty aquacrop restaurants.

- ***Supermarket tasting***—Local supermarkets may provide samples of products for customers to taste in their stores. These are often set up by distributors or processors to promote products among people who are not now consuming them. Sometimes new methods of preparation are introduced with supermarket tasting.
- ***Educational literature***—Pamphlets or booklets of recipes can be helpful in encouraging consumption. Cooks may not know how to prepare a product in certain ways. Thus, literature that provides preparation information, as well as nutrition information, may be beneficial. The foods section of newspapers often contains new recipes and new product information.

Figure 9–18. Aquaculture foods are frequently promoted in health and wellness literature.

- ***Meal functions***—Aquacrop marketers will sometimes sponsor meals for various groups to introduce them to a product. In addition, a variety of recipes can be used to let the group members know about the versatility of the product.

SUMMARY

Marketing brings producers and consumers together. This linkage requires production based on consumer demand. Without considering the consumer, the farmer may produce something for which there is no market. Many steps are involved in getting aquacrops to the consumer. These involve planning what to produce, producing it, and then getting it to the consumer in a quality, wholesome form.

Various channels are available to producers. These may involve going from the farm through the processor to the consumer. Sometimes farmers market directly from the farm to the consumer, a roadside store, or a restaurant. Recreational markets are good outlets for a few farmers. Regardless, the channel used must be appropriate to the size of the farm and the amount the market will handle, also known as economy of scale.

Profitable aquafarmers develop and carry out marketing plans. Good plans allow the farmer to study the situation, set marketing objectives, develop a strategy for marketing, and evaluate how well the objectives were achieved.

Aquacrops are marketed in several forms. These forms may be on the basis of life cycle, such as eggs, fry, fingerlings, or food fish. Transportation is required to get these to the customer in wholesome condition.

Processing is used to get the aquacrop ready for the consumer. The extent of processing depends on the forms and cuts desired by the consumer. More and more consumers want easy-to-use forms. Consumer preferences are definitely important. Promotion is used to increase consumption of the aquacrops.

QUESTIONS AND PROBLEMS FOR DISCUSSION

1. What is marketing?
2. What are the marketing functions? Briefly describe each.
3. What marketing channels are available for aquaculture products? Distinguish between each.
4. What is economy of scale? How is it important in the selection of a market?
5. What questions should the aquafarmer answer in selecting a market?
6. Why is marketing planning important?
7. What is a marketing plan? What parts should it have?
8. What situations should an aquafarmer consider in developing a marketing plan?
9. What is meant by product form? Briefly describe several product forms.
10. What is processing? Describe what happens at a processing plant.
11. What processed forms and cuts are used with fish?
12. How are aquacrops transported? What factors should be considered in the hauling of live fish products? Processed fish products?
13. What is the role of consumer preferences?
14. What is promotion? How is it important?

PART THREE

Production Alternatives in Aquaculture

Chapter 10

FRESHWATER AQUACULTURE PRODUCTION

Much of the rapid increase in aquaculture has been the result of the expanded production of freshwater aquacrops. These crops may be used for food, baitfish, or recreation. The increase is largely due to producing fish for human consumption. Several species are now aquafarmed. Research is focusing on other fish species that might be adaptable to culture. In some cases, the research is to determine the environmental requirements of the species being studied. A species may be adaptable to culture when the cultural requirements are known. Aquaculture is an area with exciting new possibilities unfolding each day!

OBJECTIVES

This chapter focuses on the production of several species of freshwater aquacrops. Upon completion, the student will be able to:

- List and describe the fundamentals of freshwater aquaculture.
- Identify common freshwater crops.
- Identify species of freshwater aquacrops.
- Describe marketing procedures for the species.

- Describe size and rate of growth.
- Explain general environmental requirements of species.
- Explain how seedstock can be obtained.
- Explain how the species are cultured.
- Explain feeding procedures.
- Describe disease problems.
- Explain harvesting procedures.

FUNDAMENTALS OF FRESHWATER AQUACULTURE

Most animal aquacrops have similar fundamental requirements for culture. These, however, may vary from one species to another. Some knowledge of the requirements of specific species is essential. The following section describes several fundamental requirements for aquaculture.

Appropriate Species

The species of aquacrop to be cultured must be suited to the environment in which it is to be grown. Most species have preferences for climates within a

Figure 10-1. Floating cages are used to grow some species of aquacrops. These are being used in a large vat. Cages are more frequently used in large bodies of water where seining is impractical.

certain range of temperatures. In some cases, aquacrops may be grown in environments that are carefully controlled.

Aquacrops have different water temperature requirements. Some need "warm water," and some need "cold water," and others prefer "cool water." The temperature ranges tend to overlap. A species may survive in less than its ideal temperature, but it may also die in extreme temperatures to which it is not adapted. Best growth occurs when the species are matched to their particular requirements.

Aquacrops are adapted to water of varying salinity (the amount of dissolved salt that is in the water). Sea water varies from 33 to 37 ppt (parts per thousand) salinity. Freshwater is often defined as water with less than 3 ppt salinity. It is found in most rivers, creeks, wells, and reservoirs filled by runoff rain. Brackish water has salinity between that of freshwater and saltwater. Brackish water often occurs where freshwater streams empty into saltwater seas. The freshwater dilutes the saltiness of the sea water.

The temperament of a species has some influence on suitability. Species that attack each other may not be adapted to confinement systems.

The principle here is: Select only a species that is adapted to the production system to be used.

Available Water

Since water is the environment in which aquacrops grow, having the right water is a must. Natural supplies of water often reflect the climate of the area. For example, ponds reflect the temperature of the atmosphere. Warmer climates will have bodies of water with warmer temperatures, and vice versa.

The presence of water with certain characteristics can make possible the production of a species in a climate that is too cold or too hot for the species to naturally grow. Geothermal (naturally heated) supplies of water in cooler climates can be used to grow fish that would not normally live in the climate. Cold spring water in moderately warm climates can allow the production of cold water aquacrops where they would not normally be suited. In addition to natural water temperatures, a few sites have industries that produce heated effluent suitable for aquacrops.

The aspiring aquafarmer must look at the available water to see that it is appropriate for the aquacrop under consideration.

The principle here is: Select only a species that is adapted to the water that is available.

Nutrition Requirements

Aquacrops must have the right nutrition to grow efficiently. Streams and lakes have some natural food supply. This is used by wild fish and other life as food. In aquaculture, the levels of stocking in ponds and other growing facilities are greater than the ability of the water to provide natural foods to the aquacrops. Thus, aquafarmers must provide feed for their crops.

Not just any feed will do. Aquacrops must have the appropriate nutrients for their growth. Research has shown the dietary requirements of a few aquacrops. Certainly, providing the proper nutrition is essential in aquafarming. (General nutritional requirements were presented in Chapter 5.)

The principle here is: Feed an aquacrop on the basis of its nutritional needs.

Good Health

Health is the absence of disease. Diseases reduce the rate of growth of aquacrops. In some cases, the crop may be killed by a disease. Aquacrops must be healthy to grow efficiently. Each species has different health needs. (See Chapter 7 for details on health care.)

Good health is a part of good managment. The wise aquafarmer knows that species vary in their reaction to stress and changes in the environment. The growth environment must be managed so that unfavorable conditions do not develop.

The principle here is: Insure good health by following management practices that are appropriate to the species being produced.

Water Management

The aquafarmer can manage water to maximize its suitability for aquaculture. Dissolved oxygen should be maintained at the appropriate level for a species. Ammonia and nitrites may need to be monitored and regulated. Water pH and minerals may need to be controlled. Other water factors may need to be considered, depending on the species.

Water management is more critical with high intensity aquaculture. Fish that are stocked at high rates use more oxygen and excrete more wastes than those at low stocking densities. Weather and other organisms growing in the water may also affect how a farmer responds to the water management needs of an aquacrop. (More detail on water management was presented in Chapter 6.)

The principle here is: Maintain quality water by using production practices

that minimize the possibilities of problems and make water more suitable for a species.

Available Market

The demand for a species has a lot to do with making a profit. The prospective aquafarmer should answer this question: "Will someone buy what I produce at a price high enough for me to make a profit?" If a species grows well but is not wanted by consumers, it is not an appropriate aquacrop. Success in freshwater aquaculture depends upon the selection of an appropriate species from a marketing perspective.

Aquacrops may be marketed in a variety of ways. Some farmers sell directly to consumers; others sell to processors. Since processing plants want a year-round, uniform supply of fish, most will enter into contracts with producers. The contracts will specify various conditions, including the price that will be paid to the farmers.

The principle here is: Select a species for which there is a market.

AQUACROPS GROWN

Many species of aquacrops can be grown in freshwater. Several have become established as aquacrops; others are still in the exploratory stage. More research is needed on how to grow all of the species more efficiently. Success in aquaculture depends on the selection of an appropriate species.

The freshwater aquacrops that are reasonably well established in the U.S. are trout and catfish. Two others receiving high interest are tilapia and striped bass, particularly hybrid striped bass. Crawfish production is fairly well established in some areas.

In addition to the freshwater species included here, other species that are receiving some attention are yellow perch, carp, walleye, sturgeon, and freshwater prawns.

Trout

Trout have been cultured longer than any other freshwater species in the U.S. They were first raised for food in New York and Connecticut in the early 1900s. Earlier efforts had focused on trout to re-stock streams where natural supplies had been about depleted. Commercial trout farming began to become important in the 1950s.

Today, trout rank second in the volume of fish produced in the U.S. and first

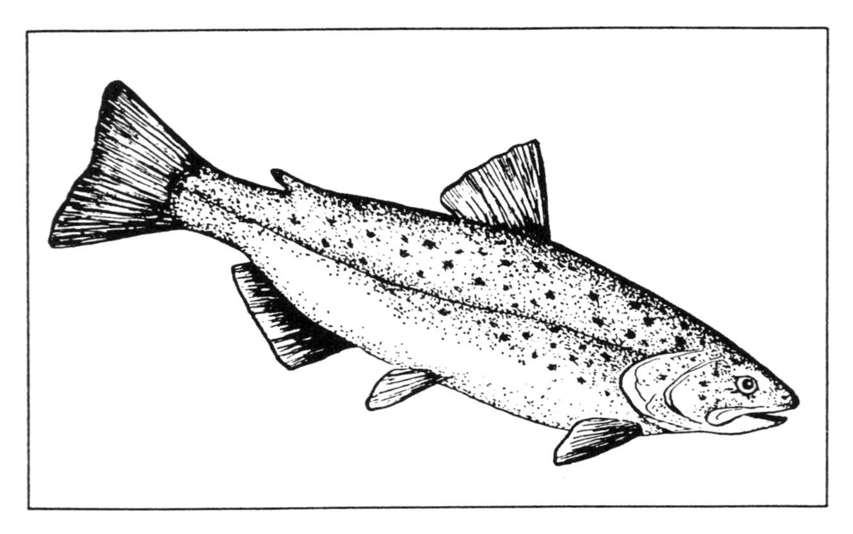

Figure 10–2. The rainbow trout is the most widely produced species of trout. It has excellent consumer appeal. The adult is typically no longer than 20 inches. Harvest size is 7 – 14 inches long.

in Canada. Idaho is the leading state in trout production. The major consumption areas are east of the Mississippi River; therefore, those produced in Idaho must be transported long distances to reach the consumers.

Marketing

Trout can be marketed at several stages in the production process. Some farms specialize in producing for these stages. The marketing may occur in the following ways:

1. Broodfish, usually of improved stocks, may be marketed to hatcheries.
2. Hatcheries may sell eyed-eggs to fry and fingerling producers.
3. Fingerling producers may raise fry and fingerlings, usually for sale to food fish farmers.
4. Food fish producers may raise fry or fingerlings to the desired food fish size and then sell them to processors or directly to consumers.
5. Fish producers may raise them for sale to fee-lakes and other recreational operations.
6. Fee-lake owners may stock them in their operations.
7. Live haulers may buy them for sale to processors, fee-lake operations, retail outlets, and others.

8. Integrated producers may grow various combinations of any of the above, such as the food fish producer who also has a hatchery and produces fingerlings.

Trout are popular food fish. Restaurants frequently offer special dishes featuring trout, which are often served as whole fish. The best quality, most flavorful food is from trout that have received a ration high in protein from animal sources.

Species Grown

Although several species of trout have been tried, the choice is primarily between two species: rainbow trout (*Salmo gairdneri*) and brown trout (*Salmo trutta*). Of these, the rainbow species is more widely grown. In addition to being adapted to culture, the rainbow trout has outstanding consumer appeal.

Identification

Rainbow trout vary considerably in color, depending on their habitat. Various shades of green and blue to brown are found in the dorsal (top) area. A pale pink to brilliant red band usually extends from the gills to the tail. The sides are silvery, with white on the ventral (belly) area. Small, black spots are found on the back and sides of trout. Spots are also present on the dorsal, adipose, caudal, and anal fins. The spots are always black.

Size

Adult rainbow trout are typically no more than 20 inches long. The largest known rainbow trout was considerably longer and weighed 52 pounds. Harvest-size food fish range from 7 to 14 inches long and weigh ½ - 1 pound. This size can usually be produced on a farm with favorable conditions in 7 - 14 months. Fingerlings in the 1- to 6-inch range can be produced in 1 - 8 months.

Environment Required

Trout are cold water fish. They naturally grow in the streams of the northern half of the U.S.

Trout grow best in water with a temperature range of 50 - 68°F. The ideal temperature is considered to be 64°F. Below 50°F. and above 68°F., the rate of growth slows considerably. Water temperatures in the mid-70's and above are lethal if trout are exposed to them for more than a short while.

Rainbow trout thrive best in flowing water with a high level of oxygen. The minimum dissolved oxygen level is 7 ppm. A pH of 6.5 – 9.0 is best. The trout will grow (but not spawn) at low stocking levels in ponds that have water that is 10 feet or more deep.

Water for trout should be cold, clear, free of pollution, protected from summer heat, and have plenty of oxygen.

Seedstock

Trout have been cultured longer than any freshwater fish in the U.S. Because of this history, improved varieties of rainbow trout have been developed. An example is the super trout, which is also known as the "Donaldson trout."

In nature, trout spawn in fast-flowing streams with gravel bottoms. Spawning is usually in the spring but may occur in the fall. Producers of eggs typically raise or buy improved varieties of broodfish. Natural reproduction is not used in trout farming. The eggs may be artificially fertilized and hatched in flowing water.

Growers of food-size trout may buy eyed-eggs and hatch them or buy fry or fingerlings that have been grown out by a trout farmer. Sometimes a trout farmer will maintain broodstock and go through the complete cycle from broodfish to food fish.

Culture

Most trout are cultured in flowing spring or well water. Raceways are typically used, but cages and closed systems are sometimes used.

The standard rate of water flow in a raceway involves a complete exchange of the water every hour. Raceways with a 1 – 3% slope will allow water to flow at a proper rate. The water control structure at the end of each raceway segment should provide a drop to help reaerate the water through splashing. The general rule is that 1.8 pounds of marketable trout can be produced in 1 gallon of water per minute of water flow.

Stocking rates vary from 300 to 750 fingerlings (2 – 5 inches long) per acre of water. The rate depends on the rate of water flow and the level of management to be used in growing the trout crop.

Feeding

Trout must have a high-protein diet. In nature, they are carnivorous, eating insects, fish, fish eggs, and other small animals. Commercially prepared feeds are specially designed for the needs of trout. These feeds typically have a protein

Figure 10–3. Closed systems are sometimes used to grow trout. Plenty of water must be available. (Courtesy, Master Mix Feeds)

level of 42%. Trout should be fed twice daily. On small farms, the feed may be thrown out into the water by hand. Larger farms may use power blower feeders or self-feeders.

Diseases

Bacteria, viruses, and protozoa may cause diseases. Environmental diseases may result from high levels of nitrogen forms, low dissolved oxygen, and contamination of the water that has been polluted by industries and other farming operations. Good management is a key in trout disease control. Bringing only clean, healthy fingerlings onto a farm is an important step in preventing disease outbreaks.

Harvesting

Trout are harvested by seining, trapping, netting, or draining the raceway or pond. A little less than two years is needed to grow a trout from eggs to food-size fish.

Catfish

Catfish are by far the leading freshwater aquacrop produced in the U.S. The value of the crop has increased annually for over two decades. The culture of catfish began to emerge in the late 1950s. In 1960, there were only about 600 acres of catfish farming. By the early 1990s, this had reached 150,000 acres. Nearly 370 million pounds are annually produced. An entire industry, supported by fish equipment manufacturers, feed mills, processing plants, product promotion, and other areas, has emerged. The leading states in catfish farming are Mississippi, Arkansas, Alabama, Louisiana, California, Missouri, Texas, and Oklahoma.

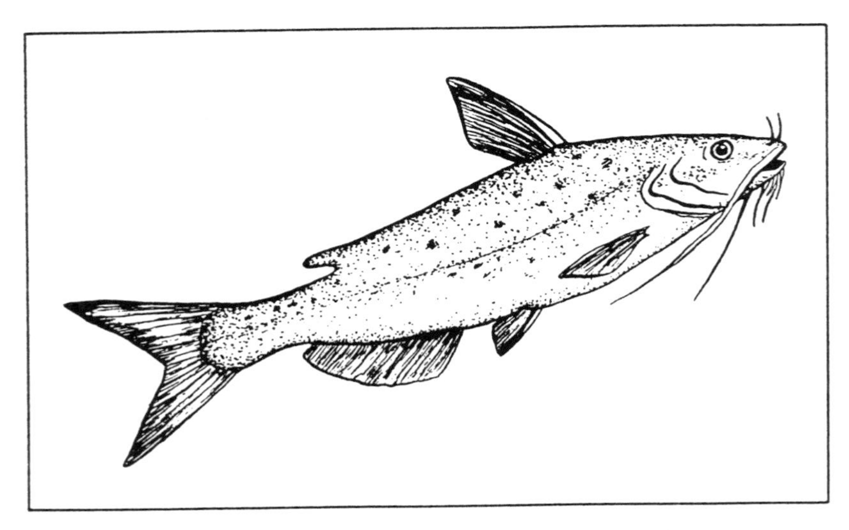

Figure 10–4. The channel catfish is currently the largest freshwater aquacrop in the U.S. The flavor is excellent and a variety of delicious dishes can be prepared. The most popular food fish size is 1 – 3 pounds.

Marketing

Catfish are similar to trout in ways of marketing. One advantage that now exists is the vast industry that has developed. Furthermore, the industry has formed organizations to promote the consumption of catfish.

Catfish farming has expanded rapidly because of the quality of the product that has been produced. Catfish are good sources of protein in the human diet. The flesh has an excellent mild flavor and can be prepared in a variety of delicious dishes. Catfish may be sold to the consumer whole, cut into steaks or fillets, or into other easy-to-cook forms.

Marketing the catfish crop may occur at a number of stages of production, as follows:

Figure 10–5. Catfish being harvested as food fish. (Courtesy, MFC Services)

1. Broodfish may be sold to hatchery operators.
2. Hatcheries produce eggs and often fry and fingerlings.
3. Fingerling producers obtain eggs or fry from hatcheries and grow them to the size needed for stocking in growing ponds.
4. Food fish growers stock ponds with fingerlings and grow them to food size for sale to processors, fee-lakes operators, and live haulers. (This is by far the major area of catfish production.)
5. Recreational fee-lake operators obtain fish from growers who may be producing food fish or fish especially for the fee-lake market.
6. Live haulers buy fish from growers for re-sale to fee-lakes, fish markets, and others.
7. Many of the above marketing stages / outlets are combined so that farmers carry out more than one phase.

Species Grown

Nearly 50 species of catfish are found in the U.S. Of these, only a few have been farmed. The predominant species in aquafarming is the channel catfish (*Ictalurus punctatus*). Other species receiving some attention have been the blue catfish (*Ictalurus furcatus*) and the white catfish (*Ictalurus catus*).

Identification

One of the distinguishing features of catfish is the presence of barbels about the mouth. These are sometimes referred to as "whiskers." (They are actually feelers that help the fish in its natural habitat to maintain its position in the water and to locate food.)

The channel catfish has a deeply forked tail and rounded anal fin. The dorsal part of the body has a bluish color that changes to silver on the sides and white on the belly (ventral side). Younger channel catfish have spots, but these are usually gone by the time the fish weigh 5 pounds.

Size

Adult channel catfish range from 11 to 30 inches long and weigh up to 15 pounds. On rare occasions channel catfish may reach weights of 50 - 60 pounds. The desired food fish size is 1 - 3 pounds. This size can be obtained in less than two years under good growing conditions. In the first year, the eggs are hatched and grown to fingerlings. In the second year, the fingerlings are stocked in growing ponds where they should readily reach market size.

The size of the fish is important in determining how the fish will be processed and the forms and cuts that will be made. Small fish of ¾ - 1 pound may be left whole. Slightly larger catfish are processed into valuable fillets. Other sizes may be cut into steaks.

Catfish dress out at 55 - 60% of their live weight. This means that the skin, head, and viscera are 40 - 45% of the weight of a live fish. A fish weighing 1 pound would yield a dressed fish weighing slightly over ½ pound.

Environment Required

Channel catfish are warm water fish. They grow best in water that is 75 - 85°F. When the water temperature is below 60°F. or above 95°F., channel catfish virtually stop eating. And when they do not eat, they do not grow! Sites where there is a minimum of time when the water temperature is below 70°F. are preferred for catfish culture. Short winters and long summers are most desirable.

Channel catfish will grow in ponds, streams, raceways, and other water facilities. The water should be free of pollution and sufficiently oxygenated. Catfish will survive for a short time in water as low as 1 ppm dissolved oxygen, but the level must be at least 4 ppm for them to grow efficiently. The most desirable pH range for the water is 6.3 - 7.5; however, 5.0 - 8.5 may be used under certain conditions.

Seedstock

Broodstock were initially obtained from rivers and lakes as wild fish. In recent years, efforts have been made to improve the quality of the broodfish. Through selection, only the best broodstock have been used to produce eggs. Farmers may maintain their own broodfish and produce seedstock or buy their fingerlings from fingerling producers.

Catfish naturally spawn in the spring when the weather begins to warm. In the wild, they seek out hollow logs, stumps, or other secluded places in the water. The female lays the eggs and the male swims over them, depositing sperm on the egg mass, which is also known as a spawn. One egg mass may contain thousands of eggs. A rule of thumb is that a female catfish will produce 2,000 eggs per pound of body weight. In nature, the male catfish tends the spawn until it hatches. He swims immediately above the egg mass, fanning the water with his fins and occasionally bumping the mass.

On catfish farms, the broodfish may be kept in broodfish ponds until spawning season. They may then be moved in pairs (one male and one female in each pair) to pens. Here they are provided with spawning containers (nests) of milk cans or other similar places to hide while spawning. The spawning containers are regularly checked during the spawning season. The eggs may be allowed to hatch naturally in the pond or may be transferred to an artificial hatchery. An artificial hatchery uses some of the same motions as the male fish uses in nature. The water is kept moving and at a temperature of 70 – 85°F. Hatching occurs in 6 – 10 days, depending on the water temperature. The warmer the water, the shorter the hatching time.

The newly hatched fry have an egg sac attached. This sac provides nutrition for a few days. When it is gone, the fry are ready to start eating. Feed must be provided. Normally a fine meal feed of 28 – 32% protein is fed every two to four hours around the clock. Fry may be reared in ponds or in troughs. Some producers prefer trough rearing because they have better control over the conditions to which the fragile fry are exposed.

Over a few summer months, the fry grow into fingerlings. Most farmers prefer to stock growing ponds with 6-inch fingerlings. Nearly a year after the fry were hatched, the fingerlings are stocked into growing ponds. When stocked in the spring, they will be food fish size by fall if properly fed and managed.

Culture

Most channel catfish are produced in ponds filled with fresh well water. They may also be grown in raceways, tanks, and cages.

Fingerling catfish 5 – 8 inches long are stocked in ponds at a rate depending

Figure 10–6. Removing an egg mass from a spawning container to be carried in a tub to a hatchery. (Courtesy, Louisiana Agricultural Experiment Station – Louisiana Cooperative Extension Service, Louisiana State University Agricultural Center)

Figure 10–7. Two large spawns are shown here. The darker spawn on the left is closer to hatching. (Courtesy, Delta Pride Catfish, Inc.)

Figure 10–8. Hatching troughs are used to hatch eggs artificially. (Courtesy, Aquacenter, Inc.)

Figure 10–9. Fingerlings being moved to a growing pond.

on the level of management. It is possible to stock 4,000 – 5,000 fingerlings per acre if careful management is to be followed. Lower stocking rates of 1,000 – 2,000 should be used if the grower does not plan to provide aeration of the water. The natural food supply typically available in catfish ponds will support only a few hundred fish per acre. Catfish do not usually have other fish stocked in the same pond with them. (Producing one kind of crop in a water facility is known as *monoculture*.)

Water management is critical to success in catfish farming. The oxygen level must be maintained. Round-the-clock monitoring of catfish ponds is essential in the warmer months of the year. This involves testing for dissolved oxygen. Aerators that splash the water, throw water into the air, or inject oxygen into the water may be used to add dissolved oxygen. When the oxygen gets too low, the fish "gulp" at the surface in an attempt to get oxygen from the air. They will die if the oxygen problem is not solved quickly.

Feeding

Channel catfish must receive a ration that meets their nutritional needs. The natural food supply in pond water is far below what a crop of catfish needs. Catfish are typically fed a ration that is 32% protein. The major ingredients in the feed are soybean meal, corn, and fish meal. Some of the protein in any ration should be provided by the fish meal. Catfish grow better with some animal protein.

Feed for catfish is commercially available from a number of suppliers. It may be manufactured as pellets that float, sink, or have neutral buoyancy. (Neutral buoyancy means that the pellets neither float nor sink. They can remain in a stable position in the water.) Most farmers prefer the floating pellets. The fish must come to the surface of the water to feed. This allows the farmer to observe their behavior. Catfish that do not feed in warm weather are likely to have some problem such as oxygen deficiency or a disease.

Large catfish farms use truck-mounted power blowers to blow the carefully measured amount of feed out into the ponds. Small farmers with only a few acres may throw the feed into the water by hand. A few farmers have used automatic self-feeders. Catfish should be fed once or twice daily and at the same time each day. Many producers feed only once per day, but research has shown that two

Figure 10–10. A tractor-pulled feeder that blows feed out into a pond. (Courtesy, Master Systems, Inc.)

feedings increase the rate of growth. The total weight of the feed fed is the same with two feedings as it would be with one feeding.

The amount to feed is based on the estimated weight of the fish as related to the temperature and other factors. Fish should be fed only the amount they will consume in a few minutes and no more. In warm weather, catfish under 1 pound are fed at the rate of 3% of body weight. If the fish weigh over 1 pound each, the feeding rate is dropped to 2% of the body weight. For example, if a pond was stocked with 4,000 fish that average ½ pound in weight, the pond contains 2,000 pounds of catfish. The total amount to feed a day is 3% of 2,000, or 60 pounds. Overfeeding should be avoided. Uneaten feed contributes to water problems.

Diseases

Several different diseases may attack catfish. Using plenty of good water can help eliminate many disease problems. Farmers should stock only healthy fish. Any new fish should be held in isolation for a few days to see if any diseases develop. Catfish may be stressed when handled, such as when harvested and hauled. Thus, following good practices can reduce the possibility of stress and the diseases that may result.

The most common catfish diseases are those caused by bacteria, viruses, fungi, algae, parasites, nutritional deficiencies, and environmental problems. Only those treatments which have been approved for use on food fish should be used. Food fish which have been improperly treated may be ruled unfit for human consumption.

Harvesting

Catfish are typically harvested at a weight of 1 - 3 pounds from ponds by seining. In the past, harvesting usually involved lowering the water level, but in recent years whole ponds have been seined without any water being removed. This has been made possible by the development of power seine haulers that can be pulled by farm tractors. Once the fish have been concentrated into a small area of the pond, they are scooped into a brailing bag for lifting to a haul truck.

Before catfish in a pond are harvested, samples should be checked for off-flavor. This merely involves quickly cooking a few samples and then tasting them. The trained taster can determine if the fish have the right flavor. Fish with off-flavor should never be harvested. Processors will not accept them.

Crawfish

The production of crawfish is fairly well established in some areas. California, Texas, Oregon, Washington, Wisconsin, Louisiana, South Carolina, and Mississippi have some crawfish production. Louisiana is by far the leading state.

Crawfish are harvested from the wild as well as cultured. Production in the U.S. is approximately 100 million pounds annually, with the majority being cultured crawfish.

Some 300 species of crawfish are found in the U.S. Each has its own local environment adaptations. The crawfish industry is not nearly as highly developed as the catfish or trout industry.

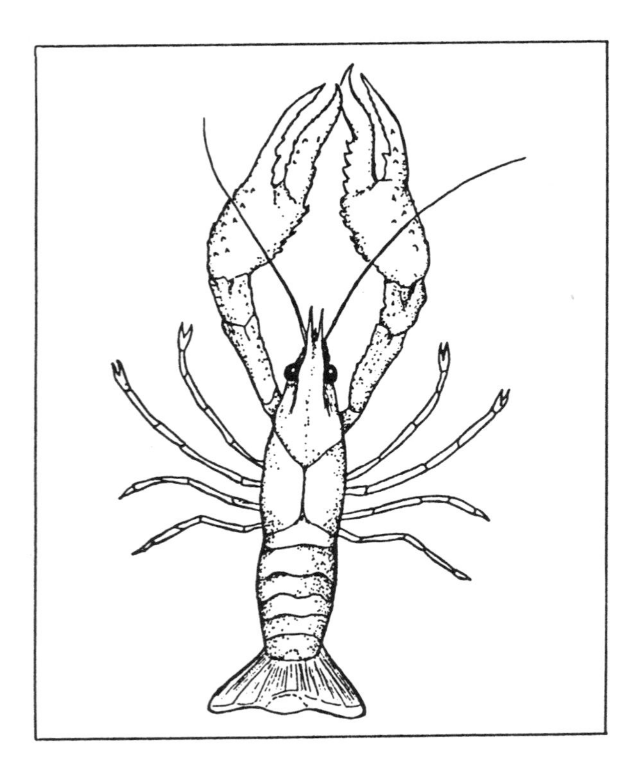

Figure 10–11. Crawfish are popular in some areas. The red crawfish is more popular in the South, while the white crawfish is popular in the northern states. Nearly a year is required for a crawfish to reach harvest size.

Marketing

Crawfish are typically marketed by the farmers to consumers, restaurants, and wholesalers. The markets tend to be local, though some regional distribution occurs during crawfish season. The minimum size for marketing is a weight of about 1/2 ounce. A preferred size is 1 - 1 1/2 ounces. Six to 14 months is required for crawfish to reach this size.

Figure 10–12. Crawfish being graded in preparation for shipment. (Courtesy, MFC Services)

Crawfish may be graded on the basis of the number required to weigh a pound. The count ranges per pound are:

- Large or Number 1: 15 or fewer crawfish per pound
- Medium or Number 2: 16 – 25 crawfish per pound
- Small or Number 3: more than 25 crawfish per pound

The larger crawfish are marketed live to dealers, restaurants, and others. Some of the large crawfish may go into export trade. The small crawfish may go to a peeling plant for processing.

A new marketing niche is developing for soft-shelled crawfish. This is the stage between hard shells that occurs when crawfish molt.

Crawfish may be sold in a variety of forms. Most are sold live, but boiled and peeled-tail forms are sold to dealers and restaurants. Mechanical processing is needed to expand the market for peeled forms.

Species Grown

The predominant cultured species are the red crawfish (*Procambarus clarkii*) and the white crawfish (*Procambarus blandingi acutus*). The red species may also be referred to as the "red swamp crawfish" or "crawdad." This species is native to the Mississippi and Ohio River valleys. It has been introduced into several western states, including California and Nevada. The red crawfish thrives in the climate of south Louisiana and is the major species grown.

The white crawfish is also known as the white river crawfish. It is found in the Mississippi River valley, in the Great Lakes area, along the Atlantic Coast, into Florida. It has a much lighter color than the red crawfish. It is better adapted to cooler climates. In nature, the white crawfish is more likely to be found in rivers than the red crawfish.

Identification

The crawfish is a crustacean similar to the shrimp. Its body is covered with an exoskeleton. This hard structure or shell protects the soft tissues of the body from injury.

The crawfish has five legs on each side of the body, with the two front legs forming pincers (sharp claws also known as *chelas*). If the crawfish loses a leg, or pincer, it grows back. There are two long antennae, or feelers, and two short ones on the head. Underneath the abdomen, the crawfish has swimmerets. The color may vary, depending on the species, but the red crawfish is a deep red.

Crawfish molt as they grow. The hard shell of a crawfish does not expand as the crawfish grows. The shell is shed to allow the body to get larger. Very young crawfish molt every 5 - 10 days, while older crawfish that are still growing may molt every 20 - 30 days. The new shell hardens about 12 hours after the old shell is shed. A crawfish may molt about 11 times before it reaches maturity.

Size

Crawfish seldom reach a length greater than 6 inches. The desired weight for harvesting is about 10 crawfish per pound.

Environment Required

Crawfish are naturally found in shallow, weedy swamps and ponds. They will grow in a wide range of soil types, but the soil must retain subsurface water that is available to crawfish when the area is drained. They typically prefer warm water with a temperature of 65 - 85°F., but they will grow in a wide range of water temperatures. Crawfish will go dormant in water below 45°F. Above 88°F., they burrow into the earth. White crawfish tend to grow better at cooler temperatures.

Most agricultural chemicals are very toxic to crawfish. Fields that have had insecticides applied to crops grown on them are not suitable for crawfish until all residues are gone.

Good-quality water is needed. The presence of some salt will not pose a problem. Water with too much iron or sodium bicarbonate is not suitable. Water in the pH range of 5.8 - 8.2 has been found satisfactory.

Crawfish burrow into the earth. This is a part of their life cycle. They typically come out of their burrows at nightfall and daybreak. Their burrowing may weaken pond dams or other earthen structures.

Seedstock

Adult crawfish may be obtained from the wild or from crawfish growers. Crawfish mate in the late spring. During mating, the male deposits sperm in a receptacle on the female. The female typically digs a burrow near the edge of the water and goes inside where she lays the eggs in the summer. As the eggs are released by the female, sperm is released from the receptacle to fertilize the eggs. The eggs are attached to the underside of the tail of the female crawfish until hatching. Red crawfish hatch in 14 - 21 days, while white crawfish hatch in 17 - 29 days. Each female may hatch 400 young, with 700 being the maximum. Red crawfish may hatch a few more than the white crawfish. Survival of the young is much better if open water is available.

Culture

Water management is the most critical part of crawfish culture. The level of the water in a growing facility has a lot to do with the stages the crawfish go through.

Four kinds of grow-out facilities may be used, as follows:

1. *Permanent ponds*—These are specially constructed ponds for crawfish production with a maximum water level of 24 inches. Rice may be planted in them for forage. These ponds have no trees or bushes in them and may grow 1,200 pounds of crawfish per acre a year.
2. *Rice ponds*—These are facilities that are rotated for rice production. The crawfish eat the rice forage remaining after the rice crop has been harvested. These ponds can produce about the same pounds or more as permanent ponds.
3. *Wooded ponds*—These tend to occur naturally in low areas. Trees and bushes may be in them and may interfere with management practices. Wooded ponds produce 400 - 600 pounds of crawfish per acre a year.
4. *Marsh ponds*—These are created by damming off low areas on land that

is of little value. An acre of marsh pond may produce 300 - 500 pounds of crawfish a year.

Crawfish are often grown in permanent ponds or in rice ponds that can hold water that is 18 - 24 inches deep. A good size for a pond is 20 acres. The levee around the pond should be wide enough for tractors or small trucks to travel for easy access to the pond. Since most crawfish are harvested with traps, a smooth bottom to the pond is not essential.

Figure 10–13. Construction of permanent crawfish ponds requires a levee of about 24 inches. Approximately 1,200 pounds of crawfish may be grown per acre each year. (Courtesy, MFC Services)

When construction of the pond is completed, the bottom should be disked or rototilled before the pond is filled with water. Water from sources that might contain trash fish should be filtered. Once the pond is filled in May or June, 25 - 50 pounds of both male and female brood crawfish should be stocked per acre. After stocking, the crawfish will begin burrowing. At this time, the water level should be drawn down to just a few inches. The area will later be flooded to a depth of 18 inches or so in September to encourage maximum growth during the fall and early winter months.

New systems of growing soft-shelled crawfish in water trays in greenhouses are being studied. The current systems are labor-intensive, but do result in the growth of crawfish. Production of soft-shelled crawfish is greater at water temperatures of about 80°F.

Feeding

Some sort of forage is usually planted in the water. Some farmers plant grain sorghum or millet in crawfish ponds. Rice is most common. The rice is planted when the water level is about 4 inches deep (after the draw down) and seeded at the rate of 100 pounds per acre. After it has been seeded, the remaining water is removed. Sometimes fertilizer is applied when the rice is planted. Since the purpose of the rice is to feed the crawfish, producing a high yield of rice is not a concern. (In some cases, rice and crawfish are double-cropped.)

As the rice grows, water is added back to the field to a depth of about 4 inches for the remainder of the summer. More water is added to the crawfish ponds in September and October. This flooding causes the crawfish to come out of their burrows and release their eggs. The crawfish feed on the green rice, other organic matter, and microorganisms that are in the water. As the water level is increased, rice stems and leaves begin to decay. This ties up oxygen in the water. Sometimes aeration of the water is needed to keep the level of dissolved oxygen above the 3 ppm minimum.

The crawfish feed on the rice throughout the winter. In heavily populated ponds, all of the forage may be gone by late winter or early spring. In this case, some supplemental feeding may be needed. Commercially prepared crawfish feeds are available. Other farmers use range pellets that have been prepared for use with beef cattle.

Diseases

Crawfish are subject to several diseases. Nutritional diseases result when the crawfish do not receive the proper diet. Plenty of edible plants must be available, or the crawfish will have tails that are not filled out. In some cases, inadequate forage will cause their livers to be brown or black. Well-nourished crawfish have full tails and yellow livers.

Very few diseases cause problems in crawfish. Sometimes bacterial diseases may attack crawfish, causing some decay. These can readily be controlled by treating with a 3 ppm potassium permanganate solution.

Crawfish may also be subject to damage from pollution in the water, such as agricultural chemicals, and to attack by predators. Common predators are trash fish that get into the growing pond, as well as birds, raccoons, frogs, snakes, turtles, and water beetles.

Harvesting

Harvesting can begin in the late fall or early winter after stocking in the previous summer. The original broodstock are harvested first. The minimum size to harvest is a length of 3 inches.

Most crawfish are harvested with traps. The traps are constructed of screen wire or ¾-inch chicken wire that allows smaller, young crawfish to remain in the growing pond. About 10 traps are used for each acre. The traps are baited with fish heads, internal organs of beef animals, or dead trash fish. The traps are checked daily, with the crawfish being removed for marketing. The harvesting season ends by May or June. (At this time, the pond is prepared for the next crop of crawfish.)

Crawfish typically have hard shells when harvested. A current trend is to harvest soft-shelled crawfish. This occurs in the 12-hour period between the time the hard shell is molted and the new shell is in place. With soft-shelled crawfish, the shell is eaten with the meat. Research on getting large numbers of crawfish to molt at the same time so that large harvests can be made is underway. Some producers have reaped a good profit with these, but too many producers have flooded the market. Also, the soft-shelled crawfish must be watched constantly because the hard-shelled crawfish will attack and destroy those without shells.

Markets should be lined up before the crawfish are harvested.

Tilapia

The first tilapia were cultured some 4,000 years ago in Africa. An Egyptian tomb originating some 2,500 years ago depicts a tilapia. From the ancient days, tilapia production has expanded in some parts of the world to provide protein. It has been valued for many years in Africa and the Far East. Tilapia are grown in Israel, where they are known as St. Peter's fish. Today they are grown in over 100 countries. Interest in tilapia production is fairly new in the U.S. In fact, tilapia is viewed as a fish crop that has yet to prove itself.

Some states have laws forbidding the possession and release of tilapia into streams. This is because the tilapia is thought to take over streams and lakes into which it is introduced. This is true only in warmer climates, as it cannot live in water below a temperature of 50 - 55°F.

Marketing

Tilapia have been promoted in fairly narrow markets in the U.S. These are

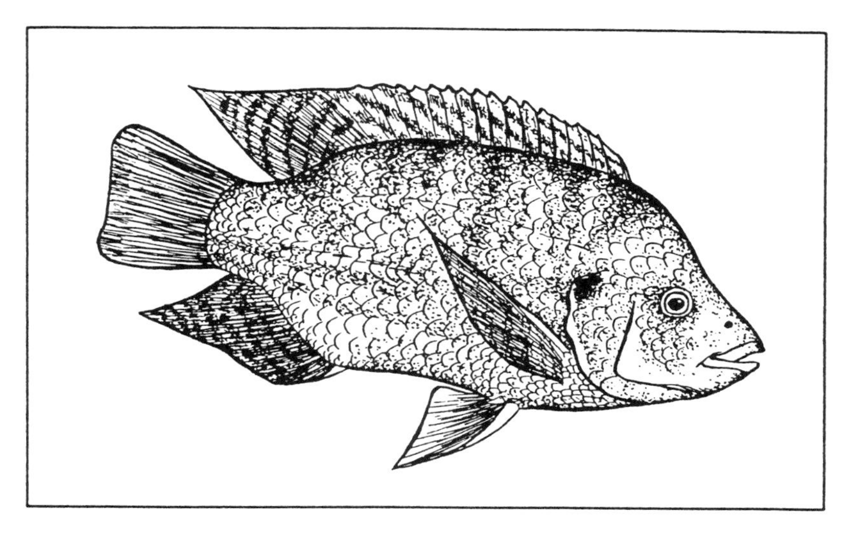

Figure 10–14. Tilapia are receiving considerable attention as a new aquacrop. They grow very well in warm water. They are typically harvested at a weight of 2 pounds.

typically ethnic markets in the large cities. Efforts are underway to expand the demand for tilapia into other markets.

Tilapia are now grown in a few locations, particularly in the southern U.S. Locations in northern areas with heated water are trying to get the industry started.

The tilapia now being distributed undergo very little processing before sale to the restaurant or the fish market. The farmer will likely stop feeding about 24 hours before harvesting, chill the whole fish, and ship it under refrigeration to the buyer. No preparation is done to the fish; the head, internal organs, and skin are left intact. The buyer is responsible for all preparation.

Species Grown

Many species of tilapia exist. The java tilapia (*Tilapia mossambica*) and the blue tilapia (*Tilapia aurea*) are the most common species.

Identification

The tilapia is a scale-covered fin fish resembling the North American sunfish. It is a thin, flat fish with depth from dorsal to ventral side.

The java tilapia is greenish-gray with three faint vertical bars. It has a dull gold edge to the tail. The fins may sometimes have red edges but not with the same intensity as the blue tilapia.

In cooler water, the blue tilapia is bluish-gray with black vertical bars. A blue bar may appear across the cheek in high water temperatures. Crimson borders may appear on the dorsal and caudal fins of healthy blue tilapia. Some tilapia are very colorful fish.

Size

When mature, the tilapia weighs a maximum of 5 pounds or a little more. It can grow rapidly, reaching a weight of ½ pound in 6 months.

Environment Required

Tilapia prefer shallow, fertile ponds and lakes where the water temperature never goes below 50°F. They are tolerant of poor-quality water and will survive in water with a dissolved oxygen content as low as 1 ppm. Tilapia need a water pH of neutral to basic (pH of around 7); they do not do well in acid water. Tilapia will grow well in small ponds and in confinement tanks at heavy population densities. In nature, tilapia feed on a wide range of plankton, soft green plants, and invertebrates. Young tilapia typically favor zooplankton. Due to their huge appetites for many aquatic plants, tilapia are sometimes used to control aquatic weeds.

Seedstock

Adult tilapia may be obtained from wild sources, selected from food fish stocks, or imported into the U.S. Tilapia are fairly easy to breed in captivity. Some aquafarmers buy fingerlings from fingerling producers; others have broodfish and hatcheries. The fingerling producers operate hatcheries and fingerling grow-out ponds.

Tilapia females may reach sexual maturity at 2 – 3 months of age and at the small size of 2½ – 4 inches in length. They can be very productive during the warm season when the water temperature is 75 – 85°F. The number of eggs produced depends on the size and age of the female. Frequency of spawning is also a factor in the number of eggs produced, with fewer number of eggs being produced when the female spawns frequently. Tilapia produce 100 – 1,500 eggs per spawn.

Tilapia species vary in how they care for their spawns. Some species spawn and care for their eggs and young in a nest. Other species spawn into a nest, but the females hatch the eggs in their mouths. (They are known as mouth brooders.)

With the mouth brooders, females lay eggs in a nest over which the male deposits sperm. The eggs and sperms are sucked into the females' mouths for fertilization and incubation. Hatchery operators may collect the females every 10 days or so and check their mouths for eggs. The eggs may be removed for hatching in special jars. Other hatcheries allow the females to incubate the eggs in their mouths and then collect the fry when they are released by the females. Eggs usually hatch in three to five days.

Other tilapia use nests for hatching their young. The spawning pair form a nest by making a depression in the sediment at the bottom of the pond or stream. The eggs are laid in the nest and the male deposits sperm above them. The pair guards the nest until hatching. Some producers use aquaria for producing seed-stock. The tilapia form a nest in the sand in the bottom of an aquarium and spawn much as in the wild. Aquaria spawning is best in water of 75 - 85°F.

Culture

Tilapia are commercially grown in ponds, tanks, and other facilities. They are very adaptable to a wide range of cultural situations. The major requirement is that the water temperature be above 50°F., with the preferred temperature 75 - 85°F. Thus, unless the water is heated, tilapia must be grown in tropical and

Figure 10–15. Tilapia may be grown in highly intensive tank systems. The quality and quantity of the available water are limiting factors in such production systems. (Courtesy, *Aquaculture Magazine*)

semi-tropical climates. In tanks with flowing water, tilapia can be stocked at heavy densities. They will tolerate low oxygen, high ammonia, and other conditions that would be lethal to many species of fish.

In ponds, tilapia will utilize a wide range of natural foods. Tilapia normally spawn easily, and crops can be started about any time of the year in warmer climates. Since tilapia reproduce profusely, one problem in ponds is that of population control. Researchers are working on methods to stock all-male populations. In some cases, certain parental stock will produce nearly 100% male fish. Hand sorting fish by sex is time-consuming but not too difficult. Some researchers are using hormones that reverse the sex of female tilapia. This is not yet an acceptable practice to most consumers. Still another approach involves producing hybrids. Breeding has produced some hybrids that are all of the same sex and others that will not reproduce. Some research has involved stocking predatory fish with the tilapia and allowing the predatory fish to keep the population down by consuming them. In a few places, tilapia are being cultured in greenhouses and cages. Greenhouse culture is possible in cold climates. Cages are useful only when the natural temperature of the water is appropriate for tilapia.

In a good growing situation, tilapia will reach ½ pound in 6 months. A weight of 2 pounds can be achieved in 12 months or so from the date of hatching.

Feeding

In ponds, tilapia feed on plants, insects, and other natural organisms. Supplemental commercial feed may be added. Fertilizing ponds with a complete mixed fertilizer can stimulate plankton, algae, and aquatic plant growth. Commercially available feeds are used to supplement natural growth in ponds or to feed intensive production systems. Feed with 28 – 32% protein is essential in intensive farming.

Diseases

Tilapia are resistant to many diseases. They are tolerant of poor-quality water. Very low oxygen levels and water temperature below 50°F. are the two principal causes of loss. Sometimes tilapia become stunted. This is a result of overpopulation for the amount of feed available. Providing sufficient feed will usually overcome this problem.

Harvesting

How tilapia are harvested depends on the system in which they are grown. Where possible, cast nets are used. Seining can be used in ponds. Draining methods may be difficult because the fish may try to burrow into the mud. In intensive systems of raceways, the fish may be dipped with nets from the water. With ponds and streams, hooks and lines can be used for a few fish, but this method is not practical for commercial aquafarms.

Many tilapia are harvested at a weight of about 2 pounds. This size is large enough to make nice, portion-cut fillets. In some areas, small tilapia are harvested and cooked whole without any processing.

Striped Bass

Striped bass have recently begun to receive attention as an aquacrop. They have been important sport fish for quite a while. They produce a high-quality food for restaurants, supermarkets, and other outlets.

Marketing

Some attention has focused on developing a market for the striped bass. Restaurant testing has shown that the fish are highly acceptable to consumers. They are considered to be mild-flavored fish with good market appeal.

Current producers are marketing to niche outlets. Ethnic groups in the larger cities of the Midwest and East like the striped bass. A marketing strategy may involve producers shipping whole fish (no processing) on ice to distributors in

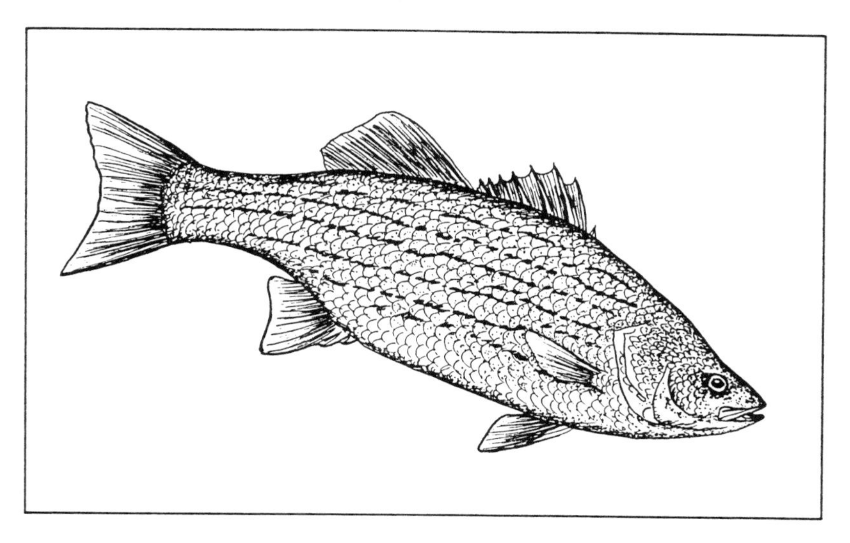

Figure 10–16. Hybrid striped bass are thought to have considerable potential for intensive production. Market-size fish weighing 1 – 2 pounds can be grown in about 15 months.

large cities. For example, the China Town area of New York City is targeted by some growers. In these restaurants, the chefs dress the fish and prepare them for customers. Prices paid must reflect the cost of production as well as the transportation.

Market-size hybrid striped bass can be grown in 15 - 18 months.

Species Grown

Attention has focused on a hybrid species of the striped bass. This hybrid was developed by crossing a female striped bass (*Morone saxatilis*) with a male white bass (*Morone chrysops*).

Identification

Striped bass are silvery-white to grayish with black stripes that extend to the tail from behind the gill cover. Hybrid striped bass have much the same coloring except that the stripes tend to be broken or jagged in pattern. The body is covered with scales and has a somewhat forked tail.

Size

Striped bass can grow rather large, weighing 30 pounds or more. At the end of the first growing season, a hybrid striped bass may weigh ½ pound. Marketable size is considered 1¼ pounds. In 18 months, the hybrid striped bass will grow to a weight of 2 pounds. Typical brood females weigh 15 - 30 pounds.

Environment Required

Striped bass grow well in a variety of situations. They can grow in saltwater, brackish water, or freshwater. Most commercial production is in freshwater.

Hybrid striped bass grow best at a water temperature of 77 - 88°F. They will survive temperatures as low as 40°F., but feed consumption is reduced and growth is slowed below 60°F.

Water quality is important. Dissolved oxygen should be no lower than 4 ppm for good growth; however, hybrid striped bass will survive for short periods of time at an oxygen level as low as 1 ppm. The best pH is 7.0 - 8.5, but hybrid striped bass are tolerant of a pH range of 6 - 10. Water alkalinity of 100 mg/L or above is desirable, but the fish will survive in extreme water alkalinity. Transferring the fish from high alkalinity / hardness water to low alkalinity / hardness water can cause death. As hybrid striped bass age, they become more tolerant of increased ammonia levels in the water.

Seedstock

Reproducing hybrid striped bass requires greater technology than some other aquacrops. Fingerling and food fish producers often rely on specialized hatcheries for their stock. A generally used procedure is described here.

Broodstock ready to spawn are taken from the pond to a hatchery where spawning is induced. This may involve injecting the female fish with human chorionic gonadotropin to induce ovulation some 25 - 40 hours later. The eggs are hand-stripped from the female and artificially fertilized with sperm induced from a male. The eggs are incubated in jars at the rate of 100,000 - 250,000 per jar. At a water temperature of 60 - 70°F., the eggs hatch in 40 - 48 hours. Fry are moved to ponds in five days at a per acre stocking rate of 200,000 - 300,000. Survival depends on the availability of plenty of zooplankton. Many hatcheries will lose 50 - 75% of the fry in the first 45 days. At 45 days, the fingerlings are 1½ - 2 inches long and can be stocked into growing ponds or moved to ponds at a stocking rate of 10,000 per acre to grow the 6- to 8-inch size fingerling. Maximum growth occurs at a water temperature of 82°F.

Culture

Hybrid striped bass can be grown in a variety of culture systems, including ponds, cages, raceways, and tanks.

Figure 10–17. A highly automated round tank system is used here to intensively produce hybrid striped bass. (Courtesy, Chore-Time Equipment)

With ponds, stocking rates range from 1,000 to 1,500 fingerlings per acre. Experienced farmers who follow good management practices may stock at much higher rates of 3,000 - 4,000 per acre. When fingerlings are stocked in the spring and properly managed, they will reach harvest size of 1½ pounds by the fall. Survival rates of 80% or more are the result of common good management, such as aeration, feeding, and proper handling.

Raceways and tanks with high water volume exchange are receiving considerable attention in hybrid striped bass farming. Circular pools or tanks are very good with intensive culture. Plenty of good-quality water must be available. A few producers have access to well water in the 80°F. range, which is very good for hybrid striped bass farming. With the very intensive systems, a power failure can pose particular problems.

Feeding

Hybrid striped bass fingerlings need a feed that is 38 - 52% protein, such as salmon or trout feed.

Food fish grow better on a feed with 38 - 44% protein. A few manufacturers are making a high-protein striped bass feed. If it is not available, trout feed or catfish feed may be used. One drawback with catfish feed is that it has a protein level of about 28%, which is somewhat below that needed by hybrid striped bass.

Some producers use automated feeders that provide feed every hour or so. Others may hand feed one or two times a day. The rate of feeding is 1 - 3% of the fish weight in feed each day for fingerlings. As the fish grow, the rate should probably be reduced to 1% of the weight of fish each day. The rate of feeding also depends on the temperature of the water. Water problems resulting from overfeeding are more likely to occur in warm water.

Diseases

The diseases of hybrid striped bass are continuing to be investigated. The common diseases that affect other cultured fish, caused by bacteria, viruses, or protozoa, also affect hybrid striped bass. Research has shown that water salinity is helpful in the control of diseases in this fish. There are no approved drugs for use on hybrid striped bass.

The fish are very sensitive to being handled. Minor abrasions can lead to infections. Changing the water can cause problems, particularly if the hybrid striped bass are moved to soft water.

Harvesting

Hybrid striped bass are typically harvested at a weight of 1 pound or more. Harvesting methods depend on the kind of water facility used. In ponds, they may be seined. In tanks and raceways, they may be netted. With cages, harvesting is simplified.

With hybrid striped bass, most producers withhold feed beginning about 24 hours before harvest. This allows the fish time to digest the feed that is in their digestive system. When the fish are shipped on ice without processing, withholding feed is very important.

Figure 10–18. High-quality hybrid striped bass being packed for shipment as whole fish. (Courtesy, *Aquaculture Magazine*)

SUMMARY

Several species of aquacrops are adapted to farming in freshwater. The major freshwater crops are trout, catfish, and crawfish. Two aquacrops thought to have considerable potential are tilapia and hybrid striped bass.

When aquafarmers are considering which aquacrops to produce, they should keep in mind some fundamental principles. These are:

1. Select only those species that are adapted to the production system to be used.
2. Select only those species that are adapted to the water that is available.
3. Feed an aquacrop on the basis of its nutritional needs.
4. Insure good health by following management practices that are appropriate to the species being produced.
5. Maintain quality water by using production practices that minimize the possibilities of problems and make water more suitable for the species being produced.
6. Select a species for which there is a market.

QUESTIONS AND PROBLEMS FOR DISCUSSION

1. What are six important fundamentals of freshwater aquaculture? Briefly describe each.
2. What are three established freshwater crops? Name two that are being expanded. (Include both the common and the scientific names of each.)
3. Compare the five freshwater species included in this chapter on the following:
   a. Marketing
   b. Identification
   c. Size
   d. Environment required
   e. Seedstock
   f. Culture
   g. Feeding
   h. Diseases
   i. Harvesting
4. Select the species that is (are) best suited to your community. Explain why you made the selection(s).

Chapter 11

SALTWATER AND BRACKISH WATER AQUACULTURE PRODUCTION

Aquaculture production in saltwater and in brackish water involves many species. The types of animals cultured in saltwater and in brackish water include fish, molluscs, gastropods, and crustaceans.

Because saltwater and brackish water species comprise over half of the imports of fish and seafood into the U.S., the potential economic advantage of producing these species is excellent. For example, imported shrimp account for over 60% of the U.S. supply. While ocean catches of shrimp and other seafood remain constant or decrease, the demand for these products has increased. Aquaculture has the potential to fill some of the void.

The challenge is to efficiently produce an aquacrop that will economically compete with the wild catch. Of course, aquaculture should be able to provide a product that is fresher at any time during the year and thus demand a superior price for a superior product.

This chapter presents the primary species of aquacrops cultured in saltwater and brackish water in and around the U.S. and a summary of the production of several secondary species. The routine management, feeding, disease prevention, and harvesting of the aquacrops are discussed.

OBJECTIVES

This chapter stresses the principles of saltwater and brackish water aquaculture. Upon completion of this chapter, the student will be able to:

- List and describe the fundamentals of saltwater and brackish water aquaculture.
- Describe the common aquacrops produced in saltwater and brackish water.
- Select sites and facilities for saltwater and brackish water aquaculture.
- Describe the culture of shrimp.
- Describe the culture of salmon.
- Describe the culture of oysters.
- Briefly describe the culture of secondary aquacrops, including clams, mussels, red abalone, striped mullet, milkfish, lobsters, blue crabs, pompanos, and red drums.

FUNDAMENTALS OF SALTWATER AND BRACKISH WATER AQUACULTURE

Saltwater and brackish water aquaculture, sometimes referred to as *mariculture*, involves the production of aquacrops in marine or estuarine waters. "Marine waters" refers to the oceans or gulfs that are primarily saltwater. "Estuarine waters" refers to bays, inlets, and mouths of rivers, where saltwater and freshwater are mixed.

Salinity

"Salinity of water" can be defined as the amount of salt in the water, which is usually sufficient for aquafarmers. The precise definition is a little more complicated. A measure of salinity technically refers to the total amount of solid material in 1 kilogram of water when all carbonates and organic material have been completely oxidized. With this accomplished, the following formula will provide the salinity: salinity = 0.03 + 1.805 × chlorinity.

"Chlorinity" refers to the total amount of chlorine, bromine, and iodine, in grams, contained in 1 kilogram of water. A precise definition of "salinity" is not usually a concern to aquafarmers, who care less about how to define the word and more about how to measure it and manage it for a particular aquacrop.

The two most common methods of measuring salinity for aquaculture production are measuring the density and measuring the refractive index of the water.

The measurement of salinity by density is the least expensive method. A hydrometer is used to measure density, and the reading is converted to salinity with the use of a conversion table. Readings from hydrometers are very accurate.

The easiest method of measuring salinity is through refractometry, although a refractometer costs more than a hydrometer. A refractometer resembles a pocket telescope and requires only one drop of water for the measurement. Once the drop of water is placed on the refractometer, the salinity can be read from the scale provided on most refractometers. Some will require the reading to be converted with the use of a conversion table.

The salinity of water is usually reported in parts per thousand (ppt). A refractometer is usually accurate to within 0.5 ppt, while a hydrometer is more accurate. The most accurate method of measuring salinity, taking a measure of the conductivity, requires expensive equipment and large amounts of water and is not suitable for most aquaculture operations. In aquaculture production, the producer tries to maintain salinity within a certain range, which can be measured easily with one of the less expensive methods.

Although various classifications exist, freshwater is typically considered to be less than 0.5 ppt. Saltwater is usually defined as being 16.5 ppt or greater. The range in between is referred to as brackish water. As a comparison, the open ocean averages about 35.0 ppt salinity.

The species discussed in this chapter vary widely in the amount and range of salinity needed in the water in which they are raised. Salmon, for example, are *anadromous*, which means they spend most of their lives in saltwater, but return to freshwater to spawn. Several of the species that are found naturally in saltwater or brackish water, such as the red drum and striped mullet, have the potential for production in freshwater. Conversely, some freshwater species, for example, some tilapia and the rainbow trout, have shown tolerance for moderate levels of salinity and could be reared in saltwater.

Production Considerations

One major consideration for the production of saltwater and brackish water species is that of finding a suitable location for production. Very few people own parts of the ocean or bays that are necessary for this type of production. As a general rule, the water used for saltwater and brackish water aquaculture, the oceans, gulfs, bays, and river mouths, is considered to be public. Many states, however, have procedures in place for leasing certain areas for aquaculture production.

If manufactured facilities are used for saltwater and brackish water aquaculture, special considerations must be taken. The higher the salinity of the water, the more corrosive it becomes to metal parts. Any metal parts used with saltwater must be protected from the water.

Different aspects of water quality are introduced in saltwater and brackish water facilities. If water of the same salinity is added to closed recirculating systems or to static systems such as ponds, the salinity of the facility will constantly increase as the water evaporates but the salts do not. Freshwater must be added periodically to lower the salinity. If the water coming into a facility is too low in salinity, the salinity may be raised by adding artificial sea salts, but this can be very expensive if large amounts are needed. If the water used is drawn from a well, its salinity tends to be more stable than if it is drawn from surface water.

Many states have regulations that are designed to protect natural fishing stocks from depletion but that also affect the aquafarmer who wishes to culture saltwater and brackish water species. Many states have minimum size requirements for the catch or sale of certain species, which may cause problems for the aquafarmer trying to stock a production facility. Sometimes it is illegal for the farmer to possess the size of aquacrop needed to stock the facility. In most cases, permits can be obtained either to catch fry for stocking or to buy suitable fry from hatcheries.

Another problem may be the closing of seasons. Many states have closed seasons, which are designed to protect species during times of spawning and to insure wholesome food products. Normally these laws would not affect the aquafarmer, except that in some cases the laws also prohibit the sale of the species when the seasons are closed, thus causing the aquafarmer to miss a prime marketing opportunity when demand is high and supplies of commercially landed species are gone. In Georgia, for example, oyster farmers can sell their oysters only during the time when oysters can be taken from the public oyster grounds. Fortunately, several states have recognized aquaculture as agriculture and have allowed aquafarmers to obtain special permits to excuse them from regulations intended for others.

COMMON SALTWATER AND BRACKISH WATER SPECIES

The three most important types of aquaculture in saltwater and brackish water involve the production of shrimp, salmon, and molluscs. Several species of these three types of aquacrops are produced in U.S. coastal waters. A description of the particular species and their importance to aquaculture follows.

Shrimp

The shrimp is the most important and extensively farmed crustacean in the

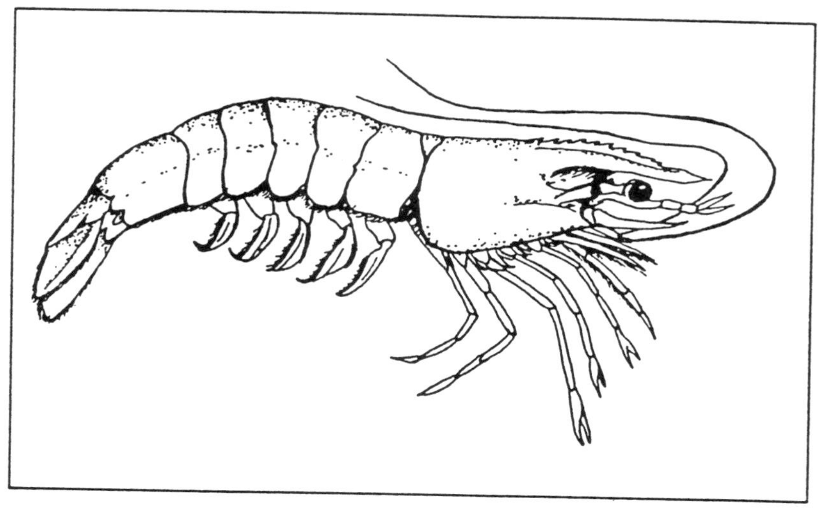

Figure 11–1. The shrimp is one of the most widely cultured saltwater and brackish water species.

world. Two factors cause shrimp to be ideal for intensive cultivation: (1) they grow rapidly in intensive production facilities and (2) there is a great demand for shrimp around the world, especially in the U.S.

The most common shrimp grown in the U.S. are penaeid shrimp, meaning they all come from the genus *Penaeus*. The three species most commonly grown are the *Penaeus aztecus* (brown shrimp), the *P. setiferus* (white shrimp), and the *P. duorarum* (pink shrimp).

Salmon

Although the Atlantic salmon (*Salmo salar*) is widely cultured in northern Europe and in a few areas of the U.S., the predominant salmon culture in the U.S. is that of the Pacific salmon, genus *Oncorhynchus*.

The species of Pacific salmon cultured in the U.S. include *Oncorhynchus tshawytscha* (chinook salmon), *O. kisutch* (coho salmon), *O. nerka* (sockeye salmon), *O. keta* (chum salmon), and *O. gorbuscha* (pink salmon). The culture of these salmon is based mostly in the Pacific Northwest in Alaska, Washington, Oregon, and northern California.

The culture of salmon has very stiff competition from commercial landings, so the production must be cost-efficient. In fact, more salmon are exported from the U.S. than any other fish or seafood, mostly because of the large commercial landings.

Methods of salmon production include salmon farming and salmon ranching (which takes advantage of the salmon's natural instinct to return to their place of birth to spawn).

Oysters

Of the three different types of molluscs commonly cultured in the U.S., the oyster is easily the most important. Oysters are in great demand as high-priced seafood around the world, with excellent market potential for species produced by aquaculture.

The oyster most commonly cultured in the U.S. is the *Crassostrea virginica* (American oyster). It is farmed in the coastal waters of the Atlantic and Pacific oceans and in the Gulf of Mexico.

Secondary Species

Several other species of aquacrops are suitable for culture in saltwater and brackish water, although their culture in the U.S. is not as widespread as that of shrimp, salmon, and oysters. Some of the culture of these species is still in the experimental stages. With other species, techniques that have been developed and proven successful in other countries should work in the U.S. In some cases, markets still need to be developed in the U.S. in order to make the culture of these secondary species profitable.

Some of these species include *Mercenaria mercenaria* (hard clam), *Mya arenaria* (soft clam), *Mytilus edulis* (common edible mussel), *Homarus americanus* (American lobster), *Callinectes sapidus* (blue crab), *Chanos chanos* (milkfish), *Micropogon undulatus* (Atlantic croaker), *Sciaenops ocellata* (red drum), *Trachinotus carolinus* (pompano), *Mugil cephalus* (striped mullet), and *Haliotis rufescens* (red abalone).

SELECTION OF SITES AND FACILITIES

The relationship of the site to the ocean or gulf and to the type of facility used is extremely important to the success of any saltwater or brackish water aquaculture operation. The species being produced, the facility used, and the availability of the land or water to be used exclusively for aquaculture all determine if a site is suitable for production. The various zones used in saltwater and brackish water are the shore, the intertidal, the sublittoral, and the seabed.

Shore

The shore is the land that is adjacent to the sea. It is used as the site for most hatcheries to reduce transportation costs. The primary advantage of using the shore is that the tides do not affect the water level of the facility, only the sea water outlet and inlet. The primary disadvantages are that sea water must be pumped into the facility and that freshwater runoff into the facility may be a problem in saltwater ponds.

Planning facilities on the shore is relatively simple in that the tides do not affect the facilities. Ponds and tanks are the primary facilities used in the shore zone.

Figure 11–2. A concrete enclosure located on the shore. Note the netting around the drain pipe to keep the aquacrop from escaping.

Intertidal

The intertidal zone is the area that is covered with water during high tide but not during low tide. The primary advantage of using the intertidal zone is that pumps do not have to be used to pump water into the facility or to pump water out to harvest or treat the pond bottom. Another advantage is the easy access provided to the facilities at low tide. The primary disadvantage is that the tides have to be studied and careful planning is required to build the facility so that the tide will be a help and not a hindrance. Ponds should be built so that the water level can be adjusted easily during high tide and so that the ponds can be drained completely during low tide.

The intertidal zone is used primarily to build ponds that hold the water in but have sluice gates to permit the flow of sea water in and the flow of freshwater out. It is also used for some culture of molluscs on the bottom, but this requires careful management.

Sublittoral

The sublittoral zone includes relatively shallow (usually 5 - 100 feet deep) inshore areas such as bays and lagoons. The sublittoral zone has the advantage of always having water. Its primary disadvantages are the reduced access afforded by land and the amount of freshwater that may be introduced into the bays during heavy rains (not a concern for brackish water aquacrops).

When the aquafarmer is selecting a facility for use in the sublittoral zone, the difference between the water level at high tide and at low tide is very important. Ideally, the difference would be as small as possible, while allowing adequate movement of water at low tide (stocking densities must be calculated at the low tide level).

The sublittoral is the zone of choice for net enclosures used in the production of many saltwater and brackish water species, including salmon and shrimp. The nets are placed at the outlet end of the bay to keep the aquacrop in but to allow the movement of water both in and out. Because the nets do not have to be placed around the land edges, the enclosures are inexpensive relative to the amount of water available to the contained aquacrop. The major concern is to keep the nets clean so that water movement is not restricted.

Sublittoral facilities also are good choices for floating cages for fish production and rafts for mollusc production. Of course, the fish cages require more intensive management, including daily feeding. When fish cages are used, some regular means of access is necessary. Once molluscs are placed on their rafts or other structure, they require almost no supervision.

Figure 11–3. An inlet for a sublittoral facility. The screen is to keep predators and trash fish from entering the facility with the incoming sea water.

Seabed

The seabed zone is the ground which is covered by the sea at all times. As a general rule, the seabed offers a constant rate of salinity, its primary advantage. The primary disadvantage is that it is subject to pollutants and limited to the production of molluscs or crustaceans (in cages moored to the bed).

Before the seabed is selected as a site, the area under consideration must be checked for dissolved oxygen levels and pollutants. Water currents must also be adequate to provide for the removal of wastes and to bring in food for the aquacrop.

The seabed is the primary choice for mollusc aquaculture, as long as it is not too silty. Also, pens moored to the seabed or to a dock often provide a suitable environment for shrimp.

CULTURE OF SHRIMP

Shrimp are widely cultured in Asia, but the practice is fairly new to the U.S. In historic aquaculture operations, shrimp were cultured almost by accident in brackish water ponds and estuaries as part of a polyculture with fish. The monoculture of shrimp is a fairly recent occurrence, even in Asia.

Several factors point to an increase in the production of shrimp through aquaculture in the U.S. Although the most abundant supply of shrimp in the world is in the Gulf of Mexico, shrimpers (commercial fishing operations) cannot supply even half of the shrimp demanded by the market. Also, the shrimp grow well in intensive culture operations and have few problems with disease. Methods of obtaining seedstock, hatching, facilities, management concerns, diseases, and harvesting are discussed as follows.

Obtaining Seedstock

The major problem in culturing shrimp is that of finding a reliable source of broodstock. Although shrimp are easily spawned in captivity and reared through the larval and juvenile stages, they do not usually produce eggs or copulate under captive conditions in ponds. Recently, however, a few nurseries have been successful in spawning captive shrimp in the U.S. (particularly in Hawaii) and in Ecuador. Although the numbers of seedstock produced are too small to contribute much to the stocking of growing facilities at present, in the near future, hatcheries will be providing much of the seedstock for shrimp aquaculture operations.

The two primary methods for stocking growing facilities other than the new methods of controlled spawning are (1) catching larval shrimp for stocking and (2) catching gravid (pregnant) females and moving them to hatcheries where their offspring can be hatched and grown to stocking size.

Catching larval shrimp for stocking is a difficult venture. The two primary ways to catch larval shrimp are by (1) seining with very fine mesh seines and (2) by building tidal ponds with sluice gates that allow larval shrimp to enter the pond with the water from the rising tide. Both ways have serious problems. Catching larval shrimp with the use of seines almost always results in a high mortality rate, so many more shrimp must be caught than are actually needed. Seining by hand is only feasible during certain times of the year when the larval shrimp are most numerous in tidal streams. If tidal ponds are constructed so that the tides carry larval shrimp into the ponds, larval predator species usually come into the ponds with the tide as well, which leads to loss of shrimp as the predators mature.

Capturing gravid females and moving them to hatcheries may be even harder than capturing larval shrimp for stocking. The mortality rate for shrimp caught by seining with a boat is usually 85% or greater, no matter how carefully they are handled after being caught. Using cast nets results in fewer deaths but is not very efficient. Even then, not only must the females have eggs, but they also must be captured with spermatophores already attached or they will not spawn. Once gravid females have been supplied to hatcheries, the process of hatching and rearing the shrimp through the larval and juvenile stages is usually fairly successful.

The demand for shrimp and its potential as a significant aquacrop in the U.S. has led to much experimental work in supplying seedstock. New methods being tried are methods to drive larval shrimp into tidal ponds or estuaries and methods to lure or draw the shrimp into the ponds. In other countries, aquafarmers have found that larval shrimp attach themselves to seaweeds to keep from being washed back out to sea, so aquafarmers put down "lures" of grasses, which the shrimp cling to, then they pull up the grass and shake the shrimp into a dip net. The idea behind these experiments is to try to find a way that will provide many more shrimp than simply opening a sluice gate and letting the tide bring them into the pond.

The advances in hatchery techniques, such as those being developed at the Oceanic Institute in Hawaii, will make the problem of obtaining seedstock much less of a problem in the future. These hatcheries will be able to provide a reliable source of healthy seedstock for aquafarmers who wish to produce shrimp. This assurance should cause the shrimp aquaculture industry to grow rapidly in the next decade.

Hatching

The hatchery techniques used in the U.S. have largely been adapted from the Japanese, who have been very successful in hatching shrimp. Once gravid females have been captured, they are placed in tanks in sea water, which is replaced every one or two days. The best salinity for spawning is about 32 – 35%, with the optimum temperature range between 72° and 91°F.

After hatching, the baby shrimp, called *nauplii*, go through several molts within the first three days, when they begin eating microscopic plants and animals and are called larvae. The larvae move rapidly through the growth stages and are considered postlarvae after 10 – 12 days. The postlarvae are kept in tanks until they are between 30 and 40 days old and then are usually moved to nursery ponds.

The postlarvae remain in the nursery ponds for about 50 days, or until they reach a weight of about 1 gram (.035 ounce). The juveniles, called seedlings, are then stocked in the food production ponds at a rate of 30,000 – 40,000 per acre.

During these stages, the shrimp can withstand a wide range of salinity, from 15 to 37%, although the optimum range is 27 to 32%.

Larval shrimp must have an abundant supply of phytoplankton or other food, such as brine shrimp. Some hatcheries supply the phytoplankton by providing a high turnover of sea water in the facilities. Others supplement by adding cultured phytoplankton or brine shrimp to the diets. As the shrimp are moved into nursery and grow-out ponds, their diet may be supplemented with other foods such as ground fish, chicken blood meal, and similar animal products. Commercially prepared feeds are also available. Some producers feed the shrimp by fertilizing the pond with some type of organic fertilizer to encourage phytoplankton growth.

Food Production Facilities

Shrimp are most commonly reared in ponds with an abundant supply of sea water nearby, although tanks and other facilities can be used. The facility must allow for good movement of sea water to maintain a high level of phytoplankton. If intertidal ponds are used, the sluice gates must be large enough to keep the water pressure from becoming too great on the dam. Also, care must be taken

Figure 11–4. A modern shrimp hatchery where the shrimp are reared in fiberglass tanks. Note the polyethylene structure of the facility. Common in greenhouses, the covering is designed to regulate the temperature and reduce production costs. (Courtesy, Gary Fornshell, Mississippi State University)

to find a site with enough clay in the soil to hold water, which is sometimes difficult in coastal areas.

An alternative to intertidal ponds is to take advantage of a sublittoral impoundment of water which allows for the placing of a net or a sluice gate at the mouth. However, sublittoral impoundments with the right depth for shrimp production and the ability to keep freshwater out in periods of heavy rainfall are hard to find.

Management Concerns

One major concern in producing shrimp is to maintain the desired range of salinity in the water. As the shrimp get older, the range of salinity should be maintained between 22 and 37%. Older shrimp are not tolerant of lower salinities, so freshwater must be diverted away from the pond or growing facility.

Shrimp also grow best in the 70° – 90°F. temperature range, although they can tolerate higher and lower temperatures. Some mortality may occur, especially in white shrimp, if the air temperature drops quickly below 40°F.

Shrimp can usually grow to market size in about 150 days after stocking. The average facility in the U.S. produces about 1,000 pounds per acre, but much more will be able to be produced as advanced methods of stocking are developed. The potential for yields of 4,000 – 5,000 pounds per acre under ideal growing conditions is good. These numbers commonly are reached and exceeded by Asian aquafarmers who produce shrimp.

The species of shrimp that seems to grow best in pond culture is the white shrimp. The brown shrimp grows more slowly than the white but is more tolerant of colder water. The pink shrimp does not do well in most pond cultures due to its preference for sandy bottoms. All three species seem to grow well in tank culture, although tank culture is still in the experimental stages in terms of making a profit for the producers.

Diseases

Because shrimp farming is a relatively new method of production in the U.S., little is known about the diseases of shrimp grown in intensive culture. Some of the diseases to which shrimp are susceptible in nature will undoubtedly become more serious when shrimp in intensive culture systems get them. A few of the reported diseases are discussed here.

Several bacteria species of the genera *Vibrio* and *Pseudomonas* may invade the body fluids and exoskeleton of shrimp. The bacteria do not harm the quality of the meat, but they may make the shrimp unsightly and hard to sell in the shell.

The most dangerous disease for larval shrimp is contracted from fungi of the genus *Lagenidium*. This disease is very contagious and fatal. It can be controlled with malachite green.

The cotton shrimp or milk shrimp disease is caused by some type of protozoan organism. Other protozoa may cause problems, but the extent of these is not yet known.

Shrimp may also be invaded by cestodes, nematodes, and trematodes, although they are probably an intermediate host and have not suffered severe problems due to these parasites.

Harvesting

Shrimp are usually harvested by hand, either with seines or dip nets. In a tidal pond, the sluice gate is opened to drain the water, and the shrimp are caught at the gate or seined from the lower water levels. For this reason, it is important that the pond not have any significant holes on the bottom where the shrimp can congregate and avoid harvesting.

In some areas, the market for fresh shrimp may be such that the facility may need to be partially harvested every day or several times a week. This is a particular concern of aquafarmers who sell cooked and uncooked shrimp at roadside stands. This need must be considered in the planning and construction of the facility.

CULTURE OF SALMON

The Pacific salmon species from the genus *Oncorhynchus* make up a very important part of the fishery of the U.S. A significant market has developed for the salmon, both in the U.S. and in other countries. More salmon are exported from the U.S. than any other fish or seafood, almost $1 billion each year. The exports, however, are usually of canned salmon, while a significant amount of fresh salmon is imported to U.S. markets in the winter months. Aquaculture should help reduce the need for these imports by providing a year-round product for U.S. consumers.

The U.S., Japan, Canada, and the Commonwealth of Independent States (CIS) are the major countries with Pacific salmon fisheries. The Atlantic salmon is cultured extensively in Norway, with a few operations in the U.S. and Canada.

Government-supported salmon hatcheries have been producing seedstock of salmon for the commercial and sport fishing industries for many years on the Pacific coast of both the U.S. and Canada. Most of the fish produced from these hatcheries are caught either at sea or on their journey back to their place of birth to spawn.

Obtaining Seedstock

Obtaining seedstock for salmon production is not a problem. Salmon that are ripe for spawning are easily caught as they make their way back to their place of hatching to spawn (called their natal stream). Then they are moved to hatcheries with tanks or to artificial spawning channels built into natural streams or rivers.

Some artificial channels are built in areas that are near enough to natural spawning grounds to attract females ready to spawn. Other artificial channels are planted with fertilized eggs from hatcheries. These channels typically have a bed of medium to fine gravel (approx. 2 – 8 inches in circumference), a steady supply of running water (with dissolved oxygen levels near saturation), and a water temperature of between 50° and 60°F. The temperature and flow of water may be managed with water from a dam or a main channel.

The time of year when broodstock can be caught varies with the species. The chinook, for example, will enter the river as soon as they reach it and begin to ascend to their natal stream. They will rest in deep pools until fall, their preferred spawning time. The coho will gather in the ocean close to the mouth of the river through the summer, building their fat reserves. In the fall, they will begin their journey upstream.

Hatching

In nature, the female digs a hole, fills it with eggs, and lets the male fertilize the eggs by depositing the sperm, called *milt*. The female then covers the hole with soil and gravel to complete the process. A female will usually have enough eggs to fill three or more holes, called *redds*, with a total of about 4,000 eggs.

In artificial hatcheries, the ripe female is usually killed and the eggs are removed surgically. This poses no real threat on the salmon population since the salmon generally die immediately after spawning. The eggs are fertilized with sperm taken from the male, which may or may not be killed in the process. The egg mass is placed in trays in troughs of flowing water or in incubators with water dripping over them.

The fish hatch out of the eggs after five to seven days with the yolk sac still attached. This sac provides the nutrition for the next few days. After the yolk sac is absorbed, the fish are called fry. Once they reach 2 inches in length, they are called fingerlings. At a length of 6 inches, they become smolts.

Chum and pink salmon can be moved to a stream or estuary very soon after they absorb the yolk sac. In nature, they spend very little time in freshwater after spawning. Chinook, coho, and sockeye salmon, on the other hand, must be kept in freshwater for some time. In nature, the chinook, coho, and sockeye may stay in freshwater for up to two years before moving to the open sea.

The main problem with hatchery spawning of salmon is that the average survival rate of adults has been less than 1%. In other words, less than 1% of the fish hatched from the eggs have returned to spawn in the same location. The rate for natural spawns is between 1 and 5%.

Techniques to improve the survival rate of artificially hatched salmon have been researched and tested. One method that seems to improve the rate is to place some type of gravel in the hatchery tank so that the fry spend less energy holding their position. Another is to decrease the stocking rate so the flow of water can be reduced to a level more like the natural spawning streams. Another method is to move the fish from the hatchery to gravel-bottomed channels at the eyed stage. The use of these and new methods under research should bring the adult survival rate to a level similar to that of natural spawn.

In the past, 50% of the salmon returning to spawn were allowed to spawn in order to provide a steady supply of salmon for the commercial fishery. With the improved techniques of salmon hatcheries and a better survival rate, less than 50% should be required to keep the number of salmon available at an appropriate level.

Feeding

Chinook, coho, and sockeye fry that are being kept in freshwater are usually fed a diet of brine shrimp, ground meat or ground fish, or a combination of these. For fish that are being grown for food production, the diet is similar to that of trout (discussed in Chapter 5). Salmon nutrition has not been studied as much as trout nutrition, so not all of the specific requirements are known. As carnivorous fish, salmon require a high-protein diet. The protein is obtained through either meat or fish products or through a commercially prepared diet that is high in protein.

The primary source of protein in many salmon diets is from the carcasses of salmon that are spent from spawning. These carcasses are ground into meal and mixed with a vitamin supplement. Other species of fish and animals may also be used to make the feed. Several commercial nutrient and vitamin supplements are available.

Food Production Facilities

Food production of Pacific salmon commonly takes one of two forms, salmon ranching and salmon farming. In salmon ranching, the aquafarmer usually operates a hatchery, releases the fingerlings or smolts to the sea, and waits until the fish are ready to return. The wait may be from two to five years, depending on the species. Coho and pink usually return within two years, sockeye within three years, chum in three to four years, and chinook within four to five years.

Salmon farming usually involves the use of net pens or sublittoral enclosures. When the smolts or fingerlings are about 6 inches long, they are moved to the enclosures from the hatcheries. Salmon farming in sublittoral enclosures usually involves either the chinook or coho species, since these two are the most tolerant of freshwater. With floating net pens moored at sea, however, any species can be cultured.

Salmon farmers sometimes have a hard time making a profit. The price of salmon may vary widely due to the uncertainty of the wild catch. Even when the commercial catch is low, the lower costs of salmon ranching make it much more profitable than salmon farming. The one aspect of salmon farming that makes it potentially profitable is the ability of the salmon farmers to provide a reliable source of fresh salmon on a year-round basis. The salmon runs are so unpredictable that salmon ranchers cannot do this at present. Hatcheries have been trying to establish runs on rivers in years when runs are not prevalent, but the success of these is still not known on a long-term basis.

Management Concerns

The primary management concern of salmon ranchers is to provide a suitable hatchery environment that will result in a high adult survival rate. The adults can then be harvested upon their return to their natal stream.

Salmon farmers must be concerned with the same management factors as aquafarmers growing fish in any enclosure. In net pens or sublittoral enclosures, the dissolved oxygen, nutrition, temperature, wastes, and salinity all are important factors because the fish cannot move to more suitable waters if one of these factors is not present in sufficient amounts.

In net pens, nutrition is usually the major management concern. Since the fish cannot move about and catch their food, the aquafarmer must provide a nutritionally balanced ration. The site for the pens must be selected carefully with adequate salinity and temperature for maximum growth. Wastes and dissolved oxygen are usually not a problem because natural water movement will keep the water oxygenated and wastes moved away from the fish.

In sublittoral enclosures, the stocking rates are usually too large for the facilities to provide enough food for the fish, so their diet must be supplemented. Salinity and temperature are also important factors, especially the runoff of freshwater into the facilities from substantial rains. The stocking rate must be at such a level that wastes do not build up and that dissolved oxygen does not become a problem. Of course, mechanical aeration is possible, but it normally is not used in salmon production. The tides flowing into and out of sublittoral facilities will help oxygenate the water and remove some of the nitrates from fish wastes. Sites must be selected in areas far enough north so that the water temperature does not exceed 70°F.

Salmon also perform fairly well in raceways, but the costs of pumping the water through them usually keeps raceways from being as profitable as net pens or sublittoral enclosures. The salmon take a longer time to reach market size than other fish commonly grown in raceways, such as trout.

The coho and chinook salmon have shown good adaptability to freshwater lakes and have been introduced into the Great Lakes with some success. Their culture in these areas is usually as a game fish or predator fish of unwanted species. No substantial culture of salmon as food fish in freshwater is being carried out at present.

Diseases and Predators

Salmon that are artificially spawned or intensively raised are susceptible to a number of bacterial and viral diseases. These include bacterial cold water disease, columnaris disease, bacterial gill disease, fish tuberculosis, and several others.

Two methods commonly used to combat disease are drug treatment and genetic improvement. Several approved drugs are available as treatments for the bacterial diseases, but the viral diseases are largely untreatable. Some feed supplements have been developed as vaccinations against certain diseases. Providing the fish with a nutritionally balanced ration may also play a major role in reducing the stress on the fish and, subsequently, their susceptibility to disease. Several salmon breeding programs have been successful in developing strains that are resistant to a number of bacterial and viral diseases.

Two species of phytoplankton are harmful to salmon and can be particular problems with fish raised in net pens. *Chaetoceros convolutus* has spikes that hook together to form long chains of the algae. These sometimes block and damage the gills of salmon, causing them to quit feeding. *Heterosigma akashiwo* produces a mucous-like substance that can be toxic to fish, especially larger individuals.

The seal and the sea lion are two of the most common predators of adult salmon. If salmon are grown in sublittoral enclosures in areas where seals and sea lions are abundant, the facilities must be monitored closely. Bears may also be a problem but usually are not if the facilities are near populous areas.

The predators of young salmon fingerlings and smolts include numerous carnivorous fish and animals, too many to list here. While in tanks and ponds, the fish can be protected with netting and supervision. Very little can be done to protect them once they are released into the streams or estuaries.

Harvesting

Salmon farmers may harvest the fish from their facilities when the fish are anywhere between ½ and 1½ pounds. At this size, the flavor of salmon compares favorably with that of many other table fish, including trout.

Figure 11–5. Harvesting salmon from net pens. (Courtesy, Maine Agricultural Experiment Station)

Salmon in net pens are very easy to harvest. The net pens are simply opened and the fish are removed and placed into the boat with dip nets, or the net pens are emptied into the boat.

When salmon are harvested from sublittoral enclosures or from streams, seines are used, as with most fish in ponds or other facilities.

CULTURE OF OYSTERS

The American oyster is found naturally along the Atlantic and Gulf of Mexico coasts of the U.S. and has been successfully introduced along the Pacific Coast as well.

The culture of oysters in the U.S. is over 100 years old. Because of the oyster's value as a seafood, it has been a popular species for culture. Many aquafarmers have tried to increase their ability to provide quality oysters, those without irregularities in the shell, for the raw oyster market. Others have simply taken advantage of areas with little natural oyster production but good environment by transplanting oysters to these areas.

Oyster hatcheries are one option for aquafarmers who do not want to participate in the food production of oysters. The number of people who want oyster seed is increasing rapidly, including the oyster farmers and the commercial oyster fishers who want to improve their beds. In many coastal areas, the hatcheries are unable to meet the demand for oyster seed.

Obtaining Seedstock

In nature, the oyster spawns when the water temperature ranges from 68° to 80°F. The spawning season lasts for several months. The *Crassostrea virginica*, or American oyster, is an open water spawner, in that fertilization occurs outside the shell. The female releases eggs into the water, which are then fertilized by sperm from the male. As was noted in Chapter 5, the same oyster may produce sperm one year and eggs the next.

Although accurate counts of the eggs and sperm are not available, it is commonly believed that the female will release 50 - 100 million or more eggs each spawn and that the male will release more than a billion sperm! Of course, many of the eggs are never fertilized and many that are fertilized become part of some predator's meal. As a result, only 6 - 10 adult oysters survive from every million eggs released into the water.

Although some hatcheries still collect fertilized eggs from the wild, most try to hatch their own by using sexually mature oysters held in water that is at optimum temperature. Oysters can be induced to spawn by gradually increasing the water temperature to 75° - 77°F.

The fertilized eggs are then removed from the water with a fine mesh sieve and transferred to well-aerated, filtered sea water for hatching. Selecting larvae that demonstrate a rapid growth rate very early in their lives often involves several cullings of the larvae.

Hatching

Eggs hatch within a few hours after fertilization occurs. From a few days to

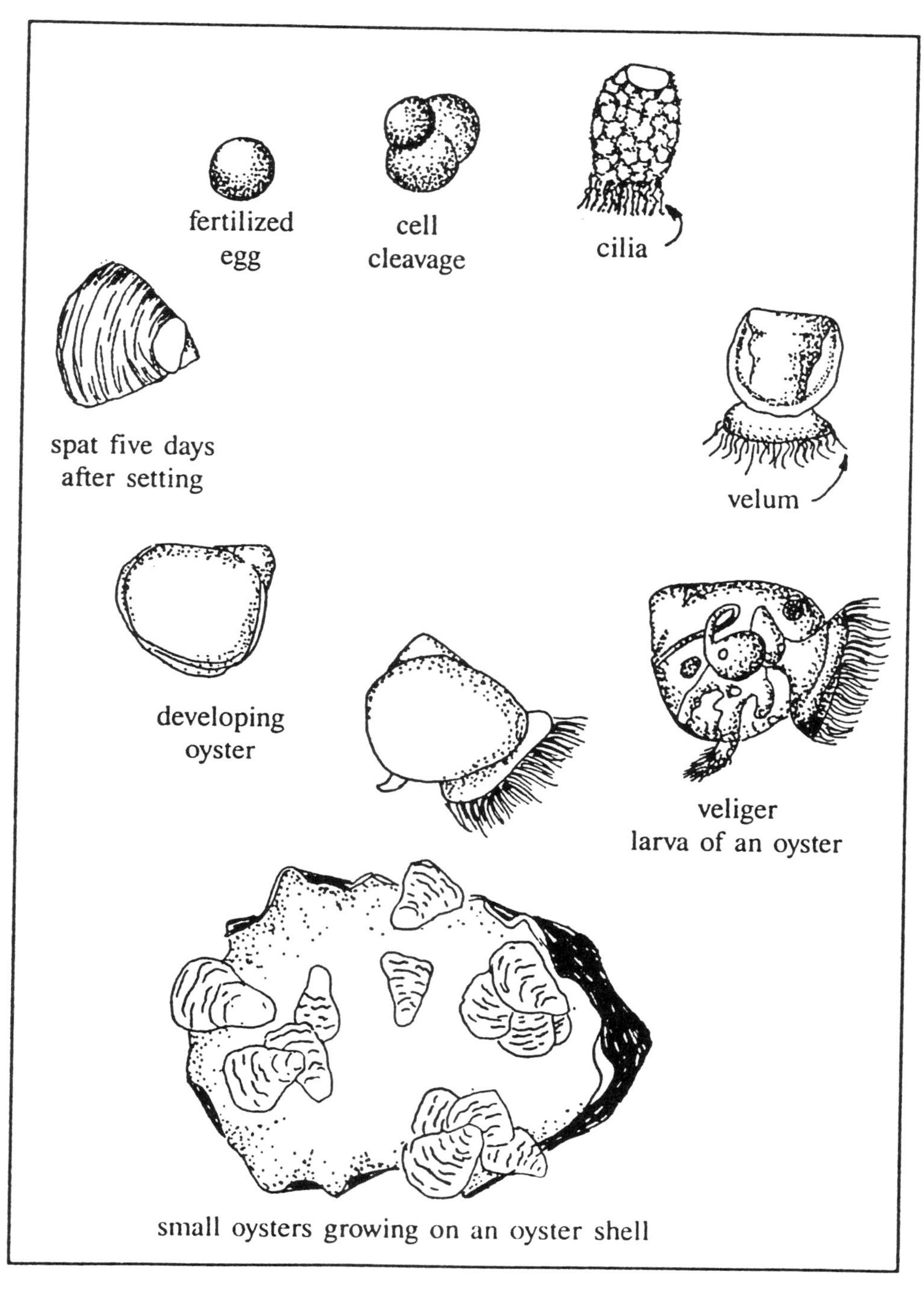

Figure 11–6. Life cycle of the American oyster, *Crassostrea virginica*. (Source: *Oystering on the Mississippi Coast*, Mississippi Department of Wildlife Conservation / Bureau of Marine Resources, Book IV — Marine Discovery Series)

two weeks, the larvae are free swimmers. This is a delicate time in nature, for the larvae may be washed out to sea on even a small current, as they are no larger than a grain of sand.

Within the first two weeks, the oysters develop their two shells and are ready to "settle" into their permanent homes. The oysters, now called *spat*, go to the bottom to look for something to attach themselves to, called *cultch*, from which they do not move for the rest of their lives. Because they do not move, oysters (and other molluscs) are called sessile organisms.

In nature, other oysters, old oyster shells, other shells, and sometimes gravel provide the cultch. In a cultured operation, the aquafarmer spreads clam shells

Figure 11–7. Seed oysters are being placed into the water in this cage, where they will grow until they reach planting size. (Courtesy, Maine Agricultural Experiment Station)

or oysters along the bottom of the facility to provide suitable cultch. The oysters settle on the cultch and attach themselves for life by secreting a concrete-like substance that attaches one-half of their shell. After a few days, the cultch are usually transferred to the area where they will grow until harvest. This may be to an estuary, the open ocean, rafts, or cages.

If the oysters are being seeded into the open water, the timing of the placement of cultch material is very important. If it is placed in the water too early, algae or other materials may form on the cultch and make it unsuitable for the spat, which will then die. If the cultch is not put out before the spat begin to settle, they may not find a suitable place and will then settle into the silt and probably die.

Production without cultch is possible, although not as common. The spat may be kept in tanks or in a pond or an estuary on trays, although cultch may still be used. Some aquafarmers will keep the oysters indoors throughout the production. This requires a constant source of phytoplankton-rich water.

Feeding

During the hatching process, larvae are fed phytoplankton from flowing sea water or from cultured phytoplankton added to the facility. Once the oysters settle, they are provided with sea water that contains phytoplankton, the oysters' primary source of nutrition.

Oysters are filter feeders. They filter sea water through their shells, and the sea water provides them with both oxygen and food. As a result, oysters have little control over what types of organisms are available for them to eat. However, they can reject some of the food particles.

The primary concern is that their habitat have some phytoplankton but not so much as to choke them. If a strong phytoplankton bloom causes the water to be too thick with the microorganisms, the oysters cannot handle all of the food and will close their shells, sometimes suffocating or starving to death.

Food Production Facilities

By far the most likely food production for oysters is the seabed. The oysters are sown into established oyster beds, planted after becoming attached to some cultch material, placed just off the seabed on French trays, or attached to ropes (usually clam shells suspended on wires), which are suspended from rafts. It will take the oysters from two to four years to grow to market size.

The primary concern is selecting a facility that has adequate movement of water, provides phytoplankton, has a bottom that is not too silty, and is free from pollution. The adequate movement of water supplies new oxygen and removes wastes, as well as providing a new source of food for the oysters.

The bottom cannot be silty, as the silt will clog up the gills of the oysters. If the bottom becomes too silty, the oysters will close their shells for a while and let the silt settle. Eventually, however, they must breathe, so they open their shells again. If the bottom is still too silty, they will usually suffocate.

Providing water that is pollution-free is an absolute necessity in oyster production. Because oysters get their food from moving the water through their systems, they usually concentrate any pollution in their meat. Although the oysters may be unharmed, they may be unsuitable for human consumption. Population increases with their accompanying wastes and industrial pollution are the two primary reasons for the decline in the harvest of wild oysters. Many beds that were once abundant either have been killed off or are so badly polluted that the oysters are no longer worth harvesting.

One form of oyster production involves harvesting oysters from polluted waters and then purging them in clean water for several days, which allows the oysters to expel the pollutants and then be suitable for the retail market. Because the oysters will move clean water through their systems just as they did the polluted water, they can cleanse themselves of most pollutants in one to two weeks.

In these purging systems, the oysters may be moved to raceways that pump running water through the tanks or to the seabed in an unpolluted area. In either case, many oysters that were not suitable can be added to the supply of edible oysters.

Management Concerns

Once the oysters have settled, they require very little management. Many oyster farmers have full-time jobs in other professions and raise oysters as a source of supplemental income. The low management requirements of oysters make them a good choice as a second occupation in aquaculture.

As was mentioned earlier, the primary work for the aquafarmer comes in the early stages in oyster production. Once a suitable site has been selected and the oysters have been sown, very little is left for the aquafarmer to do except wait for the oysters to grow to market size. Market-size oysters are usually at least 3 inches in width. The meat of a mature oyster is usually about 15% of its body weight, although higher percentages have been obtained with some culture operations.

Aquafarmers do have some concerns in the two to four years it takes oysters to reach market size, however. Damage from predators, diseases, and pollution must be regularly monitored, so that the crop is not wiped out suddenly due to one of these problems. With pollution, removing the source of the pollution is usually the only way to control the problem. Once the source is removed and the water returns to normal, the oysters will usually cleanse themselves and continue their growth. Damage from diseases and predators is discussed in the next section.

The oyster crop may require periodic thinning, if the oysters settled well and are growing well. Most oyster farmers thin the crop by using scuba equipment to dive down and remove a certain percentage by hand. The oysters that are thinned out are usually transplanted to another location.

Diseases and Predators

The most common oyster disease is oyster fungus disease (caused by *Labyrinthomyxa marina*). Oyster fungus disease is found especially in high concentrations of oysters. It is usually not a problem in cool water or in water of low salinity. The oyster slows its growth, then stops growing, then eventually dies. There is no known cure for oyster fungus disease.

The oyster has many predators, including fish, starfish, and gastropods. Perhaps the most serious are the oyster drills, two species of gastropod molluscs, *Urosalpinx cinerea* and *Eupleura caudata*. Oyster drills drill holes in the shells and consume the oysters inside. They may be controlled by trapping, divers, and

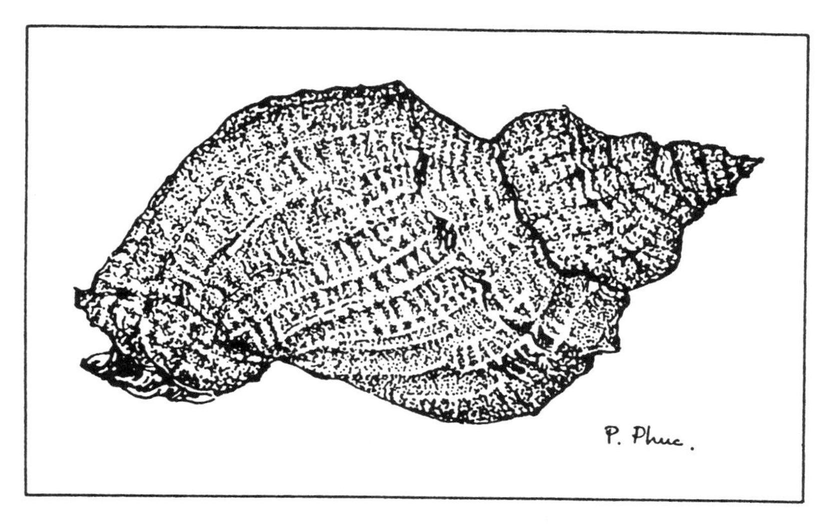

Figure 11–8. A common oyster drill. (Source: *Oystering on the Mississippi Coast*, Mississippi Department of Wildlife Conservation / Bureau of Marine Resources, Book IV — Marine Discovery Series)

chemicals. They only eat oysters cultured on the bottom; they are not a problem to oysters raised in baskets or with rope culture.

The starfish, *Asterias forbesi*, can eat up to five oysters per day. The starfish opens the shell partially by applying constant pulling pressure on the oyster. When the oyster finally opens up just a bit to breathe, the starfish can increase the opening. The starfish then extends its stomach into the shell and consumes the meat inside. Starfish are commonly controlled by lime treatments or by divers. Like oyster drills, starfish are not a problem with rope or basket culture. One method of control that will not work with starfish is cutting them up into pieces. Almost every piece will regenerate and make a new starfish, thereby increasing the population rather than decreasing it.

Leopard rays also eat oysters. They have very strong jaws that can crush the oyster shells and then the rays just pick out the meat from the shells. Placing sticks in the ground with sharpened ends pointing upward will usually control them. When the rays drop onto the oyster beds, they impale themselves on the sharpened ends of the sticks.

Oyster farmers must also be wary of human poachers, especially when the oysters near market size. Sometimes groups of oyster farmers organize into patrol groups in regular shifts to discourage thefts by humans.

Harvesting

Oysters cultured on the bottom are typically harvested by either tonging, dredging, or diving. Tonging refers to using a scissor-like device with a rake-like head containing teeth (tongs) that extend into the bottom of the water facility, moving the two sides along the bottom and dislodging the oysters, which are pushed into the basket formed by the two sides of the head. The full tongs are then moved to the surface either by hand or by power equipment. Hand tonging is very taxing physically and requires a strong person to operate the tongs.

Dredging involves moving an open basket along the bottom of the water facility with a bar with teeth on the bottom of it (dredge) to dislodge the oysters. The length of the teeth may be different, depending on the type of bottom of the facility being harvested. The dredge is pulled by a boat with ropes attached to it. A power winch is used to pick up the dredge when it is full.

Diving refers to using scuba equipment to dive down to the oyster bed with a basket, filling up the basket by hand, and then using a winch to pull the basket up to the boat or dock. Some states have regulations prohibiting diving in natural waters, but in aquaculture production facilities, it is permissible. Diving allows the producer to select only the largest oysters and leave the rest for harvesting at a later date.

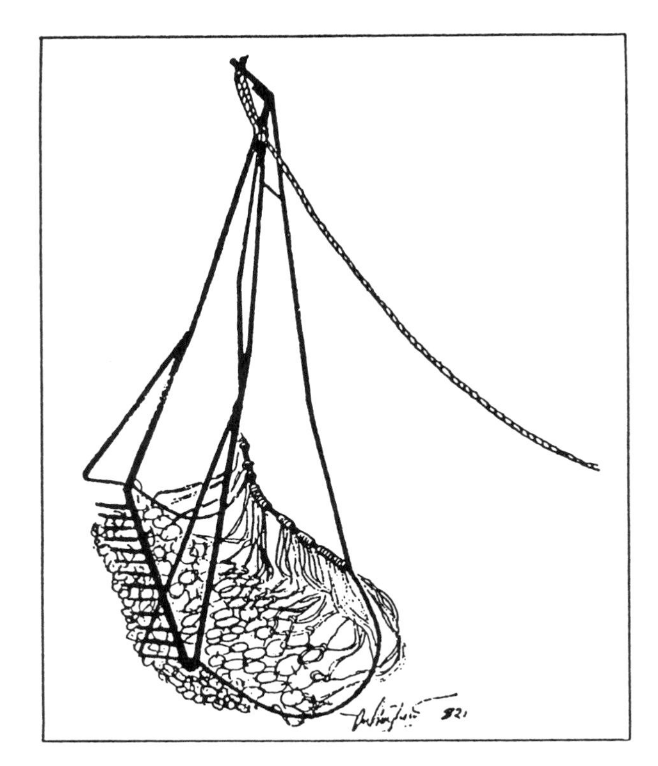

Figure 11–9. An oyster dredge. (Source: *Oystering on the Mississippi Coast*, Mississippi Department of Wildlife Conservation / Bureau of Marine Resources, Book IV — Marine Discovery Series)

Divers must be very careful not to get cut by the sharp shells of the oysters. A condition commonly called oyster finger may result. The cuts from oysters contain some substance that can cause the cuts to become very infected, to remain sore for much longer, and to leave much worse scars than normal cuts. Experienced oyster fishers' hands will often be covered with the telltale scars from oyster finger.

CULTURE OF SECONDARY SALTWATER AND BRACKISH WATER SPECIES

Clams

The two species of clams cultured in the U.S. are *Mercenaria mercenaria* (hard clam or quahog) and *Mya arenaria* (soft clam). The hard clam is found naturally all along the Atlantic Coast, while the soft clam's natural range is from North Carolina south along the Atlantic Coast. Aquafarmers have tended to concentrate on the culture of hard clams much more than soft clams, but the culture methods are virtually the same.

The clam is not widely cultured around the world. The U.S. has the most advanced culture, but much of the work with clams is still experimental. Clams are considered to have strong potential as an aquacrop because they will spawn in captivity and have a sizeable market, especially in New York and in the New England states.

Clams normally spawn only in the summer. Hatching techniques used by oyster hatcheries have proven to be very successful with clams, however. By holding sexually mature pairs in tanks and gradually increasing the temperature to between 70° and 80°F., hatcheries can get the clams to spawn at any time during the year. Today, commercial hatcheries provide nearly all of the seedstock used in clam aquaculture.

The larval clams are tolerant of temperature changes but require a salinity of at least 20 ppt with an optimum level of 27 - 30 ppt. Larval clams are fed phytoplankton (microalgae) diets, often cultured by the aquafarmer. A diet of dried microalgae (single-celled algae) has been shown to provide equal growth rates as cultured, living diets of microalgae.

Clams are cultured for market with the same techniques used for oysters. The seed clams are distributed across the beds where they settle and grow to market size. Some operations put the seed in boxes filled with sand and covered with netting to ward off predators. The clams will reach a marketable size of 2 - 2½ inches in about two years in the southern U.S. and in five to seven years in the northeastern U.S.

The aquafarmer usually needs to check periodically for pollution and for predators and to thin the clams occasionally to provide them with more room to grow. As with oysters, the clams are susceptible to human predation as they near market size.

Since clams from polluted waters can be purged by placing them in clean water, some aquafarmers culture clams in this manner. The technique is the same as that used with oysters.

Mussels

Mussels have not been cultured for very long in the U.S., but they are commonly cultured in Europe and other parts of the world. They are easily cultured, requiring little attention after being placed in the growing facility. The most common species cultured is the *Mytilus edulis*, called the common edible mussel.

The culture of mussels in the U.S. has grown very rapidly in the last few years, more than that of any other shellfish. The primary growth has been in Maine and other Northeast Atlantic states.

Mussels are considered a crop with great potential for aquaculture growth in the next few years, for several reasons. First, mussels are very easy to raise. Second, the market for mussels in the U.S. has increased; thus, aquafarmers can obtain a better price for their crop than in the past. Third, several states have developed programs for leasing offshore sites to aquafarmers for mussel production. Maine, for example has over 600 acres leased to aquafarmers engaged in mussel production. Fourth, mussels grow faster and yield more meat than other shellfish.

Most seedstock for mussel production is obtained from wild beds. The larval mussels may be seined from the water, or some structure may be placed in the water by the aquafarmer and then removed with the mussels that have attached themselves to it.

Mussels attach to the substratum by means of a thread-like *byssus* (cluster of filaments, which are chemically similar to silk), which they secrete. They will attach to almost any structure, including other mussels, other species of shellfish, and stones. Unlike oysters, however, mussels do not usually attach themselves to one place for life. They can discard the byssus and move about by crawling, or they can secrete a gas bubble and float to a new location. When mussels are being transplanted, it is important that their new environment be very much like their old one in terms of temperature, light, and salinity, or they may discard their byssus and go looking for a new home.

Although several operations have experimented with rope culture similar to that used with oysters, bottom culture has been the most profitable. The collected seed are spread over leased areas in the seabed, where they usually attach and remain.

Mussels require water with a salinity of at least 25 ppt. The locations that are the most suitable for mussel growth may not be the best for setting seed. Mussels grow best in locations where the tide changes dramatically, providing the mussels with a strong current of new food. The strong currents, however, may carry many of the larval mussels out to sea before they have time to set. If an area of wide tide changes is selected as a production site, the mussels should be set by means of rope culture or pole culture before being placed in the water.

If a good site is selected and the beds are periodically thinned, the mussels will usually grow to a market size of 2 – 3 inches each in one year. It may take two years, depending on the size of the seedstock and the climate. In the wild, it usually takes five years for mussels to reach market size. Very little growth occurs in winter months.

Red Abalone

The only gastropod culture of any significance in the U.S. is of the red abalone, *Haliotis rufescens*. The red abalone is the largest abalone in the world, reaching a weight of over 3 pounds. The abalone is considered a good aquaculture species because of the declining commercial catch and its low management requirements.

Red abalone, like other snails, are very slow-moving and easy to catch. As a result, they have been overfished to the point that California now has closed seasons and size limits on abalone catches. The commercial catch of abalone, once over 5 million pounds per year, is now below 1 million pounds per year.

Red abalone are found naturally along the western coast of the U.S. and Mexico, from southern Oregon to Central Baja, California. They prefer well-oxygenated water with a stable salinity and a temperature range between 45° and 65°F.

The largest abalone hatchery in the world is located in California, producing over 4 million abalone a year. This operation reports a profit margin of 50%, almost unheard of in any industry, and expects recent expansion to increase the profit margin within the next few years.

Red abalone will spawn throughout the year, but the peak spawning season is in the spring—April and May. When constantly provided with food, red abalone will spawn every 75 - 90 days. Also, red abalone will reach sexual maturity and be ready to spawn within three to four months after spawning.

The broodstock are usually harvested from the sea and transported to the hatchery. The male spawns first; the presence of sperm induces the female to shed her eggs. Because of their prolific spawning rate, red abalone are a good choice for hatcheries.

In the postlarval stages, red abalone are fed microalgae called *diatoms*. The diatoms are usually cultured by the aquafarmer in the tanks in which the larvae are placed. At first, the diatoms grow faster than the postlarvae can eat them, so the tanks are usually covered to inhibit growth of the diatoms. After about 50 days, the tanks are uncovered to encourage growth because of the increased grazing rate of the juveniles.

At six months of age, the abalone are switched to weaning tanks where kelp (seaweed) is provided along with the diatoms for forage. The kelp is the most expensive part of abalone culture, as it is mostly water and the food conversion rate is low (10 - 15%). In the weaning tanks, some structure is provided to give the abalone more surface on which to grow. This is usually accomplished by placing sections of PVC pipe in upright positions in the tanks. After they attach to the structure, the structure can be moved to another location.

After the abalone are one-year-old, they are moved to grow-out tanks and provided with a diet of giant kelp. They reach harvest size in four to five years. In high-density raceways with high water flow and filtered water, yields of 200,000 pounds of abalone per surface acre of water are possible. About 30% of the body weight is meat.

In all structures used to produce abalone, a constant flow of sea water is necessary. For the grow-out tanks, for example, a flow rate of around 1,000 gallons / minute is recommended.

Striped Mullet

The striped mullet, *Mugil cephalus*, is one of the few fish commonly cultured in brackish water. In nature, the striped mullet is found in tropical and semi-tropical waters around the world. The mullet is not considered a food fish in much of the U.S., except in Hawaii and some southeastern states, but it has potential as such. One reason may be that the wild striped mullet sometimes carries a fluke that is harmful to humans, but this should not be a problem in cultured fish.

The striped mullet can be induced to spawn in captivity, leading to an increase in the potential for aquaculture of this species. The rearing of mullet fry to fingerling size, however, has been a major stumbling block of attempted mullet operations. At present, the seedstock for mullet culture is usually obtained from commercial harvests. The mullet may grow as large as 5 pounds, but 1 – 2 pounds is usually the market size.

The striped mullet has several characteristics that make it a good fish for aquaculture production. It can withstand a wide range in salinity, from 0 ppt to 38 ppt, which means it can be cultured in almost any body of water. It also can withstand a wide range in water temperature, from 40° to 95°F. It performs well in polyculture with other fish. As a herbivore, the mullet does not require feeding; its nutrition needs can be met with a fertilization program. In addition, even though it is not popular as a food fish in most of the U.S., its flesh is considered quite good in most countries.

In other countries, the striped mullet is reared in estuarine waters, including brackish water ponds and sublittoral enclosures. The facilities are stocked either with fish obtained from commercial operations or with fry, which spend their first few months in brackish water areas, that have been caught.

The striped mullet has enormous aquaculture potential in the U.S., but little culture exists at this time. Some experimental culture has been conducted in the southeastern U.S., particularly Florida, usually with success, but commercial operations have not been interested to this point.

Milkfish

The milkfish, *Chanos chanos*, has many of the same characteristics of the striped mullet as they pertain to aquaculture production in brackish water ponds. It is herbivorous, grows rapidly, can withstand wide ranges in salinity and temperature, and is very disease-resistant.

The milkfish is a popular seafood throughout the tropical Pacific. Although most of the milkfish sold come from commercial fisheries, the supply is supplemented by aquaculture in many countries. Milkfish grow larger than mullet, sometimes reaching 50 pounds or more, but usually are marketed at 1 - 2 pounds.

The milkfish spawn at sea and the eggs are hatched there. The fry make their way to estuaries and brackish water, where they may spend from a few months to a year.

Because milkfish will not spawn in captivity, the facilities used to rear them are stocked from either commercial catches or from fry that have been captured as they move into water near shore. Milkfish production is usually referred to as ranching, because the milkfish are usually held in sublittoral enclosures, where they feed on microalgae until they are harvested. Very little management is necessary.

Milkfish fry can be reared in nurseries for stocking in ponds for grow-out. Once they are moved to grow-out facilities, either ponds or sublittoral enclosures, the primary management needed is fertilizing the facilities to promote algal growth. In other countries, such as the Philippines, the use of ponds is prevalent. In the U.S., however, very little milkfish production is carried out, and what is done is usually in sublittoral enclosures.

The potential for milkfish aquaculture is very good for Hawaii and parts of southern California. Although milkfish may grow well in waters of the southeastern U.S., these waters are not their natural habitat and few markets exist for milkfish in these states.

Lobsters

As lobsters are one of the most expensive seafood items, their culture has long been a goal of aquafarmers. Hatcheries for the *Homarus americanus* have been in production for over 100 years, but the farming of lobsters from the egg to the market-size adult has not yet been profitable. The hatcheries have been used in stocking programs to supplement the number of lobsters available from the commercial catch.

The primary drawback of lobster aquaculture is the long period necessary to grow a lobster to market size. Most lobsters require a minimum of five years to mature, with many individuals taking six to eight years. Although keeping lobsters in tanks with warm water can speed up the growth process, this is still not profitable for trying to raise lobsters from the larval stages to the market size. At

Figure 11–10. A typical Maine lobster boat. Most lobster hatcheries produce seed lobsters, which are released to supplement the wild stocks. (Courtesy, Robert C. Bayer, University of Maine)

present, the best chance for culture is to raise juvenile lobsters (two- to three-year-olds) to market size in two to three years.

One problem with intensive production is that lobsters are highly susceptible to crustacean shell disease, a form of exoskeleton necrosis. Crustacean shell disease can cause mortality if the lobster does not molt first. Molting usually sheds the disease with the exoskeleton, but the disease is highly contagious, causing problems in the intensive crowding found in aquaculture production.

Blue Crabs

Blue crabs (*Callinectes sapidus*) are primarily a fisheries product and not normally an aquacrop. Ample supplies of crabs to meet the market demands can be caught by commercial fishers, so aquaculture techniques are not needed for the general market.

One market that is being filled by aquaculture, however, is that for soft-shelled crabs, called soft crabs, those that have just recently shed their shell, or molted.

Figure 11–11. A blue crab, *Callinectes sapidus*. (Source: *Development of a Soft Crab Fishery*, Florida Sea Grant College, Report No. 31)

Historically, the crab fishers placed crabs that were about to molt into boxes or fenced pens in the estuaries until they molted. Recently, however, aquafarmers have used indoor recirculating systems for soft crab production.

The production of these soft crabs is very intensive. The water must be maintained at about 20 ppt salinity and constantly moved through the system.

Figure 11–12. A soft crab facility. Many of these facilities are small because of the intensive labor required to watch the crabs and to remove the soft-shelled crabs immediately after molting.

An upflow sand filter or a combination of an upflow sand filter and a fluidized bed filter is used to remove wastes and ammonia from the water.

Soft crab production in the U.S. has largely been concentrated in Louisiana and Florida. The potential for other southeastern and eastern states is good if the market expands and anticipated prices hold at a level that will allow the producer to make a profit.

Pompanos

Pompanos (*Trachinotus carolinus*) are saltwater fish commonly found along the Atlantic Coast from Massachusetts to Florida. Pompanos bring a premium price because they are not very abundant and they have an excellent flavor. Prices are typically $4 to $5 per pound, or up to $16 per pound for fillets.

Many experimental operations and a few commercial ventures have tried pompano production, with various degrees of success. The primary culture of pompanos in the U.S. has been in waters off the Gulf coast of Florida. Various facilities have included net pens, tanks, and sublittoral enclosures. Most operations have fed floating commercial feeds, sometimes supplemented with shrimp or frozen codfish.

The operations which have tried pompano production have developed a system of spawning the fish in captivity and rearing the fry. One company had success rearing the pompanos from eggs to a weight of about ½ pound, but its tanks were too shallow to allow for further growth. The potential for rearing the fish in net pens or larger tanks after they reach ¼ - ⅓ pound appears to be excellent.

Although the production of pompanos is expensive, the technology exists for commercial production. Several companies are now getting underway, some of which should be able to realize a profit in pompano production.

Red Drums

The culture of red drums or redfish, *Sciaenops ocellata*, is not widely practiced at present. The red drum is, however, a very popular fish in many Cajun-style restaurants and also a popular sport fish, which has led to overfishing of its primary habitat, the coastal waters of Louisiana, Mississippi, Alabama, and Florida. Due to the decline in numbers in the wild population and a substantial increase in the market, several experimental operations have tried to culture red drums, some with success.

Seedstock for these operations is generally obtained from commercial fisheries or from the incoming tide by letting it fill saltwater ponds or sublittoral

Figure 11–13. A red drum, *Sciaenops ocellata*. Note the black spot near the caudal fin, a feature of the red drum. (Courtesy, Gary Fornshell, Mississippi State University)

enclosures. Before aquaculture of this species can prosper, more reliable means of obtaining seedstock must be found.

As soon as reliable sources of seedstock can be developed, the potential for red drum aquaculture will increase substantially. Red drums appear to be relatively easy to raise in saltwater and brackish water in warm climates.

SUMMARY

Of all the types of aquaculture production presented in Part Three, saltwater and brackish water include the most varied species. Aquacrops that are cultured in saltwater and brackish water include several types of fish, molluscs, crustaceans, and gastropods. As a general rule, the species that are widely cultured became popular as food taken from wild catches. Often the early culture practices of these species were not aimed at food production but at replenishing wild stocks for commercial and recreational harvests.

Aquafarmers must efficiently produce high-quality saltwater and brackish water aquacrops to compete with the market from the wild catch. They must constantly refine their techniques and look for ways to increase production without raising costs.

The aquafarmer must know the habitat requirements of the aquacrop being produced and how to provide that habitat in a cost-efficient manner. Salinity and other water quality factors must be monitored and adjusted to maintain the optimum levels for production. Site location is very important and depends on several factors, including the supply of water, the supply of seedstock, a source of food for the aquacrop, and available markets for the finished product. The four major categories of water facilities for saltwater and brackish water aquaculture production are the shore, intertidal, sublittoral, and seabed.

Three aquacrops dominate saltwater and brackish water aquaculture around the world. Shrimp, salmon, and oysters make up over 80% of all saltwater and brackish water aquaculture.

Other saltwater and brackish water aquacrops include clams, mussels, red abalone, striped mullets, milkfish, lobsters, blue crabs, pompanos, and red drums. The culture of some of these species is still experimental, but most show potential for production. Some of them will only be cultured for improvement of wild populations.

In the U.S., saltwater and brackish water aquaculture has not progressed as rapidly as that of some freshwater species such as trout and catfish. Saltwater and brackish water aquaculture should increase in scope within the next several years. The culture practices for many species have been well established. The culture of these species is needed as a food source and for the replenishment of wild stocks.

QUESTIONS AND PROBLEMS FOR DISCUSSION

1. What is the biggest challenge facing saltwater and brackish water aquafarmers?
2. Define "salinity."
3. Describe the two most common methods of measuring water salinity.
4. If the water salinity is too high, how can it be lowered? If it is too low, how can it be raised?
5. Why must metal parts be protected from saltwater or brackish water?
6. How might government regulations affect saltwater or brackish water aquaculture?
7. What are the three most important types of saltwater and brackish water aquaculture?
8. List the secondary saltwater and brackish water aquaculture species.
9. What three factors determine if a site is suitable for saltwater or brackish water production?
10. List the primary advantages and disadvantages of shore facilities.
11. What is the intertidal zone? How is it used for aquaculture production?
12. Why is the sublittoral zone used so often for net pens and cage culture?
13. What precautions must be taken in the selection of a seabed site?
14. How is seedstock obtained for shrimp production?
15. Describe the process for hatching shrimp.

16. Briefly explain the management concerns for shrimp production.
17. Describe the spawning of salmon.
18. Briefly explain three techniques for improving the survival rate of artificially hatched salmon.
19. Differentiate between salmon farming and salmon ranching.
20. What are some management concerns of rearing salmon in net pens?
21. What diseases and predators cause problems in salmon production?
22. Where is seedstock for oyster production usually obtained?
23. What is cultch? Why is it important in oyster production?
24. How do oysters obtain nutrients from the sea water?
25. What are the primary management concerns with oyster production?
26. What predators cause problems for oyster farmers?
27. What are the two species of clams cultured in the U.S.?
28. Why are mussels considered to have great potential as an aquacrop?
29. Describe the artificial hatching of red abalone.
30. What are some characteristics of the striped mullet that make it a good choice for aquaculture?
31. What is the primary problem in lobster production?

Chapter 12

ORNAMENTAL, BAITFISH, AND RELATED AQUACULTURE PRODUCTION

Many people think of aquaculture as being only the production of aquatic crops for food. Though food production does comprise a large share of aquaculture, ornamental, baitfish, and related aquaculture are profitable enterprises for some aquafarmers.

Ornamentals and baitfish are a source of pleasure for many people. As the attractive aquarium shows us, ornamental species come in many colors, shapes, and sizes. They have varying environmental requirements, depending on their natural habitat. There are a large number of ornamental species, including a few of the same species used in food and bait production.

Any sport fisher will likely tell you how important baitfish were on a recent fishing trip, as well as how sport fishing brings pleasure to people's lives. Only a few species of baitfish are prominent.

Related aquaculture includes the production of laboratory species that are used for research and educational purposes. Though the related species are important in certain locations, the major focus of this chapter will be on the ornamental and baitfish species.

OBJECTIVES

This chapter emphasizes ornamental, baitfish, and related aquaculture production, including the species cultured and the considerations in production. Upon completion, the student will be able to:

- Explain the meaning of ornamental, baitfish, and related aquaculture.
- Describe the components of an ornamental aquaculture system.
- Explain the considerations involved in the selection of ornamental aquaculture species.
- Describe general considerations in ornamental aquaculture production.
- Identify common types of ornamentals and baitfish.
- Describe general considerations in baitfish production.

MEANING AND IMPORTANCE

Millions of dollars are spent each year in the U.S. on ornamental, baitfish, and related aquaculture species. In most cases, ornamental and baitfish species are for recreational purposes. They bring pleasure to those who have or use them. The related species are for research and educational purposes.

Figure 12–1. A large freshwater aquarium with several species of aquatic animals is shown here. (This one is used for instruction at Millsaps College, Jackson, Mississippi.)

Ornamental Aquaculture

Ornamental aquaculture includes a wide range of aquatic plants and animals that are used for non-food purposes. Most commonly, individuals think of ornamental aquaculture as the pet fish that are kept in an aquarium. However, ornamental aquaculture is much more than pet fish.

Ornamental aquaculture is concerned with aquatic species that are kept for their appearance and gracefulness. Their owners enjoy observing their behavior and watching them grow. They find caring for them and maintaining a good environment to be relaxing and an escape from the stress of routine everyday life. They often enjoy the challenge of maintaining the ecosystems in the aquaria or other containers where the ornamentals are being grown. Some individuals gain considerable benefit from the therapeutic (healthful) value of tending ornamentals.

As pets, aquatic plants and animals are more economical than other types of pets. They do not make noise, run away, or destroy property. Many species of ornamentals can be kept in a small amount of space and without a lot of training in how to maintain them. Renters do not meet with the same objections of building managers for small aquaria as they do with pet cats or dogs.

Ornamental aquaculture can consist of tropical or fancy fish; other times it includes pet or hobby aquatic animals and plants. There is much diversity in what is involved. Some species prefer freshwater; others prefer saltwater or brackish water. Providing the right environment is a challenge. The one common element is that all are maintained in a water environment—usually in an aquarium.

Ornamental aquaculture also adds to the decor of homes, offices, and businesses. An attractive aquarium stocked with fish, plants, and other aquatic species can serve as the focal point in many places. Schools use ornamentals for educational purposes in teaching many areas of science and ecology. Offices and businesses may employ custom care services to maintain their aquaria. (These services can provide good employment for people who have the skills and interests!)

Some ornamentals can survive in simple aquatic systems, such as the common fish bowl. Others require sophisticated water management systems in aquaria. Beginners should start with the species that are easy to grow; as individuals acquire the necessary knowledge, they can advance to species that require more skill. Mixing species can help maintain a quality ecosystem (community of organisms). (Mixing species is also known as polyculture.)

Ornamental plants and animals and the supplies needed to maintain them are sold through pet stores, discount stores, and other retail outlets. All of these provide career opportunities for people with the necessary skills. Some pet stores may have a third or more of their business in ornamental aquaculture. The value

Figure 12–2. Beginners can buy complete aquarium kits, such as the 10-gallon outfit shown here.

of ornamental species, equipment, and supplies runs into hundreds of millions of dollars each year in the U.S. Growers are needed to rear the ornamental aquatic species for the retail market. Florida is the leading state in ornamental production. California and several other states have some production.

Ornamentals may be cultured or captured in the wild and imported into the U.S. Most of the ornamentals imported into the U.S. are from, in order of decreasing value, Singapore, Thailand, the Philippines, Indonesia, and Hong Kong. The South American countries of Colombia and Peru provide some ornamentals for the U.S. Ornamentals are usually sold in small quantities but have a higher price mark-up than baitfish species. The value of the imported ornamentals was $44 million in 1990. Since they pass through several wholesalers and brokers, the retail value is much greater.

The U.S. also exports ornamentals to other countries. In 1990, the value of exports was nearly $12 million. The top countries receiving ornamentals from the U.S. were Canada, Mexico, Hong Kong, Japan, and Taiwan.

Baitfish Species

Baitfish species are produced so that they can be used in catching other species, primarily food fish. Sport and commercial fishers use baitfish. The farm value of baitfish exceeds $100 million a year in the U.S. Baitfish have been produced in the U.S. since the 1920s, with Arkansas, Kansas, Missouri, and Minnesota currently the leading states.

Wild baitfish species are captured much as wild edible species are captured for human food. Baitfish may be seined from a creek or a pond. Seining for wild baitfish often results in a lack of uniformity in species, size, and quality.

Some species of crustaceans, such as crawfish and shrimp, are also used as bait. These may be the by-products of harvesting for human food, or they may be seined especially for use as bait.

Baitfish producers must have a market for their crop. They may sell directly to bait-and-tackle stores or operate retail outlets on their farms or in conjunction with fee-lakes.

Baitfish have many of the same cultural requirements as food fish. Some producers grow both food fish and baitfish on their farms. A few culture both ornamentals and baitfish, particularly goldfish. Baitfish are priced lower than ornamentals and are often produced in larger volumes.

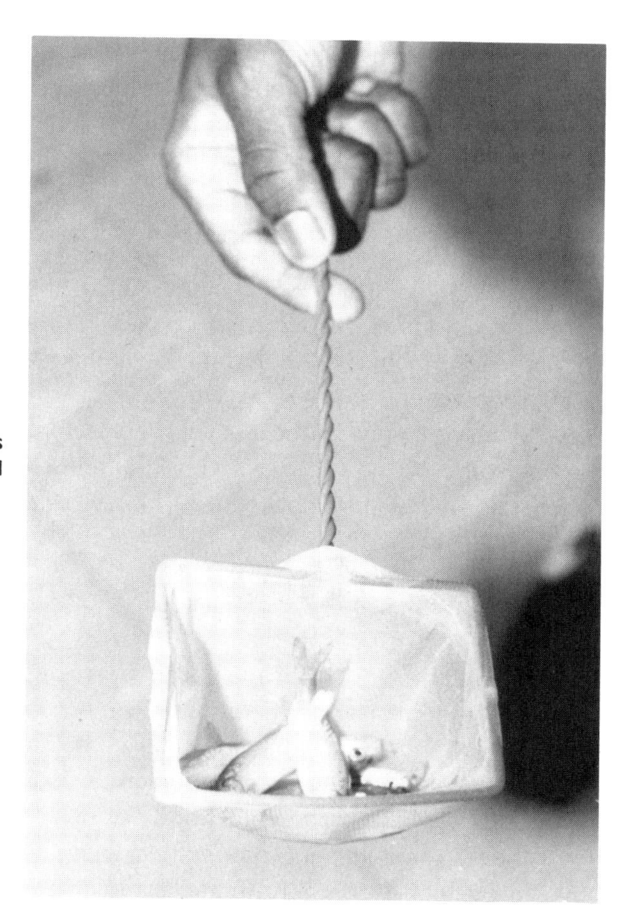

Figure 12–3. Shiner minnows ready for use in catching bass and other sport fish.

Related Species

Many educational and research programs use aquaculture species for study. Biology students in high school and college frequently use bullfrogs in their laboratory study of anatomy. Various species of fish are used in teaching morphology and anatomy.

Research programs frequently use fish to study how various environmental factors influence growth. Not only does this research focus on improving efficiency in aquaculture, but it also seeks to find answers for the betterment of human life.

One example is the Japanese medaka (*Oryzias latipes*)—a species of fish that is 1 - 1½ inches long as adults. This species is hardy in a wide range of environments and will spawn any time of the year. The females are very prolific and will produce 3,000 or more eggs in a single breeding season. The eggs are excellent for studying embryo development. The small fish are excellent for physiology and genetic research. Millions of these fish can be economically produced for research and educational purposes.

The laboratories using species in research often have facilities to grow the supply needed. A small number of specialty biological supply houses meet most of the needs of educational programs.

ORNAMENTAL CULTURE SYSTEMS

A large number of species (perhaps 100 or more) may be used for ornamental purposes. Each species has its own unique requirements for culture and growth. However, some species can share the same environment. Several general considerations apply in the culture of ornamentals. The following section addresses some of these general considerations from the perspective of the fish culturist as well as the hobbyist.

System Needs

The system needed for maintaining ornamental species is roughly the same for both the hobbyist and the commercial ornamental farmer. A system that will operate successfully must have a water container, a satisfactory source of water, and water quality maintenance equipment. The quantity and size of these depends on the scope of the ornamental hobby or business. The hobbyist may have only a few pieces of equipment to operate one aquarium; the ornamental farmer may have a large installation of water tanks and other facilities. Some growers of fancy, more expensive fish have been known to establish viable aquabusinesses in garages at their homes.

Water Containers

Various water containers and attachments are used with ornamentals. Water containers may be aquaria, tanks, ponds, or other facilities. The hobbyist will likely have an aquarium with a capacity of 15 – 40 gallons. Aquaria are typically rectangular glass containers that allow for ease in viewing the ornamental species and for cleaning the water environment. Common dimensions are 36" × 18" × 15" (capacity is 42 gallons if filled entirely with water) and 24" × 15" × 12" (capacity is 18 gallons if filled entirely with water). The commercial ornamental grower may have a number of larger aquaria. Sometimes round glass bowls, bottles, or other items are used. Plastic materials are sometimes substituted for glass. Sizes and shapes should be convenient to use and not so large that reaching all areas of the containers is difficult.

Construction materials and design are important. All glass materials are preferable; however, metal or plastic frames with the glass cemented in place are sometimes used. All-glass aquaria are held together with a durable silicone rubber cement. Scratched glass should be rejected, as this indicates a flaw that could later result in a break and water leak. Glass thickness should be at least ¼ inch for small aquaria except for the bottom, which should be ⅜ inch thick. Any new aquarium or one that has been unused for a while should be tested for water leaks. It can be partially filled in a location where leaked water will not be a problem and where it will be easy to empty.

Water containers must be positioned on strong stands or other supports. The weight of water adds up quickly. For example, the filled 18-gallon aquarium described previously weighs about 155 pounds. (The weight of an aquarium in pounds can be calculated by multiplying 0.036 [the weight of a cubic inch of water in pounds] times the number of cubic inches in the aquarium. The number of cubic inches is found by multiplying length × width × height. For example, an aquarium that is 24" × 15" × 12" contains 4,320 cubic inches. The weight is 0.036 × 4,320, or 155 pounds. And this is a small aquarium!) An aquarium should have a styrofoam or plastic cushion between the bottom and the surface on which it sits. It may also be possible to use a wooden frame that would not require a cushion.

Water containers may also have covers, hoods, and lighting attached to them. The covers are to keep fish and other aquatic animals from jumping out, to keep water from splashing out, and to prevent unwanted items from being dropped into the aquaria. Covers may be made of glass, metal, or other material. Metal

Figure 12–4. Establishing a small goldfish bowl involves using "aged" water, water container, decorative crushed stone, artificial plant material for decoration, food, and fish. (Note that the goldfish are being held in a plastic bag with water and air until the bowl is ready.)

Figure 12–5. A large aquarium operation used for research purposes.

covers should be of a material that will not rust or corrode and is non-toxic to the aquatic species. Cover glasses are often of thinner glass than the walls and bottom of the aquarium.

Fluorescent lighting is the standard in aquaria because it is cooler and more economical than regular light bulbs. It consists of tubes that enhance the growth of aquatic plants and organisms. Individual aquaria may be fitted with lights, or in the case of several aquaria in the same room, the room may be lighted appropriately. It is generally recommended that aquaria lights provide 10 watts per cubic foot of aquarium. (To determine the watts needed, calculate the number of cubic inches in the aquarium and divide by 1,728, which is the number of cubic inches in a cubic foot. This gives the number of cubic feet. Multiply the

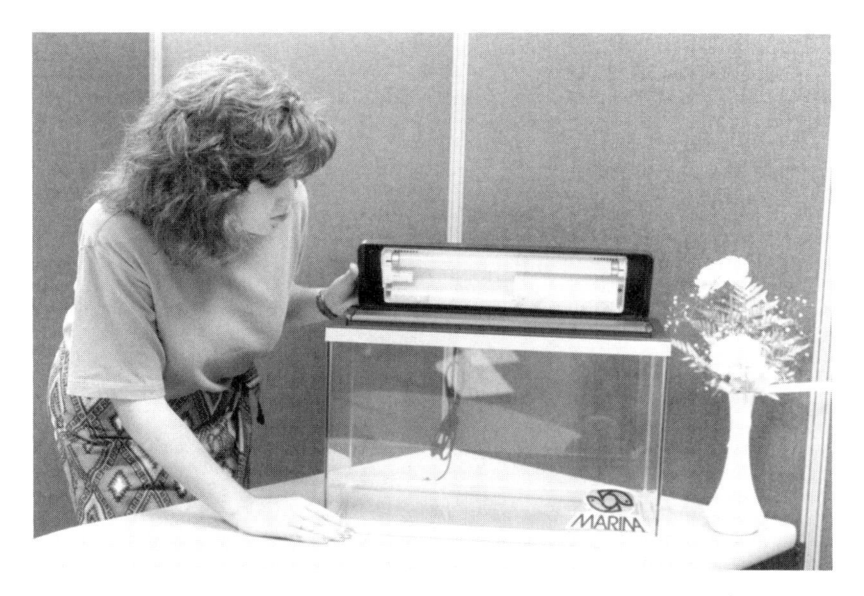

Figure 12–6. Installing a fluorescent light on the hood of an aquarium.

number of cubic feet times 10 to determine the wattage needed.) Tubes should be positioned lengthwise in the tank and about 1 inch above the water level. Commercially available hoods often have built-in lighting. (Note that some species may grow best in aquaria with low levels of light.)

Water Source

Several factors are important in assessing water needs and sources. The first consideration is the kind of ornamentals that are to be produced. They can be placed in categories on the basis of water salinity and water temperature. The ornamental producer should study the availability of water as related to the requirements of the species under consideration. A "match" of species and available water is essential in order to be successful with ornamentals.

As noted previously, salinity refers to the salt content of the water. Aquaria can be established for species that are adapted to saltwater, brackish water, or freshwater. Some species have a low tolerance of water that is different from their natural habitat; others have a high tolerance of variations in water. Other chemical properties of water may also be important in its suitability for ornamental production. These include the presence of calcium, iron, and other minerals. (See Chapter 6, "The Importance of Water.")

The species selected must be adapted to the water that is to be provided for the ornamentals. In some cases, small quantities of water can be modified to suit the species to be grown. This is an expensive process for a large-scale aquaria operation. Most aquaria are used to grow freshwater species.

Water temperature requirements must be met for the species. Some need cold water, while others must have warm water. Heaters and coolers can be used to regulate the water temperature needed by a particular species.

Many aquaria producers use tap or well water that has been prepared for use in fish culture. Municipal water systems often treat the water with chlorine or other chemicals, and this makes the water unfit for use directly in aquaria. Collecting an open container of water and allowing it to "age" (stand for a while) will result in the release of chlorine and some other additives. Some growers keep

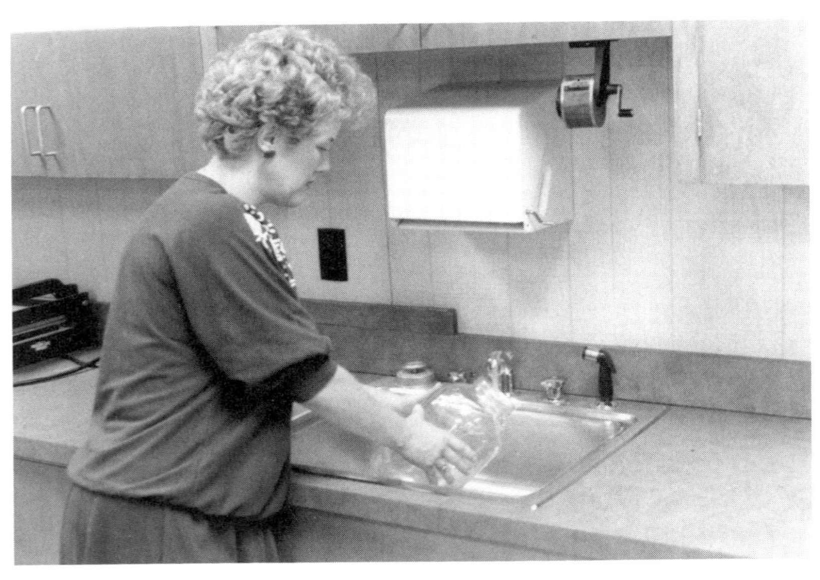

Figure 12–7. Tap water should be collected and "aged" before use in an aquarium. (Water should be caught in a jug or other container and allowed to stand for at least 24 hours in an open container. Chemicals can be bought to add to water to help prepare it for fish.)

a supply of "aged" water on hand. Water from lakes and streams, rain water, and water stored in caustic containers should not be used in aquaria.

Growers who produce ornamentals in ponds or raceways often use well water much as food fish growers do. A good example is the production of goldfish. A prolific fish that is used for both ornamental and bait purposes, the goldfish is more tolerant of a wide range of environments than most ornamentals.

Growers of ornamental species must also consider how to dispose of water. Large volumes of water usually cannot be dumped into sewage systems, creeks, or lakes. Some kind of holding and treatment facilities may be needed. The hobbyist with one aquarium produces very little used water, and this does not usually pose a problem.

Water Quality Maintenance

Water quality is maintained in aquaria by both mechanical and biological means. Several areas to be considered are aeration, temperature, and filtration.

Aeration involves keeping an adequate amount of dissolved oxygen in the water. In aquaria, this is often with devices used to inject air or pure oxygen into the water. Flexible plastic tubing carries air from an air pump to a porous material in the bottom of the aquaria known as air stone. Air bubbles from the air stone into the water and rises to the top of the water. Aquatic plants can produce some oxygen through photosynthesis. In most cases however, the amount they produce is inadequate for a heavily stocked aquarium. Without aeration, oxygen can enter the water only from the surface. This occurs at a very slow rate in still water. Splashing increases the diffusion of oxygen into water.

Ornamental species usually have definite temperature requirements. Water temperature needs to be kept constant throughout the day. Ornamentals should be protected from rapid warming and cooling. Drops of more than 2°F. or increases of more than 5°F. in one day should be avoided. Aquaria located inside most buildings need to have a water temperature higher than the temperature of the air surrounding the aquaria. Heaters with thermostats (sometimes known as thermostated heaters) suspended in the water should control the temperature within a range of 1.5°F. of their thermostat settings. Most heaters are powered by electricity. Heaters have a wattage rating. The wattage needed depends on the size of the aquarium and the amount of heat needed to keep the water within the desired range. As a rule of thumb, a 75-watt heater is adequate for an aquarium holding 18 gallons of water. A 100- to 150-watt heater is adequate for a 40-gallon aquarium. Good electrical wiring should be available so that extension cords and other dangerous makeshift wiring will not need to be used.

Adding some cold water, adding ice for a drastic change, or using a refrig-

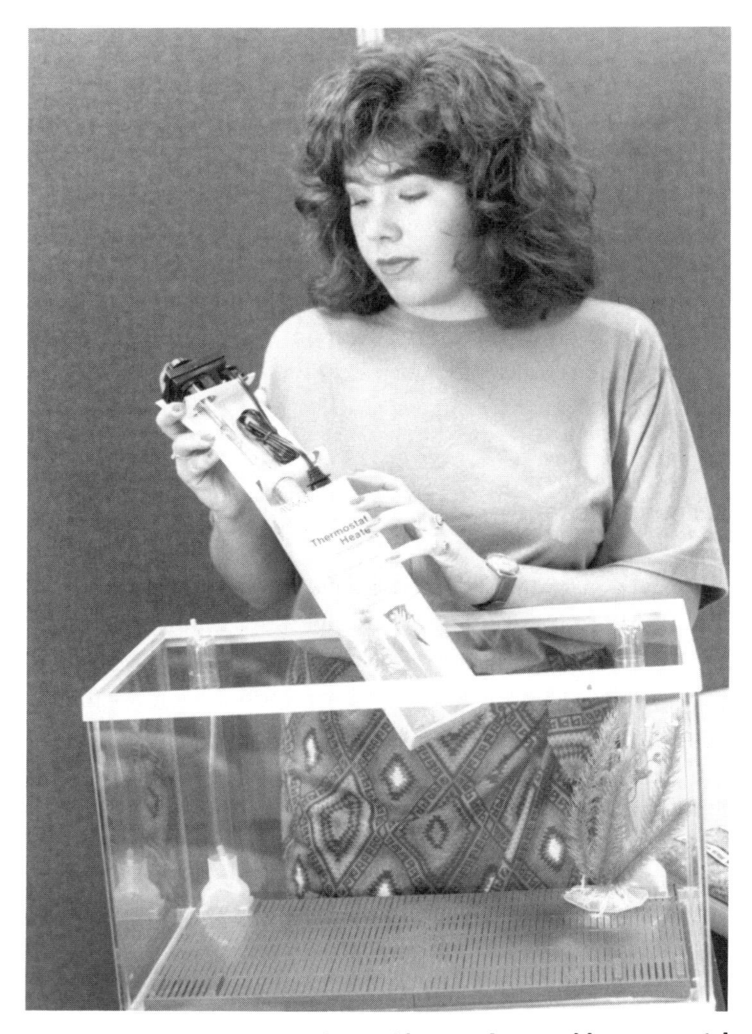

Figure 12–8. Thermostat heater kits are often used by ornamental enthusiasts. Here a heater is being prepared for installation on the side of the aquarium.

eration process will temporarily cool water that is too warm. Most of the time, water is more likely to need warming than cooling.

Water filtration, both mechanical and biological, is used to remove excess feed, feces, gases, and other substances from the water. Mechanical filtration involves passing water through some sort of material that removes solids that are suspended in the water. Biological filtration involves using bacteria to convert harmful substances to less harmful forms. Filtration kits that fit inside or attach to the outside of the aquaria are often used by hobbyists. Large producers may have filtration systems outside the aquaria to remove undesirable materials in the water.

Filtration involves pumping or flowing the water through a material that removes the harmful debris. The filter may use nylon threads, plastic mats, charcoal, or gravel. A hobbyist may use a gravel-covered bottom in the aquarium to filter the water. Charcoal is an excellent filtration material because it can absorb a large amount of unwanted rubbish. Glass wool filtration material should never be used because of the possible danger to the species and to the person who handles it. Most filtration systems use a combination of materials, since different materials filter different substances from the water.

Figure 12–9. An under-gravel filter is being prepared for placing in the bottom of an aquarium. Small crushed gravel will be placed over the entire filter before the aquarium is filled with water.

In recent years, the use of a gravel filtration system has grown in popularity with hobbyists. This usually involves covering a commercially available kit with 2 – 3 inches of fine gravel in the bottom of the aquarium. One of the nice features of this type of filtration is that it makes use of biological filtration. This means that there are many bacteria present in the gravel that consume some of the harmful materials and convert them to other forms. For example, bacteria convert ammonia to less toxic forms of nitrogen. (This was discussed in Chapter 8, "Aquaculture Facilities.")

Some aquaria hobbyists stock snails, crawfish, aquatic plants, and other species to help maintain water quality. For example, snails feed on decaying organic matter and crawfish scavenge the bottoms of aquaria for leftover food. A variety of aquatic plants may be used to remove waste materials left by the fish or other species.

SELECTING ORNAMENTAL SPECIES

Both plant and animal species may be grown in aquaria. These may be suitable to freshwater, saltwater, or brackish water. It is important to know the required environment for a species before trying to grow it in an aquarium. *The proper environment must be provided.* Some species are very difficult to grow and should be attempted only by experienced growers. In the selection of a species or combinations of species, the practical aspects of available water, species requirements, and knowledge of the grower about the species and aquaculture must be considered.

There are hundreds of ornamental species that can be cultured by the hobbyist or the commercial grower. Only general considerations with a few specific examples are included here.

Ornamental Plants

Plants are grown in aquaria by hobbyists to make the aquaria more attractive and to help in maintaining water quality. Growers of plants keep pet stores and other retail outlets stocked with various plants for hobbyists. The volume of production for the retail market is small. A few plants will quickly grow and provide all that are needed in an aquarium. Commercial ornamental fish growers may view plants as pests in the management of their production systems. For example, plants interfere with seining water to capture ornamental fish that may be grown in ponds or large tanks.

Aquaria plants may be classified as rooted, bunched, or floating. Aquarium supply departments in pet stores will have books that give considerable detail on some of the plants listed here.

Note: Some of these plants are considered pests in streams and lakes and should not be placed in them. Regulations may make the stocking of these plants a violation of the law, with fines and other penalties for doing so.

Rooted plants. Rooted plants grow best in aquaria with gravel on the bottom. They are attractive plants but can cause problems as their roots may grow into the filtration system that is in the gravel. When an aquarium is set up, the roots of the plants should be covered in the gravel and, if possible, anchored in some way to the bottom. Some examples of rooted plants are:

1. *Hygrophila*—This plant (*Hygrophila polysperma*) tends to grow well, but requires good light. It is planted in small bunches. The narrow, 2-inch long

leaves grow on stems that may be several inches long. Propagation is by cuttings or leaves. It typically grows in freshwater.

2. *Banana plant*—This plant (*Nymphoides aquatica*) produces banana-shaped tuberous roots that lie on top of the gravel in the bottom of an aquarium. The leaves may be 2 inches across and grow on stems that are 3 or 4 inches long. The plant grows best in freshwater with a temperature of 70° - 82°F. Plant-eating fish often prefer it.

Figure 12–10. The banana plant gets its common name from the banana-shaped tuberous roots.

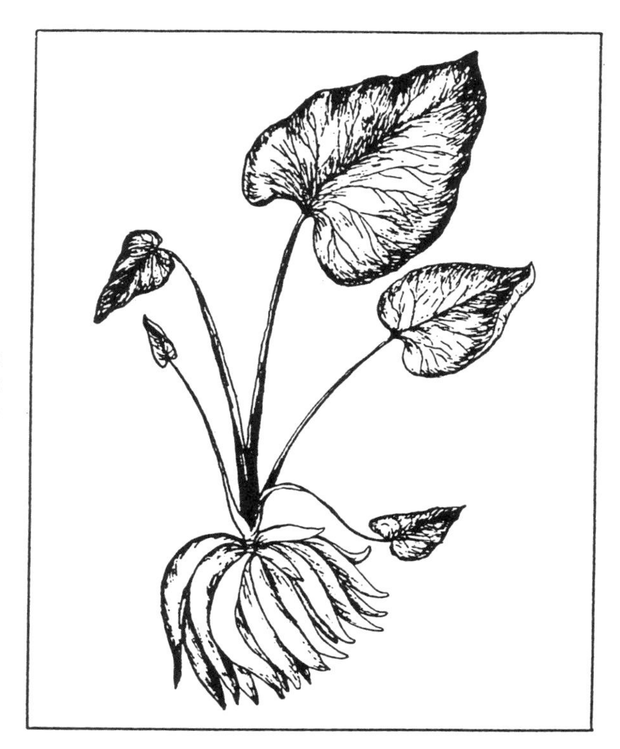

3. *Amazon sword*—The Amazon sword plant (*Echinadorus paniculatus*) is suited to larger aquaria because it may reach a height of 2 feet. A smaller variety, the pigmy Amazon sword (*Echinadorus tenellus*), may be used in smaller aquaria. The pigmy grows about 4 inches high. New plants are developed from runners and should be cut off and removed to keep from over-crowding the aquarium. Both are nice plants. Inadequate light may result in discolored leaves.

Bunched plants. Bunched plants are probably more popular in aquaria than rooted or floating plants. They are propagated from cuttings that may or may not produce roots. They typically grow quickly and need cutting back to keep them from taking over the aquarium. Some examples of bunched plants are:

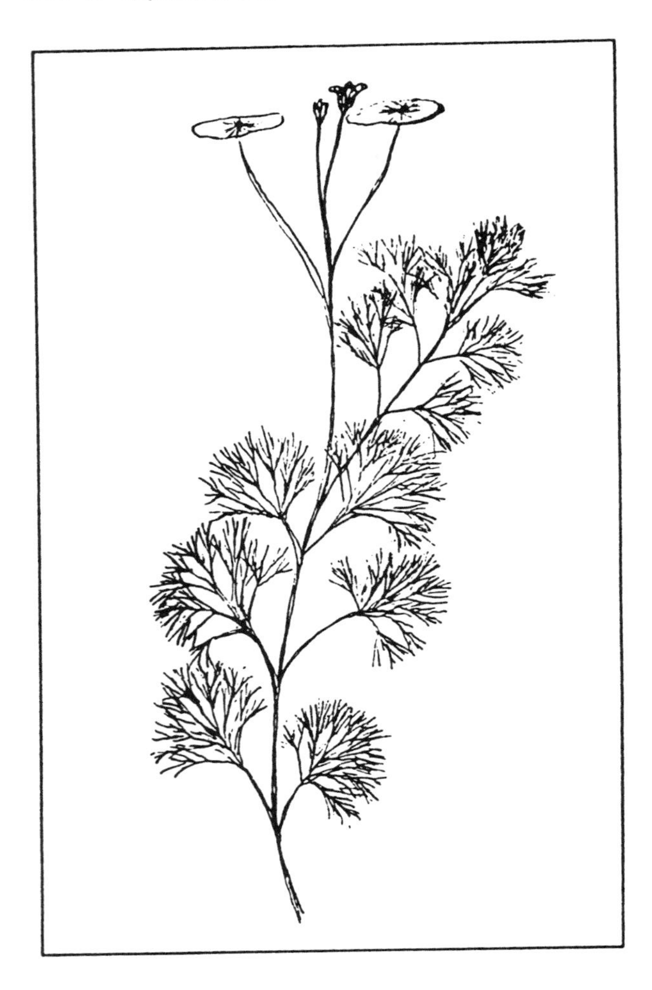

Figure 12–11. The fanwort plant has fine underwater leaves and may have flowers and floating leaves.

1. *Fanwort*—This plant (*Cabomba caroliniana*) grows fine leaves in a fan-shaped arrangement on a stem that may reach several inches in length. Sensitive to water temperature, the fanwort can be killed by sudden changes in water temperature. It prefers water in the 72°F. range. Light and aeration are necessary. Some species of ornamental fish eat the foliage, with goldfish being an example.

2. *Elodea*—Elodea (*Elodea densa*) is a hardy plant that thrives in water 72° - 85°F. Some species of fish nibble on the foliage. Since it grows rapidly and quite long (up to 10 feet), it should be located at the back of an aquarium and pruned often.

3. *Milfoil*—Milfoil (*Myriophyllum spicatum*) is also known as foxtail. Several species of milfoil may be used in aquaria. They are similar to hornwort and fanwort. The plants will have a deep, bright green color if they receive

eight hours of light each day. Many small leaves grow along the stems and form attractive aquarium plants. Milfoil often does best in cooler water.

Floating plants. Floating plants are considered the least desirable plants in an aquarium with other species that require considerable light. They block the light from passing into the water and to the bottom of the aquarium. Floating plants are best in an aquarium where lighting is low. Examples of floating plants include:

1. *Duckweed*—This plant (*Lemna minor*) is popular in shading aquaria from strong light. Larger species of fish will eat duckweed. The tiny, ½-inch round leaves can quickly cover the surface of the water in an aquarium. Duckweed is considered a pest in pond aquaculture. *It should never be released into ponds, lakes, or streams!*
2. *Crystalwort*—This plant (*Riccia fluitans*) forms a mass of short, narrow leaves. It must be kept thinned to no more than 1 inch thick at the surface of the water of an aquarium. Some species of fish spawn in the mass that is formed.

Ornamental Fish and Other Animals

Many species of fish are available for aquaria. Decisions about which to grow depend on the requirements of the fish and the available water and other resources. Also, the knowledge and skill of the grower are very important. The novice should begin with species that are hardy and require only minimal skill to keep them growing. As experience is gained, more difficult species may be grown.

There are many ways of classifying ornamental fish. A few examples are:

1. *Water temperature*—Fish can be classified on the basis of water temperature. Most species have a temperature range in which they grow best. The ornamentals that are known as "tropicals" require warmer water than other species. Tropicals typically prefer water in the temperature range of 72° - 80°F.
2. *Water salinity*—Water salinity refers to the salt content of the water required for growth. Saltwater (as in the ocean) is 35 ppt salt; whereas, freshwater contains virtually no salt. Brackish water has a salt level between freshwater and saltwater. Large exhibits in museums may have separate aquaria showing aquatic species in all three water environments. Many of the favored aquaria species are best adapted to freshwater.

Figure 12–12. Beginners should carefully study the establishment of an aquarium and the requirements of the species to be grown.

3. *Reproduction*—Some ornamental fish are "livebearers," meaning that the young are born ready to swim and take food. Others are egg-layers, meaning that the female expels eggs, which are fertilized by sperm from the male. A period of incubation is required for the eggs to hatch. Many of the favored aquaria species are livebearers.

4. *Size*—Ornamental fish range from less than an inch long to a foot or more. Of course, as length increases, weight also increases. Small aquaria are best suited to the smaller species of fish. In addition, smaller fish usually eat less food and require less water and oxygen.

5. *Feeding habits*—Fish species vary in their diet requirements. Some are carnivorous; others herbivorous. The grower needs to know something about the diet requirements of fish in order to be able to provide for their needs.

6. *Behavior*—Some species of fish are very aggressive toward other species and other individuals within their species. Aggressive species must often be grown in isolation to prevent them from destroying the other fish. Many ornamental producers prefer the species that are less aggressive.

7. *Physical appearance*—"Fancy" is often used to describe ornamentals. This means that they have characteristics that make them particularly appeal-

ing to people. Variations include color, body shape, fin shape and size, and presence of barbels (feelers about the mouth, as on catfish). Personal preference has a lot to do with the selection of species on the basis of physical appearance.

8. *Scientific classification* — Various schemes of scientific classification have been used over the years. These are intended to show relationships among the species on the basis of anatomy (structure of the species) and morphology (form and structure). The current approaches to scientific classification were presented in Chapter 5, "Fundamentals of Aquaculture Biology."

Easy-to-Grow Ornamentals

Some ornamental fish are hardy and will survive in a wide range of conditions. They do not stress as easily when water temperature varies. They are not as sensitive to fouled water that is low in oxygen. A few examples are presented here.

Goldfish

The goldfish (*Carassius auratus*) is a popular, hardy species of ornamental fish, which was originally developed by the Chinese and Japanese. New varieties are continually being developed. The new varieties feature fancy fins, colors, and eye

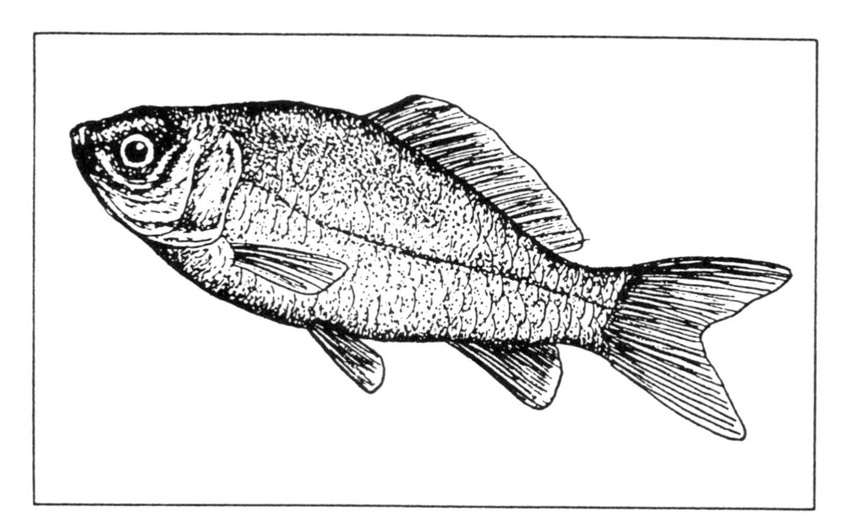

Figure 12–13. Goldfish — Though there are many varieties of goldfish, the comet is probably the most popular.

sizes. Some of the fancy varieties are the bubble-eye, black moor, comet, lionhead, veiltail, and fantail. Of course, the most popular is the common goldfish. Goldfish will grow in water that is a little cooler than that preferred by the tropicals. They will survive in backyard ponds, aquaria, and pint-size bowls or jars. The size they will reach depends on the food provided and the amount of water area available. At least 2 gallons of water is needed for a goldfish to reach a length of 2 inches and survive. Some goldfish will live as long as 15 years. Heaters are not often used in aquaria with goldfish. A variety of commercial foods recommended for goldfish should be used. In outside pools, goldfish eat insects, algae, and water plants; however, most require supplemental food. Goldfish that lose their color need some vegetation to nibble on and more light — direct sunlight may be used to try to restore color. However, all goldfish tend to lose some color as they age. Goldfish are considered to be messy fish and require water exchange in bowls and aquaria. Goldfish reproduce by laying eggs, with the eggs hatching two to four days after spawning and fertilization. The size attained by goldfish is determined by the amount of water, quality of the water, and nutrition. In ponds, they may reach a length of 2 feet! Most are marketed by the commercial grower at a length of 2 inches or less.

Barbs

The barbs include several species that are in the same family as the goldfish (*Cyprinidae*). The name "barbs" is derived from barbels about the fish's mouth;

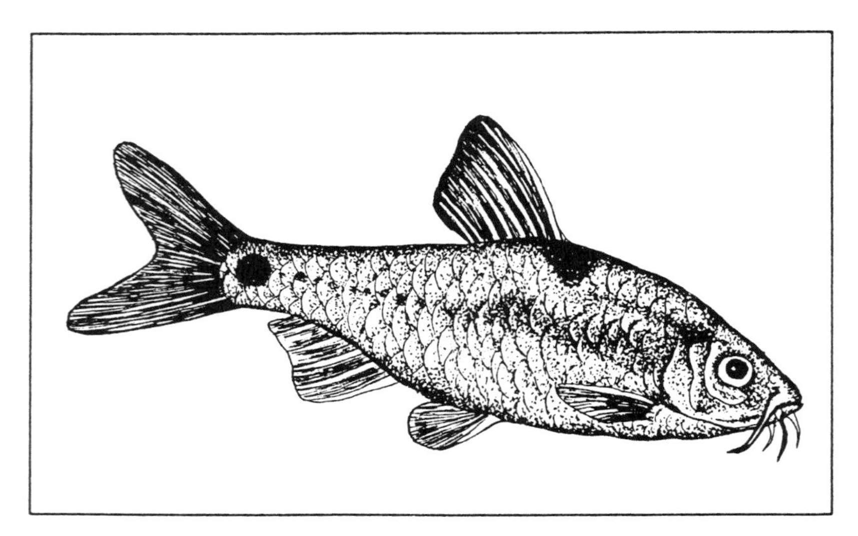

Figure 12–14. Spotted barb — This is a popular ornamental fish for beginning hobbyists.

however, some barbs have no barbels, while others have two barbels, and still others have four barbels. They prefer well-lighted aquaria. The larger barbs may be aggressive toward the smaller barbs. Barbs prefer water that is 70° - 80°F. Most of the barbs reach a maximum size of 2 - 4 inches in length. They are egg-layers with behavior similar to goldfish. Some of the species will eat their eggs after spawning. The most common species are:

1. Spotted barb (*Barbus binotatus*) — a fish that reaches a maximum length of 4½ inches and prefers water that is 75° - 78°F.
2. Rosy barb (*Barbus conchonius*) — a rosy-colored fish that reaches a maximum length of 3 inches and prefers cool water of 64° - 74°F.
3. Tinfoil barb (*Barbus schwanenfeldi*) — the largest (reaching 9 inches in length) and most active barb; it is the color of aluminum foil, except for the fins which may be orange and black.

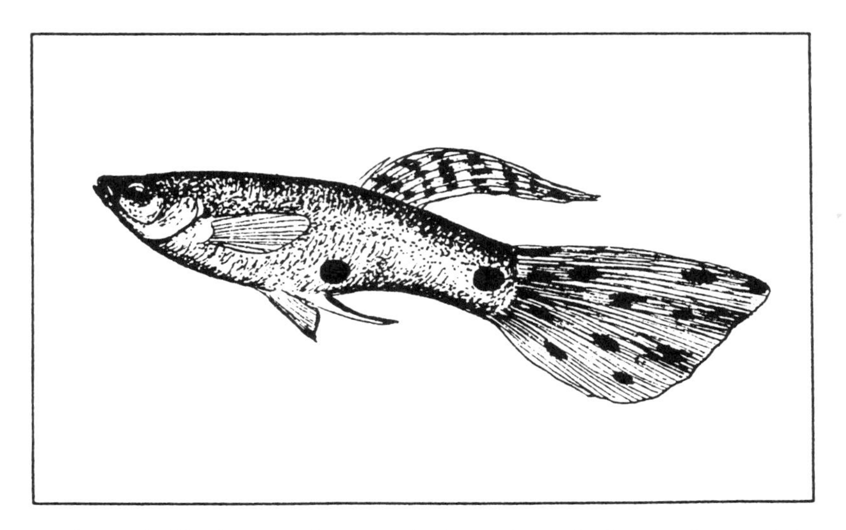

Figure 12–15. Guppy — The guppy is the most popular live-bearing ornamental fish.

Guppies

Guppies (*Poecilia reticulata*) are the most popular tropical fish. They are livebearers that reproduce profusely. A female guppy 2½ inches long may give birth to as many as 200 baby fish at a time; the average number is 40 - 50. One adult pair of guppies can fill an aquarium with young guppies in a few weeks. There are a number of varieties of guppies, with some growers developing guppies that have fancy colors and fins. Varieties include red deltatail, blue veiltail, gold flamingo, and halfback veiltail. A guppy may live about two years under good conditions. Guppies should be fed a basic diet of commercial tropical fish food. They will also eat mosquito larvae and other natural foods.

Gouramis

Gouramis are a large group of easy-to-grow ornamentals. They are known as labyrinth fish because of an air storage chamber above the gills. This chamber is called a labyrinth and is filled with air by the fish gulping at the surface of the water. The oxygen in this chamber supplements that absorbed by the gills. The air temperature should be similar to that of the water because much colder air can contribute to the development of respiratory disease in the fish. Gouramis are well-suited to aquaria and may live for five years or more. Most prefer water in the temperature range of 75°F. and need a complete commercial pet fish food. Popular ornamental gouramis include:

1. Kissing gouramis (*Helostoma temmincki*) — known for their trait of extending their thick, fleshy lips and kissing; kissing gouramis require aquaria of at least 20 gallons and should be fed twice daily, with additional vegetable matter included with one of the feedings. They may reach a length of 12 inches.
2. Blue gouramis (*Trichogaster trichopterus*) — very hardy and easy-to-raise

Figure 12–16. Kissing gouramis — Though there are many species of gouramis, the kissing gourami is probably the most interesting.

ornamentals; blue gouramis reproduce with the female laying eggs in a foamy bubble nest constructed by the male who also fertilizes the eggs. They may reach a length of 6 inches.

3. Pearl gouramis (*Trichogaster leeri*) — as pleasing pets, pearl gouramis are quiet fish that seldom attack others and do best in medium-size aquaria. They may reach a length of 4 inches.

Medium-Care Ornamentals

The fish described here require medium care in order to be successfully grown. Varieties within species may vary in the amount of care required. Water conditions, nutrition, reproduction, and temperature need careful attention. A few examples of medium-care ornamentals are presented here.

Tetras

This group of tropical fish includes several species in the *Characidae* family. The tetras comprise one of the largest families of freshwater aquaria fish. They reproduce by scattering eggs throughout the aquarium and, in general, do not care for either their eggs or their young. They tend to swim in small schools (also known as shoals). The tetras have bright colors, a peaceful temperament, and are hardy and easy to handle. They do best in aquaria with soft, slightly acid water, under low light conditions. Most prefer a water temperature of 72° - 85°F. They range from 1½ to 3 inches in length. Tetras tend to grow better if several are kept in the same aquarium so they can swim in schools of six or eight. Several species of tetras are:

1. Neon tetra (*Hyphessobrycon innesi*) — reaches a maximum length of 1½ inches and prefers dimly lit areas in the aquarium.
2. Cardinal tetra (*Cheirodon axelrodi*) — has a brilliant reddish color and is less common than the neon, with a maximum length of 1½ inches.
3. Black tetra (*Gymnocorymbus ternetzi*) — is peaceful when young but becomes more aggressive with age; its black color fades to gray as it ages.
4. Golden tetra (*Hemigrammus armstrongi*) — has a bright, golden shine and grows better in groups of four to eight fish.

Catfish

This is a large group of fish often characterized as being scavengers and

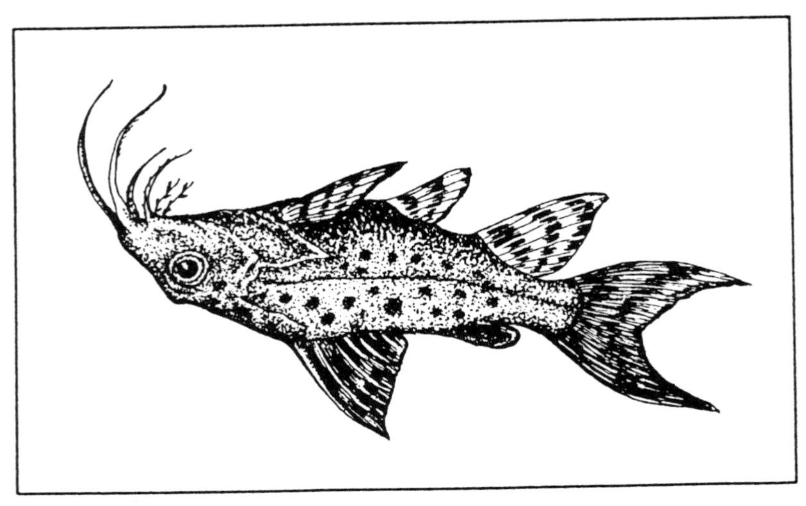

Figure 12-17. Upside-down catfish—An interesting aquarium species, the upside-down catfish has coloration reversed from other catfish. With the upside-down catfish, the back is white and the top (belly side) is dark. With other catfish, the topside (back) is dark and the underside (belly) is white.

having barbels about their mouths. Their designation as scavengers among the ornamental species bears no relationship to the cultured farm-raised channel catfish and other species grown for human consumption that receive specific rations. Most of the ornamental catfish have unique features that justify their designation as ornamentals. Most ornamental catfish prefer water in the 70° - 80°F. range. They reproduce by laying eggs. Examples include:

1. Glass catfish (*Kryptoptereus bicirrhis*)—may grow up to 4 inches long and have a clear-like body that may show rainbow colors in the right light.
2. Upside-down catfish (*Synodontis angelicus*)—has the unique characteristic of swimming inverted and prefers a dimly lit aquarium.
3. Electric catfish (*Malapterurus electricus*)—has the ability to produce an electric shock that will kill small fish and keep larger ones away—must be isolated in an aquarium.

Difficult-to-Grow Ornamentals

The ornamentals that are more difficult to grow are less common, require a higher level of knowledge and skill to culture successfully, and may sell for a higher price. Some of these fish are capable of inflicting injury when they are being tended. A few examples are presented here.

Piranhas

Piranhas (*Rooseveltiella nattereri*) are best known as fish with razor-sharp teeth that will take bites out of other fish and aquatic animals as well as the humans who tend them. Piranhas prefer water that is about 78°F. Their diet should include chunks of beef heart, kidney, or liver; raw fish; earthworms; and goldfish. They may reach a length of 10 - 12 inches. Piranhas should be isolated from other fish, including other piranhas, as they will attack, kill, and consume the fish. Not a lot is known about the culture of piranhas. Their importation and possession may be illegal in some states.

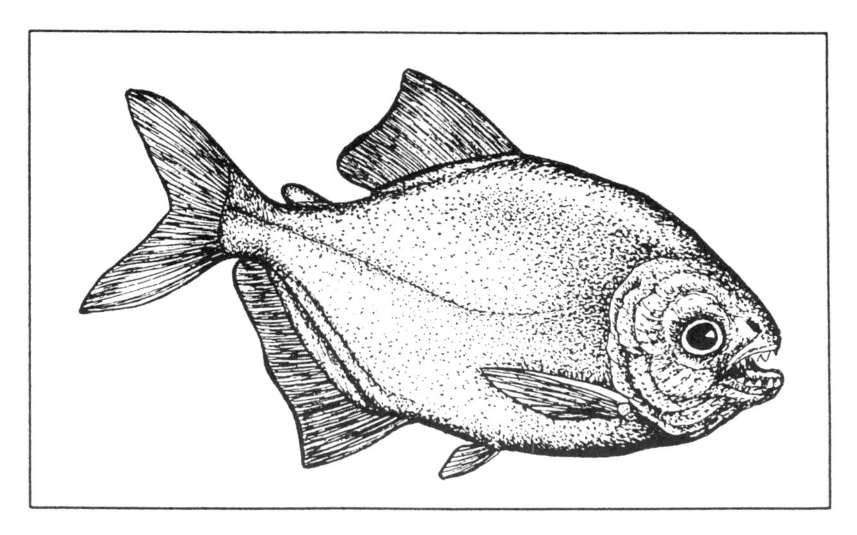

Figure 12–18. Piranha — The piranha is a rare, interesting, and dangerous aquarium species.

Hatchetfish

This group is comprised of several species of fish in the Gasteropelecidae family. Hatchetfish are peaceful fish but often do not live long in captivity. The aquaria must be kept covered as they will tend to "fly" out by using their winglike pectoral fins. Hatchetfish need long aquaria with considerable surface water. The aquaria should have a protected or shaded area, as the fish are easily frightened and may injure themselves by crashing into the glass. They do best in groups of three to six fish at a water temperature of 75° - 85°F. They should be fed floating feeds or food that remains in the upper layer of the water. Examples of hatchetfish include:

1. Marbled hatchetfish (*Carnegiella strigata*)—a species that reaches a maximum length of 2¼ inches and has irregular streaks of dark color in the pearl-like ventral side with a golden dorsal side.
2. Silver hatchetfish (*Gastropelecus levis*)—a species that is very sensitive to its environment; is slightly larger than the marbled hatchetfish.

Other Aquatic Ornamental Animals

Aquaria may also contain species of animals that are aquatic or water-loving animals. Salamanders, tadpoles, freshwater ghost shrimp, African aquatic frogs, snails, and turtles are often included in aquaria. Each has specific requirements for growth. Some species contribute to keeping a balanced aquarium. For example, the snail helps to keep the glass and decorations in an aquarium clean. Of course, it also contributes its own wastes to the water. Some species are the food of fish. For example, the ghost shrimp will be readily eaten by some fish. The interested reader should seek out books that give more detail on these species.

CONSIDERATIONS IN ORNAMENTAL AQUACULTURE PRODUCTION

The hobbyist and the producer of ornamental aquaculture must consider the same basic areas as the commercial producer of food fish, as covered in other chapters of this book. Aquaria with ornamentals should be observed at least each day, with most growers observing several times a day. Some keep written or computer logs of various factors about the water, such as temperature and behavior of the species. Several of these are presented here with particular application to the production of ornamentals.

Water Management

The water of aquaria must be managed just as carefully as that in large food fish farms. This may involve monitoring oxygen, ammonia, pH, mineral content, and other water quality and chemistry factors. Water temperature is a factor that is probably more critical with ornamentals. Since small containers of water are used, the temperature can vary suddenly. Of course, since volumes are small, it is more practical to use heaters to control water temperature. Emergency generators need to be available in case of power failure to prevent the loss of valuable ornamental species. Periodically, new water will need to be added to aquaria. (Of course, only conditioned, or aged, water should be added.) Hobbyists are reminded that at least one-fourth of the water in an aquarium should be replaced each month. This accounts for that lost to evaporation.

Nutrition

Ornamental species require nutrients for growth. Both plants and animals must have the nutrients available. The culturist must provide the nutrients that are needed, but never more than the amount that can be used. Normally, fish should be fed no more than they will eat in 10 minutes when fed one to three times a day. Commercially prepared feeds are available for many ornamental fish. These feeds tend to provide a balanced diet, but the exact nutritional needs of many ornamental species is unknown. Some variety in the diet is usually recommended.

Disease Control

Diseases can cause severe problems with ornamental fish. Sanitation, quarantine, and isolation are good practices. New fish should never be put into an established aquarium without being observed for a while for the presence of disease. One diseased fish put into a healthy aquarium can result in the loss of the entire aquarium! In a well-designed aquarium stocked with healthy species and well-maintained water, disease problems seldom occur. Entire aquaria should rarely be medicated. Diseased specimens should be placed in a separate container (small aquarium known as a "hospital tank") for treatment.

Handling

Ornamental fish may be handled when the tanks are being cleaned, reproduction is being encouraged, or they are being shipped to market. Ornamental fish are often shipped in plastic bags in styrofoam chests. The bags are partially filled with water, and pure oxygen is then injected to insure safe delivery. The styrofoam chests are usually placed in a cardboard box for air shipment. Fish that are shipped long distances may be placed in new water with new oxygen while enroute.

CULTURING BAITFISH

Some 20 different species have been grown for bait in the U.S. Most are produced in ponds, with approximately 30,000 acres of freshwater ponds devoted to baitfish production in 1991. Volume produced per acre may reach 1,000 pounds or more per acre each year. Since most baitfish are small—often no more than 2 inches long—the number of fish per pound ranges from 250 to 300 at market size. For example, an acre of pond with 1,000 pounds of 2-inch baitfish would contain approximately 285,700 fish. Four acres would produce over a million baitfish—enough to satisfy the sport fishing requirements of a lot of fishers.

A few species of baitfish account for the vast majority of the production. Golden shiners are by far the leading baitfish, followed by fathead minnows, goldfish, blue tilapia, and rudd. These species are typically grown similar to the pond production of food fish. Important areas of concern are selecting and culturing the species, managing water, feeding, controlling diseases and predators, and marketing, including harvesting, grading, hauling, and changing ownership.

Selecting and Culturing

As with food fish, baitfish species have certain characteristics that the potential grower should consider. Primary among these is the market demand, as this is the major factor in determining whether the crop that is produced can be sold for a profit. Some baitfish are used exclusively by sport fishers; others are used to grow other fish, such as goldfish, which are used as food for catfish broodfish.

The three major species of baitfish are the golden shiner, fathead minnow, and goldfish. Each has advantages and disadvantages for the baitfish producer. All three are grown in freshwater.

Golden Shiner

The golden shiner (*Notemigonus crysoleucas*) has a bright, flashy appearance that is supposedly attractive to game fish. This has made it very popular among sport fishers because they feel that a game fish is more likely to be attracted to a "flashy" baitfish. The dorsal fin on the golden shiner is pointed. The body has large, loosely attached, gold- or silver-colored scales. The fish is somewhat delicate and sensitive to handling and hot weather. Handling may result in the loss of scales. When excited, the fish may leap from tanks or other containers holding them. They are very lively when properly baited onto hooks. Most are marketed at a size of 2 - 3 inches in length; however, they may reach a maximum length of 10 inches.

Golden shiners have been cultured long enough for domesticated stocks to be developed. Wild broodstock should be avoided. Female golden shiners may become infected with an ovarian protozoan that impairs reproduction. Sexual

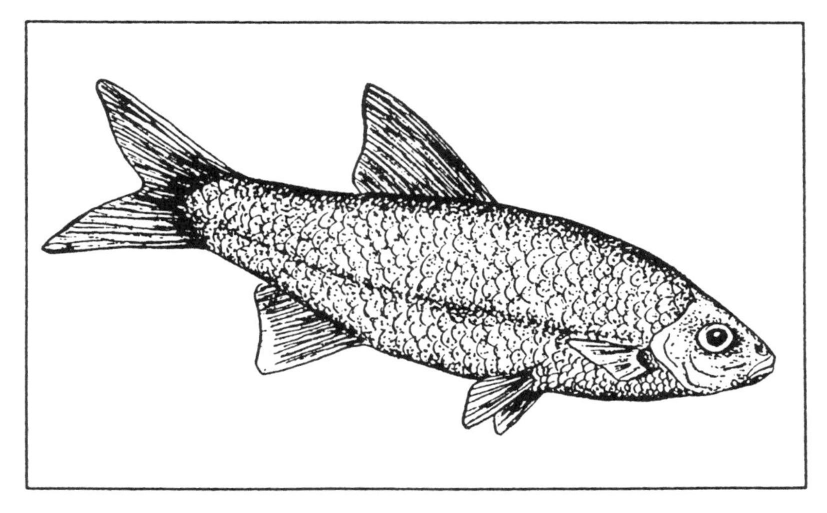

Figure 12–19. Golden shiner — The golden shiner is an active baitfish that is supposedly very attractive to sport fish.

Figure 12–20. Spawning mats used in reproducing golden shiners.

maturity is reached in one year when the fish are about 2½ inches long. Females grow faster than males. Females begin spawning in the spring when the water temperature reaches 70°F. and they continue into the summer. One female may produce up to 10,000 eggs in one season. Farmers use spawning mats to encourage the females to deposit their eggs for easy fertilization and hatching; otherwise, the females release their eggs randomly above the living plants, rocks, or other debris in the water. Hatching occurs in four to eight days after spawning at a water temperature of 75° - 80°F.

In ponds, golden shiners may be grown extensively (low population density) or intensively (high population density). With extensive culture, natural plankton bloom is a primary feed. The water may be fertilized to encourage plankton growth. With intensive culture, the golden shiners must be fed. Most growers use a high-protein, commercially manufactured feed. Fry are fed a finely ground food when they first come to the surface of the water. They will grow faster if fed several times a day. As they grow, the particle size of their food may be increased. They should not be fed more at one time than they will consume in a matter of a few minutes. A rule of thumb is to never feed more than 40 pounds of feed per acre each day.

Water quality is just as important in baitfish production as it is in the production of food fish and ornamental fish. Each species may have varying water quality needs. Producers should study the needs of species as related to the water that is available. Only species suited to the available water should be grown. Growers must regularly monitor water to insure that the oxygen level is satisfactory. An oxygen level of 3 ppm or higher is usually appropriate for baitfish. (Water quality was previously treated in Chapter 6, "The Importance of Water.")

Golden shiners are subject to various diseases and predators. The major predators are turtles, wild fish, frogs and tadpoles, snakes, and birds. The eggs are attacked by turtles, snails, crawfish, insects, wild fish, frogs and tadpoles, and certain zooplankton. Various methods will control the diseases and predators. Careful sanitation and quarantine procedures should always be practiced with new fish. Because some predators are protected by law, the aquafarmer should be careful to use only those methods of control that are legal.

Fathead Minnow

The fathead minnow (*Pimephales promelas*) is very similar to the golden shiner except that it is somewhat hardier. The body is streamlined with a rounded dorsal fin and is covered with small scales. The fathead minnow has a dull color, with the males being darker and larger in size than the females. The maximum length is about 3 inches.

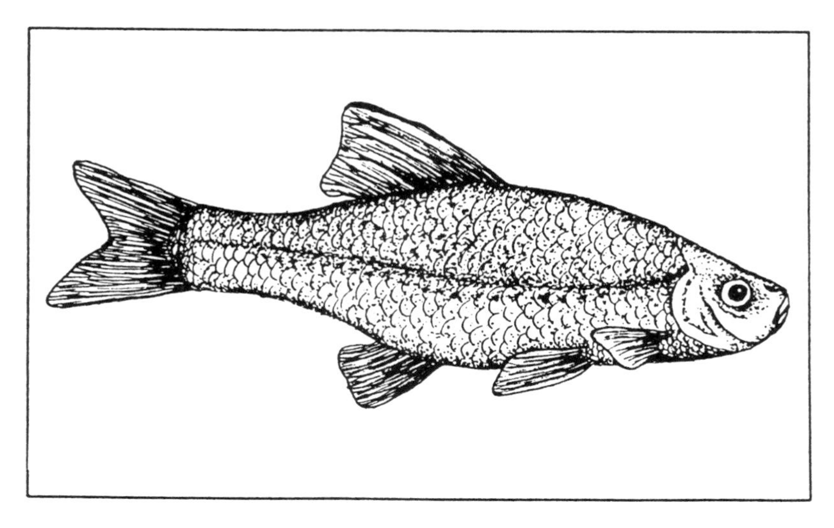

Figure 12–21. Fathead minnow — This baitfish is somewhat hardier than the golden shiner.

The spawning season begins when the water reaches 65°F. in the spring, may end during the warm summer months, and start again as the weather cools in the fall. It ceases when the water temperature goes below 65°F. The males develop numerous horn-like projections known as breeding tubercles on their heads at the time of breeding season. A pad develops just back of the head, which the males use in preparing the nesting site and caring for the eggs. Sexual maturity is reached in 1 year. The males may die after spawning. A female will produce 200 - 500 eggs at each of several spawnings that may occur during a year. After spawning, the male places the eggs on the underside of objects in shallow water. Fish culturists place bricks, boards, and tile in the water to aid egg incubation. The male guards the eggs until fry emerge in five to six days after spawning and fertilization. A spawning pond may be stocked with 2,000 brood females and 400 males per acre if the fathead young are left to grow in the pond. (A ratio of 5:1 is used; one male is stocked for each five females.) If the fry are transferred to another pond, 25,000 brood females and 5,000 males may be stocked in a pond.

Fathead minnows are fed much the same as golden shiners.

Somewhat more tolerant of handling than golden shiners, fathead minnows can be handled in warm weather. As bait, they are frequently used on casting rods. Some varieties, particularly the new rosy red, are used as feeder fish for other, larger species. Producers often obtain more per acre production of fathead minnows than of golden shiners.

The water management problems, diseases, and predators of fathead minnows are similar to those of golden shiners.

Goldfish

The production of goldfish for bait is somewhat different from that previously described for using them for ornamental purposes. Goldfish (*Carassius auratus*) are heavy-bodied fish with colors ranging from dark olive brown, gold, white, and red to black. They usually have long, graceful fins. The can grow to weights of 2 pounds if allowed plenty of space. They are hardier than golden shiners and fathead minnows. They are not as popular as baitfish because they are not as active as golden shiners. They are used on trotlines and as feed fish for bass, brood catfish, and larger aquaria fish.

Goldfish may begin to spawn when the water temperature reaches 60°F. Females will produce 2,000 – 4,000 eggs during each of several spawns in a growing season. The eggs are released randomly and attach themselves to living plants and other objects in the water. Culturists may place spawning mats in the water. Incubation lasts two to eight days, depending on the temperature of the water. Adults give the eggs and fry no protection.

Feeding is similar to that of golden shiners. Proper fertilization of the water and feeding can result in yields of 3,000 pounds or more per acre.

The water management problems and diseases and predators of goldfish are similar to those of golden shiners.

Marketing

Marketing includes several steps in getting the baitfish to the consumer. The steps include harvesting, grading, transporting, holding, and selling. Selling involves establishing a price and changing ownership.

Growers have several market alternatives: (1) the baitfish can be sold to wholesalers, who in turn sell them to retail outlets, such as bait shops; (2) growers can sell directly to local retail outlets; and (3) producers can operate retail bait sales stores themselves for selling directly to sport fishers.

The demand for baitfish depends on the interests of sport fishers. The market is very seasonal and is poor in weather that is not good for outdoor activities. The market increases near holidays during warm months of the year. Baitfish cannot be stored for long periods of time and, therefore, may be lost if the demand of sport fishers is weak when the crop is ready.

Sport fishers want baitfish that are healthy and attractive to game fish. Bait that dies quickly before reaching the fishing site is not desirable.

Harvesting. Harvesting usually involves seining the baitfish from the small ponds in which they are grown. Seines typically have a mesh of 3⁄16 inch. After the baitfish are confined with the seine, they are dipped from the pond with dip nets or boxes made of screen wire. Some are harvested with traps. Withholding feed for about 24 hours prior to harvest is usually a good idea to reduce fouling the water. In warm weather (when the surface water temperature is above 75°F.), harvesting should be done in the early morning hours when it is cooler. In small ponds, cool water can be added to lower the water temperature.

Grading. Baitfish are sorted into lots of uniform size, known as grading. Various screens and devices with bars are used to grade baitfish. The graders allow the smaller fish to pass through holes or between bars and retain the larger fish. Most bait shops want to get baitfish of uniform size and species.

Hauling. Hauling is done in tanks on trucks that are equipped to aerate the water and keep it an appropriate temperature. Baitfish should not be subjected to sudden changes in temperature or hauled in tanks at temperatures to which the fish are not adapted.

Holding. Storing baitfish until they are sold or moved requires maintaining a healthful environment. Tanks, vats, and other water facilities may be used for holding baitfish. Oxygen must be maintained in the water. A water temperature of approximately 70°F. is best. Only quality water should be used in holding facilities. Some experts recommend returning any baitfish that have been held for more than a week to the pond. Emptied storing tanks should be scrubbed with

Figure 12–22. Baitfish may be held in aerated vats for wholesaling or retailing. (Courtesy, Otterbine Aerators / Barebo, Inc.)

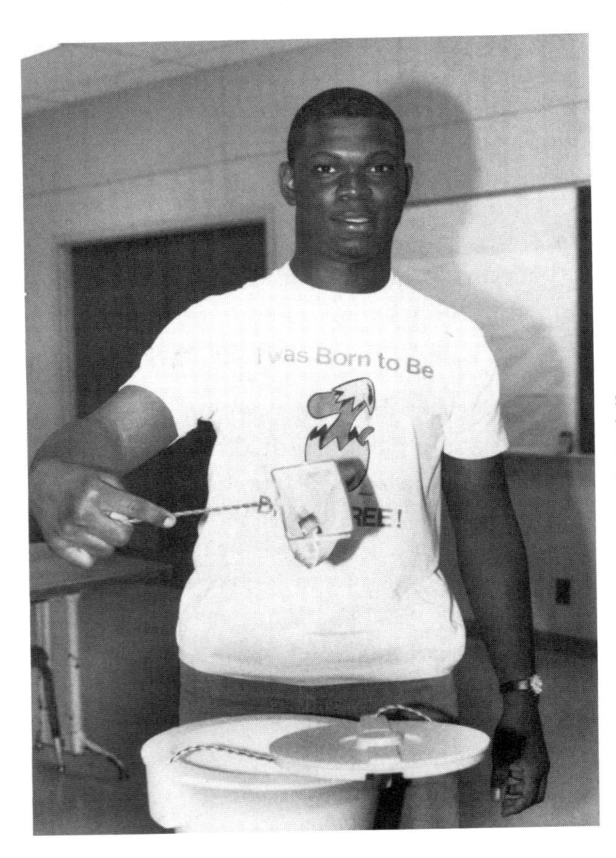

Figure 12–23. Bait species should be carefully handled to prevent injury. Minnow buckets can be used to insulate the water from sudden temperature changes.

chlorinated chemicals, such as household bleach, and rinsed well before they are used again.

Selling. Selling occurs when the buyer and the seller agree on a price, and the ownership changes when the seller is paid for the baitfish. It is a good idea to leave baitfish in the pond until a buyer is found. A rule of thumb is never harvest any fish crop unless there is a buyer. Written agreements may be used in buying and selling to spell out the qualities of the product to be exchanged and the rate of pay involved.

SUMMARY

Some aquafarmers can obtain considerable income from the production of ornamental, baitfish, and related aquacrops. As with food fish, the farmer must know the cultural requirements of the species being grown. Production practices must create environments that are as nearly ideal as possible for the crops.

Ornamentals consist of both plant and animal species, with fish being the predominant crops. Aquaria are often used to grow the fanciest, highest priced ornamentals. Some ornamentals, such as goldfish, are adapted to pond or large tank culture. Hobbyists derive considerable enjoyment from the establishment and maintenance of aquaria of mixed plant and animal species. Meeting the needs of each species requires studying the adaptations of the species. Water systems must be established and controlled to insure that the species are in a suitable environment.

Ornamentals can be classified into three categories on the basis of difficulty in growing: easy-to-grow, medium-care, and difficult-to-grow. Hundreds of species of ornamentals can be put into these categories. Easy-to-grow species include goldfish, barbs, guppies, and gouramis. Medium-care ornamentals include the tetras and catfish. Some difficult-to-grow ornamentals are the piranhas and hatchetfish. Inexperienced individuals should begin with the easy-to-grow species; as they develop skills in growing, they can advance to the more difficult species.

Baitfish are often cultured in ponds, much as food fish. The predominant species are golden shiners, fathead minnows, and goldfish. Stocking rates can be much greater than with food fish because the baitfish are harvested at a much smaller size — often no more than 2 inches long. Water management, disease and predator control, feeding, and marketing are important considerations in the production of baitfish. The grower tries to produce a baitfish that is attractive to sport fishers.

QUESTIONS AND PROBLEMS FOR DISCUSSION

1. Distinguish between ornamental and baitfish aquaculture.
2. Why are ornamentals popular?
3. What are the system needs for ornamental aquaculture? Briefly describe each of the components.
4. What is the weight of the water in an aquarium that is 36" × 18" × 15"?
5. Describe water for an aquarium.
6. How is water quality maintained in an aquarium?
7. What are three classifications of aquaria plants? Distinguish between the classifications and name examples of plants in each.
8. What are some ways of classifying ornamental fish? Briefly describe each of the ways.
9. Ornamental fish are classified into three groups on the basis of how easy they are to grow. Name the three groups and give an example of a species in each group.

10. Select one species of ornamental fish and provide some details on its cultural requirements.
11. Name three species of baitfish and describe the general considerations in the culture of each.
12. What are the functions in marketing baitfish? Briefly explain each.

Chapter 13

RECREATIONAL AQUACULTURE

Many people like to go fishing! They especially enjoy making a big catch; some enjoy talking about the big one that got away. Through recreational aquaculture, the interests people have in fishing can often be met.

Operating a successful recreational aquaculture business requires good management. Knowing some of the fundamentals can help insure success. Customers must come first. Without customers, the recreational business cannot be profitable. Just as with other areas of aquaculture, profit is the measure of success. Customers will spend their money for something they enjoy. This means that the recreational business should provide a comfortable, rewarding experience for the sport fisher.

OBJECTIVES

This chapter focuses on the operation of recreational aquaculture businesses. Upon completion, the student will be able to:

- Describe recreational aquaculture.
- Explain how to establish a recreational aquabusiness.
- Describe considerations in operating a recreational aquaculture business.
- Identify legal and risk situations.
- Use procedures to keep customers happy.

AREAS OF RECREATIONAL AQUACULTURE

Recreational aquaculture gives many people a way to escape from the routines of work and life. It helps them renew their energy and refresh their physical and mental well-being. Some say it is a diversion that allows them to relax. The enterprising aquabusiness operator will find ways to help people enjoy themselves.

Figure 13–1. Recreational aquabusiness provides the sport fisher with an opportunity to enjoy successful catches. It also provides an alternative for the grower to earn income from a profitable fee-lake or other business.

Many opportunities exist to help people and, at the same, to operate a successful aquabusiness. These include fee-lakes, bait-and-tackle stores, boat and accessory stores, tour guide services, and related areas such as food and lodging.

Fee-Lakes

Fee-lakes are small lakes or ponds where the public can pay a fee to fish. Fees may be assessed in several ways:

1. *Daily fee*—Each person who fishes pays a flat rate to fish all day. No guarantee is made that any fish will be caught. No charges are made for

the fish that are caught. This method of fee assessment is often used with low density ponds where the person fishing is not likely to catch many fish. Well-stocked ponds could result in catches of several pounds each; as a result, this procedure is probably not the best to use if someone can be present to monitor catches. The disadvantage is that some catches could be large, which would soon deplete the supply of fish. The owner tends not to get adequate pay for large catches when a daily fee procedure is used unless a large fee is charged, and this might deter patronage by the fishing public. Fees typically range from $1 to $5 a day.

2. *Season or annual fee*—Rather than making a payment each day, individuals pay a fee for a full year or fishing season. Individuals who pay the fee may fish as often and as long as they like during the season or year. Sometimes this method is used in conjunction with a fishing club. The fee should be adequate to cover expenses and provide a profit to the owner of the fee-lake. The season or annual fee may range from $25 to $150 or more, depending on the services provided.

3. *Weight-of-catch fee*—Some fee-lake operators charge on the basis of the pounds of fish caught. This method helps assure that the owner is compensated for the fish that are removed from the pond. The per pound rate depends on the kind of fish, rate needed for profit, and level that the fishing public will pay. Rates range from 50 cents to $1.50 or more per pound. Some weight-of-catch operations provide dressing services for the fish that are caught. A per pound dressing fee of 25 to 50 cents is charged, based on the weight of the fish prior to dressing. Equipment and an attendant to dress the fish must be supplied. With this method, someone must always be available to weigh the fish.

4. *Combination*—The combination approach involves both a daily fee and a weight-of-catch fee. Individuals are charged for both the opportunity to fish and the pounds of fish that are caught. This is probably the most common method. Someone must be present at all times to collect fees, weigh fish, and assist in other ways.

Catch-out ponds. Catch-out ponds involve raising the fish in the ponds where they are to be caught. The amount of fee fishing that can occur is no greater than the amount of fish that will grow in the ponds. The stocking rate, feeding, aeration, and other production factors influence the volume of fish that is produced. Catch-out ponds are not very popular because the fish supply may be depleted and returns to the owner may be less. Sport fishers like to go to fee-lakes where they can expect to make a good catch. Catch-out ponds may not always provide enough fish for this to occur.

Figure 13–2. A put-and-take operation requires getting fish from a live hauler to keep the pond stocked. (Courtesy, Otterbine Aerators / Barebo, Inc.)

Put-and-take facilities. A put-and-take operation is a pond or other water facility that is stocked with catchable-size fish that are grown elsewhere. The stock of fish is kept replenished. The operator of the put-and-take facility buys the fish for stocking from a live hauler or a fish farmer. In some cases, the owners of put-and-take may have production ponds located elsewhere to provide the needed stock. The stocking density is heavy because the owners want the fishing public to make good catches. Only quality, disease-free fish should be stocked in a put-and-take facility. The presence of sick and dead fish has a negative effect on the public. Careful water management is needed at high stocking densities. Put-and-take facilities typically involve charging the public on a weight-of-catch or combination basis.

Put-and-take operations frequently use ponds, but they may also use large tanks, raceways, or other facilities. Sometimes large, transportable tanks are moved to shopping malls or other locations. Careful and constant attention is needed in the operation of a put-and-take facility.

Bait-and-Tackle Stores

The fishing public must have certain equipment and supplies in order to fish. The selling of these may be a part of a fee-lake operation or separately established near recreational lakes. The operation of bait-and-tackle stores is much like the operation of any retail store. Convenience, good prices, quality merchandise, courteous employees, and good management are essential. The business must be conveniently located and properly advertised.

Figure 13–3. Stores located near recreational facilities often offer a wide variety of goods.

Bait-and-tackle stores typically handle a wide range of fishing gear and baits. The items stocked should be appropriate for the fish in the area. The interests of the fishing public must be considered.

Bait-and-tackle stores may offer services to the fishing public. Selling licenses to individuals that permit them to fish in public streams and lakes is frequently a part of the business. In most cases, licenses are not needed to fish in fee-lakes. The stores may also offer other services, such as dressing the catch and preparing it for shipping.

In addition to selling items for fishing, bait-and-tackle stores often sell food, fuel, books and magazines, clothing, and camping supplies. A few also have restaurant services. In areas where there is considerable sport fishing, lodging accommodations may be available near bait-and-tackle stores.

The people who work in bait-and-tackle stores must be qualified to assist sport fishers with their needs. They must be able to advise on where the fish are biting and how to catch them. Good assistance in selecting appropriate equipment can result in more sales to satisfied customers.

Boat and Accessory Stores

Boat and accessory stores provide boating and related items for the public. These stores typically sell boats, trailers, and other types of recreational equipment and supplies. In some cases, the stores may also rent equipment on an hourly, daily, or weekly basis. Storage space may be available for rent to owners of boats and other recreational equipment. Repair services for engines may also be provided.

Many fee-lakes do not allow fishing boats. Boat and accessory stores are often located near large public lakes, streams, or seashores. Considerable investment may be needed to establish and operate the business. Qualified employees are very important.

Figure 13–4. Boat and accessory stores may handle a wide range of equipment, including boat trailers.

Figure 13–5. Lakes allowing boats will need to have launch areas that are durable and safe.

Tour Guide Services

The fishing public often needs assistance in knowing where and how to fish. In some cases, there are charter boat services that take people where they can fish. Although few fee-lakes have tour guide services, the employees must be able to provide advice to the fishing public.

Tour guide services may help groups or individuals gain access to good fishing in remote areas of streams and lakes. The guides may provide only information and / or the complete equipment and supplies needed to fish. The fee is often on a daily or weekly basis.

Charter boat services are usually offered for fishing in oceans, large lakes, and other large bodies of water. The boat operator may take one individual or a small group out for a few hours or for a day or more of fishing. Tackle, bait, and other supplies are provided by the boat operator. Fees vary considerably, depending on the services and length of time involved.

ESTABLISHING A RECREATIONAL AQUABUSINESS

Establishing a successful recreational aquabusiness will require attention to a number of factors. These usually deal with what it takes to be successful in operating such a business.

- ***Determining the kind of recreational aquabusiness*** — Determining the best kind of aquabusiness to establish will involve studying the kind of recreational aquaculture that is available in the area and the demand for it by the public. A fee-lake must have customers in order to be successful. Enough people who are interested in the recreation must be in the area to provide sufficient business. Certainly, climate and its possibilities with various aquacrops is a consideration.
- ***Determining the site for a recreational aquabusiness*** — A good site is one that will attract enough customers to make the aquabusiness profitable. A location easily accessed from highways and near centers of population holds the best potential. The site should also have the natural advantages of good terrain and access to public utilities such as electricity. The location must comply with zoning regulations. The site must be available at an affordable price. Paying too much for the land can result in a debt that will be too large for the operation.
- ***Determining the development needed for the site*** — Ponds, buildings, water wells, parking areas and driveways, fences, and other construction may be needed to prepare a site for a recreational aquabusiness. Of course, the facilities should be constructed according to the intended use of the property. Cost estimates must be a part of assessing needed site work.
- ***Developing a facilities plan*** — A facilities plan describes how the site is to be prepared for the recreational aquaculture business. It includes how the earth is to be excavated, what buildings will be needed, where access roads and parking areas will be built, where ponds will be located, and other features. Ponds are typically small, ranging from 1 acre or less to a few acres. The design should provide a maximum of bank or land area for the fisher to access the pond. A small building with a weighing area, supplies area, dressing area, restrooms, concessions, and public telephone is often installed. Sometimes a small office may be included. Scales, cleaning equipment, a cash register, an ice machine, soft drink and snack concessions, and any other equipment deemed necessary, plus a first aid kit and a fire extinguisher, should be supplied.

Figure 13–6. Cleaning equipment at a fee-lake operation. (Courtesy, D & D Manufacturing Co.)

- ***Preparing a written business plan*** — A written plan requires attention to details that might be overlooked when no such plan is prepared. The written business plan contains short-range, intermediate-range, and long-range plans. Short-range plans focus on what can be done in the near future to ready the site. Intermediate plans focus on one to five years, while long-range plans are typically for more than five years. Good plans include estimates of costs and returns. A detailed site preparation plan should be prepared. Assistance from the Soil Conservation Service and other agencies may be needed. (For additional information, refer to Chapter 4, "Determining Aquaculture Requirements.")
- ***Obtaining the needed finances*** — Adequate money to support the establishment and initial operation of the recreational aquabusiness must be secured. Risks are involved. Lending agencies need some assurance that the business will be operated in a profitable manner so that loans can be repaid.

OPERATING A RECREATIONAL AQUABUSINESS

Once established, a recreational aquabusiness must be operated effectively and efficiently to try to insure its success. The owner must see that all of the following items are properly handled.

- ***Having good on-site management***—The owner or someone designated by the owner must be available to manage the recreational aquabusiness. Managers are responsible for seeing that things get done in an efficient and effective manner. Attention to day-to-day details is a must.
- ***Handling income and disbursements***—A way to handle income and the payment of charges must be established. Fees charged to customers must be properly accounted for and deposited in the bank account. Employees must be paid. Utilities, supplies for resale, fish for the pond, and other items must be paid for in a timely manner. Careful and accurate records must be kept.
- ***Having dependable employees***—Fee-lake operations typically have only a few employees. In these small businesses, it is very important that employees be qualified to do their work. Often, the owner is also the manager, who may rely on family members to help out. All employees must relate well with the public, be able to offer assistance to those who fish, and responsibly handle the details. Larger aquabusinesses dealing with boats, equipment, and other areas may need a wide range of qualified employees. Some fee-lakes are open only during certain hours or on certain days of the week. When they are open, someone must be on duty to run the operation.
- ***Advertising***—The public must be informed about a business if it is to be patronized. Advertisements in local newspapers or on local radio stations

Figure 13–7. Attractive signs promote recreational opportunities.

may be helpful. An attractive sign identifying the aquabusiness will help the public in locating it. One of the best forms of advertising is by word of mouth. Individuals who have patronized a recreational aquabusiness tell others about their experiences. Good experiences result in good comments being made about the aquabusiness. Bad experiences result in negative statements about the aquabusiness.

- ***Maintaining the facilities*** — Once constructed, facilities must be kept in good condition. Dilapidated facilities are not attractive to the public. The grounds must be kept clean and free of trash, such as cans, bottles, and paper. Buildings will need to be painted and kept presentable. Pond dams must be covered with sod and appropriately mowed. Erosion to dams or other areas must be controlled. Broken equipment or discarded fishing tackle should be removed. Trash cans should be conveniently located. Safety and health hazards must be removed.
- ***Keeping records*** — As with any business, good records must be kept on the recreational aquabusiness. These include records of income and expense. The records must be analyzed on a regular basis to determine the profit status of the operation. Some aquabusinesses use the services of an accountant to assist with preparing records and submitting reports.
- ***Submitting reports*** — Regular reports must be sent to various federal, state, and local agencies. These pertain to taxes, employee benefits, and other requirements. Most of these are due at the same time each year. A routine schedule of preparing and submitting reports can be established and followed.

DEALING WITH LEGAL AND RISK SITUATIONS

Digging a pond and putting up a sign will not cover all of the details involved in operating a recreational aquabusiness. Appropriate regulations must be observed. Ways of protecting against liability are needed.

Government Regulations

Various regulations have been enacted to protect the consumer and the environment. The operator of a recreational aquabusiness needs to be aware of and comply with these requirements.

Obtaining permits to operate a recreational aquabusiness is a time-consuming, but important activity. Major changes in the natural water reservoirs require permits from the U.S. Army Corps of Engineers and the U.S. Environmental Protection Agency. Permits may be required from state agencies responsible for wildlife conservation, particularly if exotic species of fish, such as tilapia, are to be stocked. Permits to drill water wells and to dispose of water may be needed.

From a retail business standpoint, licenses may be required for operating the aquabusiness. For example, some fee-lakes have soft drink and snack vending machines, which must have a license or decal attached, indicating that a fee has been paid to operate them.

Liability

The owners of recreational aquabusinesses are expected to provide a safe environment for the public. Owners may be held responsible for accidents or losses that occur to customers. The best procedure is to have a safe, comfortable environment. Posting warning signs about dangers is helpful but is inadequate by itself. Facilities should be kept free of hazards.

Legal action may be taken by individuals who have suffered injury. This involves going to court to try to prove that the owner was negligent. Not correcting dangers that are obvious may be sufficient grounds for negligence. In addition, safety equipment, such as life preservers and rescue equipment, should be provided.

Some fee-lake operators try to protect themselves by having liability insurance. Such insurance can be expensive and can cut into the profit of a business. Regardless, it is a good idea to have some insurance coverage. Farmers are often able to get general liability for an entire farm. Local agricultural organizations can often provide a fee-lake operator with a list of sources of liability insurance.

Safety Features

The owners of recreational aquabusinesses can follow practices that help to make their facilities safer. A few safety suggestions are listed here.

1. **Erect warning signs near places of danger.** These should state that individuals use the facilities at their own risk. Signs reminding adults that children should be supervised should also be installed.

2. **Post a notice that gives emergency telephone numbers.** The numbers of emergency officials, as well as a telephone, should be readily available. A public telephone is an added convenience of a fee-lake operation.

3. **Provide life saving equipment near water areas.** A life ring with a 100-foot rope is probably sufficient.

4. **Do not allow swimming in the same area as fee fishing.** If allowed, it should be in a separate area.

5. **Fence off ponds to keep out children, animals, and others who are not authorized to be present.** Fencing also keeps individuals from trespassing and avoiding paying the fees.

6. **Keep all boats and other equipment in safe operating condition.** Most fee-lakes do not permit boats. A few will rent boats, and only those boats that are in good condition should be available for rent.

7. **Allow only trained employees to operate equipment.** Fee-lakes that have fish cleaning equipment should provide personnel to operate the equipment. Most customers would have little training in how to use it. The customers would likely damage the equipment and possibly injure themselves. They might waste their catch or damage its suitability.

8. **Train employees in emergency procedures.** Local hospitals often have training courses to prepare people for emergencies.

Sources of Information

The area of legal regulations and liability often involves locally specific laws and regulations. Individuals are advised to get in touch with specialists of their Cooperative Extension Service for assistance.

KEEPING THE CUSTOMERS HAPPY

Treating customers with respect and providing the information they need can go a long way toward building a successful recreational aquabusiness. In addition to being friendly and considerate, workers should offer assistance that will help customers to be successful in catching the fish. When the customers have a good time, they are more likely to repeat their visit.

Keeping Things Convenient

Providing for the convenience and comfort of customers is important. Locating parking, concessions, and weighing areas convenient to the ponds helps

Figure 13–8. Educational exhibits help make a recreational aquabusiness appealing to customers. This exhibit is intended to teach the identification of sea animals.

people get more enjoyment from their outing. Providing picnic tables, chairs, and other conveniences is beneficial. Customers are more likely to repeat their visit if they enjoyed themselves, and they will tell others!

Some fee-lake operations have vending areas for soft drinks and snacks. These should be kept attractive and clean. Only fresh, wholesome products should be sold. Such services can add to the profitability of an aquabusiness.

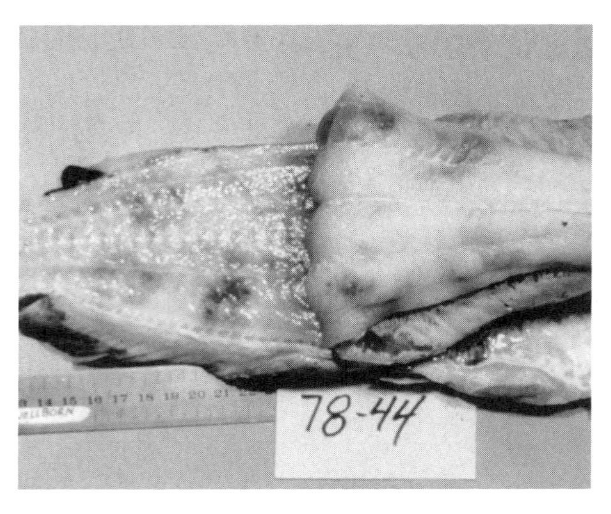

Figure 13–9. Keeping only healthy stock in fee-lakes creates customer appeal. The fish here, suffering from hemorrhages in muscles due to bacteria *(Aeromonas hydrophila)*, would not be an appealing catch to the sport fisher. (Courtesy, Eddie Harris and Pete Taylor, Mississippi Cooperative Extension Service)

Keeping the Place Attractive

Recreational customers like to visit clean, attractive places. Owners are reminded that a few simple practices can enhance attractiveness.

1. Fish weighing and cleaning areas should be kept clean.
2. Flies, dogs and cats, ants, and other insect pests should be controlled.
3. Scales should be accurate.
4. Weeds should be kept mowed.
5. Fences should be maintained.

Assisting Customers

Fee-lakes often have customers who know very little about fishing. They need assistance in how to select tackle and bait, how to catch fish, and how to handle the live fish to prevent injury. Some fee-lakes maintain a small supply of bait and equipment for sale. Many customers need instruction in how to prepare the tackle, bait the hook, remove the fish from the hook, place the fish on a stringer, and prepare the fish for cooking. Where and how to fish are frequent areas of concern.

Happy customers help to build a successful recreational aquabusiness.

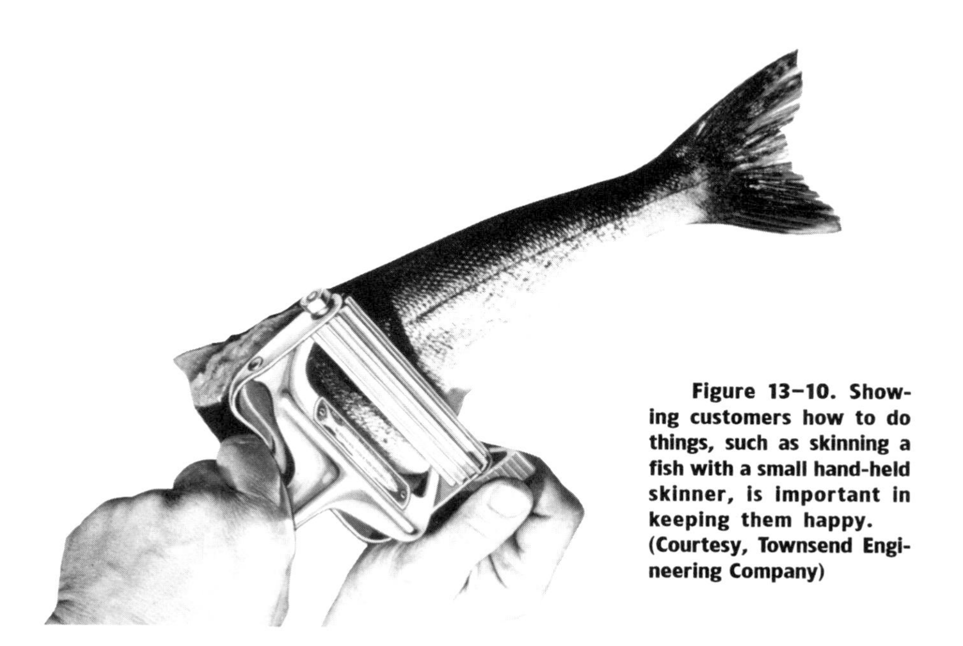

Figure 13–10. Showing customers how to do things, such as skinning a fish with a small hand-held skinner, is important in keeping them happy. (Courtesy, Townsend Engineering Company)

SUMMARY

Recreational aquaculture has considerable potential in some locations. Careful planning and management are essential ingredients for a successful fee-lake.

Fee-lakes are of two main kinds: catch-out and put-and-take. With catch-out, the fish are raised in the facilities where they are caught. Put-and-take facilities involve raising the fish elsewhere and moving them to the facilities. Fees may be assessed in several ways: weight-of-catch basis, day fee, season fee, or a combination of these.

Careful planning and operation are essential. Customer convenience and safety are of high importance. Some of the conveniences, such as soft drink vending machines, can result in larger profits.

Convenience, comfort, good catches, fair fees, courteous treatment of customers, and pleasant surroundings all go into the making of a successful fee-lake operation.

QUESTIONS AND PROBLEMS FOR DISCUSSION

1. Why are customers important to a recreational aquaculture business?
2. What are some business opportunities in recreational aquaculture?
3. How are fees assessed in fee-lakes?
4. Distinguish between a catch-out pond and a put-and-take facility.
5. What factors should be considered in the establishment of a recreational aquaculture business?
6. What are the important considerations in the operation of a recreational aquabusiness?
7. What legal and risk situations must be addressed in recreational aquabusiness?
8. List and briefly explain several safety practices that might be implemented at a fee-lake.
9. What are three areas that are important in keeping customers happy? Why are these important?
10. Assess the opportunities for a recreational aquabusiness in your local community. Identify a site and explain how you would develop it. Indicate what fish and services you would provide.

Chapter 14

PLANT AQUACULTURE PRODUCTION

Many plants grow in water. A few of these are used for food. The future could hold a significant increase in the food production and income from aquaculture plant operations. At present, little culture of aquatic plants exists in the U.S. The potential for more culture is excellent, however.

In the U.S., the primary focus of aquaculture operations has been animal production, particularly the intensive culture of animals that are in demand as food by restaurants and supermarkets. These species demand high prices, but the cost of inputs, especially feed and seedstock, is also high.

Plant aquaculture is different in both areas. The seedstock for aquaculture plants often costs very little or nothing, and the cost of growing the plants is very small. On the other hand, the market for aquaculture plants is small, and prices are not usually very good. One factor in particular makes the production of plants a prime area of expansion for aquafarmers. Many aquafarms already have the necessary facilities in place for production, and the plants may bring benefits to the animals already being cultured in the same facility.

The use of aquatic plants for human food is more widespread in other countries, particularly countries in Southeast Asia. These countries have developed plant aquaculture systems which show good potential for use in the U.S. The primary concern is developing a suitable market for the aquacrop once it has been produced.

Plant aquaculture may involve the production of plants for consumption by humans or other animals, or the production might be for specialty purposes such as ornamentals or waste water treatment. Plant aquaculture may be in saltwater, brackish water, or freshwater.

OBJECTIVES

This chapter presents the primary species of aquaculture plants produced in the U.S. and some that have potential for U.S. production. The culture of these plants, including nutrient requirements, water quality factors, and harvesting, is discussed.

The following objectives have been developed to help the student better understand the culture of the various aquaculture plants. Upon completion of this chapter, the student will be able to:

- Identify common plant species grown in aquaculture.
- Discuss the principles of plant aquaculture.
- Describe the culture of saltwater and brackish water food plants.
- Describe the culture of freshwater food plants.
- Describe the culture of aquatic plants for animal feeds.
- Describe the culture of specialty aquaculture plants.

COMMON AQUACULTURE PLANT SPECIES

The plants grown with the use of aquaculture techniques can be divided into four categories: saltwater and brackish water food plants, freshwater food plants, plants cultured for use as animal feeds, and specialty plants.

The saltwater and brackish water food plants are all algae plants. These include brown algae (kelp), red algae (nori), and green algae (aunori). The species of brown algae that are cultured are *Undaria pinnatifida* (Japanese kelp) and two species of giant kelp, *Macrocystis pyrifera* and *M. integrifolia*. The species of red algae that are cultured are *Porphyra angusta, P. kuniedai, P. pseudolinealis, P. tenera,* and *P. yezoensis*. One species of green algae, *Monostroma enteromorpha*, is cultured.

The two primary species of freshwater food plants cultured are watercress, *Nasturtium officinale*, and Chinese waterchestnut, *Eleocharis dulcis*. Other freshwater plants that have been cultured but to a much lesser extent are water spinach, *Ipomoea reptans*, and arrowhead, which includes several species of the genus *Sagittaria*.

Several species of aquatic plants are cultured for use as animal feeds. These can be divided into two groups: those used for livestock feed, particularly for cattle, and those used to feed other aquacrops, usually phytoplankton for consumption by filter feeders and larvae of several species. The primary aquacrop suitable for use as livestock feed is duckweed, which includes several species from the genera *Lemna, Spirodela, Wolffia,* and *Wolffiella*. Water spinach is sometimes used as livestock feed as well. Numerous species of algae and microalgae are cultured for use in aquaculture operations, such as for oyster or shrimp larvae or in ponds where filter feeders such as mullet are reared.

A few aquatic plants are cultured for specialty purposes. These include duckweed and water hyacinth (*Eichhornia crassipes*) for waste water treatment, certain red algae for the extraction of *phycocolloids* (gel-like substances with many uses), and others for use as ornamental plants or for making fuel.

One other use of waste water in plant production is in hydroponics. *Hydroponics* involves the culture of terrestrial plants whose roots grow in gravel or some other substrate where a shallow current of water constantly flows. Hydroponic production is not generally considered aquaculture and is not described further in this book.

PRINCIPLES OF PLANT AQUACULTURE

Plants that grow in water share many of the characteristics of terrestrial plants. Some of the characteristics, however, are unique to aquatic plants. Some of the basic principles of plant biology, functions, and culture are presented in this section.

Photosynthesis

Most plants make their own food starches through a process called photosynthesis. "Photosynthesis," simply defined, is "the process whereby the plant uses light energy, in the presence of chlorophyll, to convert carbon dioxide and water into carbohydrates, which provide energy." By-products of photosynthesis include water and oxygen, which is the source for the common statement that plants "give off" oxygen. The equation for photosynthesis is as follows:

$$6CO_2 + 12H_2O \xrightarrow[\text{chlorophyll}]{\text{light}} C_6H_{12}O_6 + 6O_2 + 6H_2O$$

Respiration

In order to utilize the carbohydrates formed by photosynthesis, the plant converts the carbohydrates to energy in the form of ATP, adenosine triphosphate. Plants accomplish this by the process of respiration. Respiration involves the conversion of carbohydrates and oxygen to energy, with by-products of carbon dioxide and water. The energy released can then be used for the various processes of the plant, including growth and reproduction. The chemical equation for respiration is as follows:

$$C_6H_{12}O_6 + 6O_2 \rightarrow 6CO_2 + 6H_2O + \text{energy}$$

Chlorophyll, which allows photosynthesis to occur, also gives the plant its green color. Several different types of chlorophyll are found in plants, including chlorophyll *a, b, c,* and *d*. Chlorophyll *a* is recognized as giving plants their green color. Other pigments in the plant include xanthophylls, phycobilins, and carotenes. Some of these cause certain plants to be reddish or brownish in color.

Nutrient Uptake

Aquatic plants obtain most of their nutrients from the water in which they are grown. This function makes aquatic plants very useful for removing ammonia and nitrite wastes from the water. Removing wastes greatly enhances the cleaning of water in waste water treatment facilities, which will be discussed later in this chapter.

Removal of wastes also should make a polyculture of plant and animal aquacrops profitable. Because the plants will remove excessive wastes that may be harmful to the animal aquacrop, higher stocking rates of animals could be used. The wastes produced by the animal aquacrop should provide sufficient nutrients to the plants so that commercial fertilizer would not need to be applied. Although little polyculture of this type has been practiced with freshwater aquacrops, the potential appears excellent.

Food Qualities

Saltwater and brackish water food plants, the various algae, are excellent sources of nutrients. The primary source of these plants to the public is through commercial health food stores, where they are recognized as health foods.

Human taste for these seaweeds is an acquired one, so not much is eaten in the U.S. Consumption is much greater in Asian countries, most notably Japan, China, and South Korea.

The algae plants are an excellent source of vitamin A, found in the beta carotene in the plants. Most of the algae have a higher percentage of vitamin A than eggs, a common source in the U.S. They are also good sources of vitamins B_2 and B_{12}, as well as vitamin C, although the amounts vary with the season. Historically, red algae has been an important part of Eskimo diets, as the major provider of vitamin C.

Compared with other plants, some algae are also high in protein, ranging from 15% to almost 50% for some red algae and green algae. Algae contain energy-providing carbohydrates. These plants also serve as an excellent source of iodine and other essential micronutrients.

Of the freshwater food plants, watercress is also high in vitamins A, B, and C, similar to the salad greens commonly eaten in this country. It is also high in iron and iodine. Chinese waterchestnuts are about 30% carbohydrates, but are low in protein, with less than 2%.

Habitat

The primary concerns for the production of aquatic plants are water temperature, salinity, and substrate material. For most algae, the desired salinity is at saltwater levels of 35 ppt or just below. Some species can survive lower salinities but will not reproduce. Brown algae will not produce gametophytes or sporophytes at salinities lower than 27 ppt, for example. Some species, in particular the green algae, have freshwater species and other species which are more tolerant of brackish water. The freshwater species mentioned in this chapter require water with a salinity of at or near zero ppt.

In general, the algae can survive a wide variety of temperatures, but different species have particular temperature ranges in which they grow best. For example, brown algae grows and reproduces best between 50° and 70°F. but can withstand cooler and warmer temperatures. Watercress requires cool, flowing water with a temperature of less than 78°F., but grows best when air temperatures are between 70° and 85°F. Chinese waterchestnuts are tropical plants in nature and grow best in warm climates and warm water, over 70°F.

Aquatic plants that attach themselves to the ground or bottom of the water facility are called *benthic*. Algae, watercress, and Chinese waterchestnut are all benthic plants. Floating plants, such as duckweed and many of the microalgae, are called *planktonic*. For benthic plants, the substrate (material which makes up the bottom of the growing facility) is very important. Many species of algae require mollusc shells or some other solid surface to which the spores can attach. Watercress requires a firm growing bed because of the large amounts of water circulated across it. Sometimes gravel or crushed rock is added for firmness to beds that are too soft.

Size

Size varies widely among the different aquaculture plants, and even within the different orders of plants, such as algae. Brown algae (such as kelp) is often called macroalgae. Giant kelp thalli (the stems) can reach 200 feet or more in length. Other species of algae, however, are unicellular (one-celled) plants that are very small. These plants can only be seen when they attach themselves to each other to form chains.

Watercress usually is harvested when the plants reach a height of 12 - 14 inches above water level, although they may grow much larger if they are not harvested. Chinese waterchestnuts are a type of sedge and grow to a maximum height of 3 - 6 feet during one growing season; they will die back if they are not harvested before the first frost.

Reproduction

Plants may be reproduced either sexually or asexually. Many of the aquatic plants discussed in this chapter reproduce by both means at different times, with some mechanical reproduction techniques also used.

Sexual reproduction involves the fusion of gametes (sex cells) followed by meiosis (cell division). The process may result in seed or in a zygote, a new plant. Asexual reproduction refers to any means of reproduction other than sexual reproduction.

In nature, algae reproduce by both sexual and asexual means. Mature plants called sporophytes release spores that become microscopic plants. The spores then produce gametes which fuse and begin the process over again. Accessory reproduction occurs when asexual spores are produced by young plants. These spores, called monospores, then become new plants.

Watercress naturally reproduces sexually, where the fusion of the gametes results in a seed. It also can reproduce by sprouting shoots, which grow into new plants. In commercial operations, however, most reproduction is mechanical, where terminal cuttings are used to start new plants. This method allows for a more uniform crop and a shorter growing period.

Chinese waterchestnuts reproduce by producing corms, parts of fleshy, underground stems (rhizomes). Each parent corm can produce hundreds of new corms during a good growing season, all of which can split off from the parent corm and become new plants. As this is the part that is also harvested for food, only some of the corms are kept for reproduction purposes.

CULTURE OF SALTWATER AND BRACKISH WATER FOOD PLANTS

The culture of the three types of seaweeds used for food — brown algae, red algae, and green algae — is not common in the U.S. Brown algae is the only one cultured in any significant amount, with most of this culture occurring off the Pacific coast of California. Red algae and green algae are abundant in certain coastal waters of the U.S., but little harvesting of these plants for human consumption occurs.

Japan leads the world in the culture of seaweeds. Most of the available waters suitable for growing these plants in Japan are already in production, but the demand still exceeds the supply. The potential exists for a market for seaweeds cultured in the U.S. to be exported to Japan, China, and other Asian countries.

Brown Algae

The culture of brown algae of the division Phaephyta is common in Japan and other Asian countries. The primary species cultured is *Undaria pinnatifida*, called "wakame" in Japan and "kelp" in Europe and the U.S.

Brown algae contains chlorophyll *a* and chlorophyll *c*, as well as alpha and beta carotene. The brownish or olive green color comes from the xanthophyll called fucoxanthin.

The culture of kelp is an involved process. Mature sporophyte plants are brought into the laboratory and kept in concrete and plastic tanks. The salinity is maintained between 30 and 33 ppt. Wooden or metal frames with cotton strings are placed in the tanks, on which the released spores attach themselves. During the summer, the spores develop gametes which fuse to become sporophytes, the plants that are eaten.

The strings, with the sporophytes attached, are tied to rafts or buoys in the open ocean. This usually occurs in September or October.

The mature sporophytes are ready to harvest in January or February. Although the whole plant is sometimes harvested, usually only part of the plant is cut off for harvesting. The plants are then dried and chopped into small pieces. The most common use is as a salad green.

The most common culture of brown algae in the U.S. is that of *Macrocystis pyrifera* and *M. integrifolia*, two species of giant kelp. These plants may reach heights of 200 feet or more.

The culture of giant kelp is more of a harvesting practice than the intensive culture of *Undaria pinnatifida*. Aquafarmers locate a bed of giant kelp and cut the kelp near the ocean floor with underwater mowers. The cut thalli are then raked to the shore. Some are eaten, but most are used for mulch or fertilizer. Some aquafarmers extract phycocolloids from the kelp, which is discussed later in this chapter.

Red Algae

Red algae is in the division Rhodophyta. The red algae of the genus *Porphyra* is the most commonly cultured seaweed in the world, although little culture occurs in the U.S. The common name is "laver" in Europe and the U.S. and "nori"

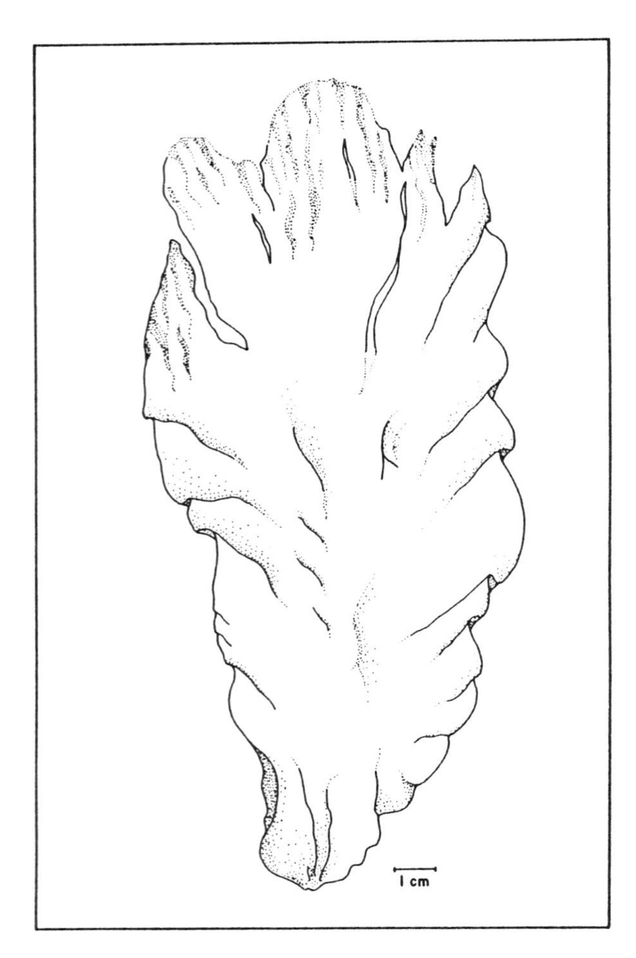

Figure 14–1. *Porphyra yezoensis*, a red algae commonly eaten by humans. (Source: J. E. Hansen, J. E. Packard, and W. T. Doyle, "Mariculture of Red Seaweeds," California Sea Grant College Program, Report No. T-CSGCP-002)

in Japan. Some reports of nori culture in Japan date back to 1570, where it is the most valuable aquacrop cultured in saltwater or brackish water.

Red algae contain chlorophyll *a* and phycobilins, which give them their red color. The phycobilins are especially suited for absorption of the green, violet, and blue light that penetrates the depths of the ocean. They allow the red algae to grow in deeper water than either brown or green algae.

In Japan, extensive culture of red algae occurs on nets placed horizontally in the open ocean. The seedstock for the nets comes from commercial laboratories. The mature plants are put in tanks with mollusc shells placed on the bottom. The released spores attach to the mollusc shells. The spores grow into microscopic plants that release monospores that grow into thalli.

Temperature is used to control the release of the monospores. When an aquafarmer brings nets into the laboratory to catch monospores for seedstock, the temperature is lowered, and the plants release the monospores. After a short time, from an hour or less to several hours, the nets are removed and placed in the open ocean, attached to poles so that they are parallel to the surface of the water.

Nets are usually placed in the water in early September. The growing season lasts from September to early April. The artificial spawning of the monospores allows for a longer growing season. Each net may be harvested three or four times before the end of the growing season; whereas, with natural spawning only two harvests are possible.

Harvesting the red algae involves cutting the thalli and leaving stock for further growth. It is then washed and cut into small pieces. After it has been washed again in a freshwater barrel, it is placed on a wooden frame and allowed to dry into mats. The dried product is very high in protein, from 30 to 50 ppt, but is expensive and usually eaten in small amounts.

Although several species of red algae are common off both the Atlantic and Pacific coasts of the U.S., very little food production occurs. Some food production of red algae occurs in Hawaii and in Alaska.

Some experimental culture is underway of another type of red algae, from the genus *Gracilaria*, found all along the eastern coast of the U.S. The yields from these studies have been exceptional, with up to 50 dry tons per acre per year produced in a highly intensive system. The systems, however, have not been energy-efficient and have required too much water to be profitable. The future does look promising, however, for some less intensive production.

Green Algae

Green algae is in the division Chlorophyta. It is the least cultured of the three primary types of algae, although some production occurs, especially in Asia. The species most commonly cultured are from the genera *Monostroma* and *Enteromorpha*. Called "aonori" in Japan, *Monostroma* is the most expensive seaweed cultured. Cultural practices in the open ocean are similar to those of red algae, but the spore-gathering process is not as advanced. Also, the production levels are not as great as with red or brown algae, because the green algae has a slower growth rate.

Some experimental culture of other green algae has shown promising results. *Chlorella*, a freshwater green algae, has been grown in the U.S. with impressive production numbers. This green algae is about 50% protein, but it does not sell well except in some specialty markets such as health food stores. The primary reason for the poor sales is that the flour made from the algae turns the food it is cooked with either black or dark green in color, so many people do not like its appearance.

CULTURE OF FRESHWATER FOOD PLANTS

Other than rice, which is generally considered an agronomic crop, very little freshwater plant aquaculture occurs in the U.S. The primary aquacrop produced in the U.S. is watercress, *Nasturtium officinale*. Most of the production takes place in Hawaii. The other important freshwater aquacrop is Chinese waterchestnuts, *Eleocharis dulcis*, a specialty crop that is not grown commercially in large quantities in the U.S. Although several other aquacrops are grown in very small quantities, none of them have been shown to be profitable as commercial products.

Watercress

Although it is the major freshwater plant aquacrop, watercress is still a very minor part of aquaculture. About 1.5 million pounds of watercress are produced in Hawaii each year, with the rest of the U.S. producing less than 100,000 pounds. The farm value of watercress produced in the U.S. each year is just over $1 million. Some people harvest wild watercress from local streams, and many people grow it in small quantities for private use.

Watercress has two primary habitat requirements. One requirement is an abundant supply of continuously flowing water, about 1 million gallons per acre per day. The other requirement is many sunny days during the growing season.

Watercress is cultured in shallow ponds or beds in 1 – 2 inches of water. The water must be relatively high in nitrates, about 4 parts per million, for optimum

Figure 14–2. A watercress cutting ready for planting. When the cutting is planted, the basal end is placed in the direction of the current.

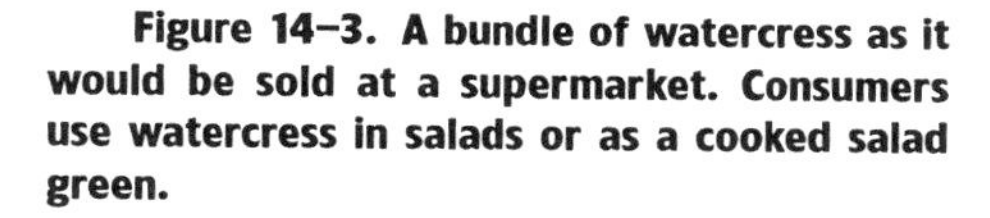

Figure 14–3. A bundle of watercress as it would be sold at a supermarket. Consumers use watercress in salads or as a cooked salad green.

growth. The nitrates must be in the water because, like many aquatic plants, watercress obtains most of its nutrients from the water.

The best growth of watercress occurs when the air temperature is between 70° and 85°F. As a result, the prime production period in Hawaii is in the winter months. It will sustain growth throughout the year, however. Watercress is also suited well to most climates in North America and will grow in all but the coldest months.

Watercress beds require a firm bottom because of the amount of running water which must flow through them. The beds have a slope of 100:1 to 200:1, which means that they are nearly level. The slope is just enough to allow for complete drainage and adequate flow-through of water. Most of the beds in Hawaii are 40 feet wide and 80 feet long, although they may vary. Concrete tile dikes are constructed around the beds to allow water to flow through and to provide foot access to the beds.

Although watercress can be planted from seeds, the normal method of propagation is by terminal shoot cuttings. Four to six shoots are placed in bunches about 1 foot apart, lengthwise to the flow of water so they are less likely to be washed away. At first, the water level is kept at about 1⁄4 inch, until the roots anchoring the plants develop properly. The water is then gradually increased to the depth of 1 - 2 inches for growing.

About 45 days after planting, the watercress is ready for harvesting, at a height of about 12 - 14 inches. The crop is harvested by hand, with the harvester grasping a bunch by hand and cutting the stems with a sickle. The stubble is left to continue to produce, and more cuttings are usually added after harvesting to replace damaged stems. In another 45 days or so, depending on the time of year, the crop will be ready to harvest again. Plants will usually yield three or four cuttings before being replaced. Obviously, the harvesting process is very labor-intensive.

The harvested watercress is vacuum-cooled and stored in a refrigerator. The shelf life is very short, usually only about one week. The consumer may use watercress as a salad green or a cooked green vegetable, much like spinach.

Chinese Waterchestnuts

In Asian countries, Chinese waterchestnut production is an important part of aquaculture and of international commerce. In the U.S., however, production is usually small-scale and not as efficient as that in Asian countries. The supply of waterchestnuts produced in the U.S. does not meet the demand, although the markets are relatively small.

Figure 14–4. A can of Chinese waterchestnuts. Consumers may buy these or fresh waterchestnuts still in the skin.

The part of the waterchestnut that is eaten is the corm, the fleshy, underground stem. It is firm and white and remains so even after being cooked.

The best location in the U.S. for Chinese waterchestnut production appears to be in the Southeast, with its warm, humid climate.

Waterchestnuts require sandy soil. Corms are planted 2 – 3 inches deep in the soil about 3 feet apart. Planting usually occurs in March. Waterchestnuts need about 1 ton of balanced fertilizer per acre, which is usually applied at three different times. The first application is at planting, with one following in May and another in August.

Waterchestnuts may be grown in raceways, flooded fields, or indoor troughs. They are a warm water plant, with their best growth occurring in the summer months.

About 220 days after planting, in late October or early November, the corms are ready for harvest. Each planted corm will have produced about 20 pounds of new corms by this time. In a good growing season, 15 – 20 tons of corms per acre can be produced.

The corms are usually harvested with rakes or small plows, covered with rubber to prevent damage to the corms. The harvesting is extremely labor-intensive. Spading forks are used to work the beds, and the soil is often screened to collect all of the corms. Although most of the corms are between 2 and 4 inches deep, some may be found as deep as 10 inches. The aquafarmer usually keeps enough corms

Figure 14–5. Harvested Chinese waterchestnut corms. Note how the corm on the right is splitting in two, the primary method of reproduction for these plants. These corms may be sold as food or retained for planting next year.

to restock the beds the next year and then sells the rest. Keeping seedstock is necessary because supplies are both hard to obtain and expensive.

After the corms have been harvested, their tough outer skin must be removed. The corms are cleaned and bleached, and then they are ready for market. Corms sold in large quantities as food vary in price from 10 to 20 cents each. In specialty ornamental sales, prices may be much higher. The waterchestnuts are used in a variety of casseroles and other dishes and as a garnish. They add a crunchy, mild taste to these dishes.

Chinese waterchestnuts are also an excellent choice for polyculture with animal aquacrops. In raceway culture, for example, the water can be used to fertilize the plants after it has moved through the animal production facility. Waterchestnuts also make an excellent biofilter for intensive indoor systems. Polyculture uses are not common practices at present; however, the potential for polyculture with waterchestnuts is excellent.

CULTURE OF AQUATIC PLANTS FOR ANIMAL FEEDS

Several types of unicellular algae have been successfully cultured for use as feed for larval stages of certain aquacrops and for herbivorous fish and molluscs.

Some algae is suitable as feed for livestock. Other aquacrops, such as duckweed and water spinach, also have been cultured for use as livestock feed. The cost of production is usually low, and the nutritional value of these aquatic plants is usually quite high. Some of the more common uses are given here.

Algae

Several green algae species of the genus *Spirulina* grow very rapidly in freshwater and will work well as feed for some animal aquacrops. The protein content of some species is 70%. *Spirulina* are found in freshwaters all over North America. When used as a feed for animal aquacrops, some mature plants are placed in the water facility and allowed to begin reproduction before the aquacrop is stocked.

Some experts believe that this green algae will someday be used more for human consumption, but very little is done at present. Although humans can digest the algae, it has a distinct flavor that most people do not like. In Mexico and some Asian countries, however, *Spirulina* is gathered and dried into patties, which are then added to other foods. It provides a nutritious supplement to other foods, but is almost never eaten by itself.

Numerous types of algae and diatoms are cultured as phytoplankton (floating or suspended plant material) for feeding to several animal aquacrops. Most of this culture is of blue-green algae from the family Cyanophyta, diatoms from the family Crysophyta, and green algae from the family Chlorophyta.

Some species of red algae and brown algae are also harvested as livestock feed supplements. In Japan and Europe this is a common practice. Off the coasts of Scotland and Ireland, some livestock producers have reduced the labor in this effort by letting their sheep and cattle graze in the intertidal zone where they can eat the algae washed up on the shore. In some areas of Scotland and Ireland, these animals feed almost exclusively on the red algae *Palmaria palmata* and the brown algae of the genus *Alaria*.

Duckweed

Several species of common duckweed from the genera *Lemna, Spirodela, Wolffia,* and *Wolffiella* are found in ponds all across North America. Duckweed is considered a weed in many aquaculture systems because its rapid growth may choke out other plants and will bother some animal species. Duckweed can usually cover the surface of a pond by the end of a growing season if it is not harvested in some way.

The duckweed plant is made of floating fronds (leaflike parts) and stems that develop a mat over the surface of freshwater ponds. The mats may get as thick as 2 inches, although the stems are usually less than 1/4 inch thick.

Duckweed is a favorite food of herbivorous fish and waterfowl. It can be harvested and used in cattle, swine, and poultry feeds. Duckweed grown in ponds that are well-fertilized is high in protein, about 40%. It also provides some essential amino acids and xanothophyll, a necessary ingredient in poultry rations.

With very little input required, duckweed is one of the least expensive aquacrops to produce. The growth rate is so rapid that weekly harvesting is common. The plants are harvested by skimming the surface of the water. The duckweed plants are dried and then added to livestock feeds.

Water Spinach

Water spinach, *Ipomoea reptans*, is more commonly cultured in Thailand, Malaysia, and Singapore than in the U.S. It is often used in polyculture with fish or freshwater prawns. It is eaten as a green vegetable in a few areas, but is much more commonly used as livestock feed.

Water spinach grows very rapidly, producing over 20 pounds of green matter per acre per day. The reason it is not cultured more extensively is that it provides very little protein and few carbohydrates. The protein content is about 2%, and the carbohydrate content is about 3%.

CULTURE OF SPECIALTY AQUACULTURE PLANTS

Phycocolloids

The most important use of seaweeds in the U.S. is the extraction of phycocolloids, gel-like substances that have many applications in science and the food industry. The two most commonly used phycocolloids are agars and carrageenans. Agar is soluble in hot water but not in cold. Some carrageenans are soluble in cold water, and all are soluble in hot water.

High-quality agar comes primarily from red algae from the genus *Gelidium*. Its most important use is as a culture medium. Almost every hospital, college, research institute, and public health agency uses agar to culture bacteria, viruses, and small plants. Agar used as a culture medium requires a high degree of purification before it is suitable.

Agar from the genera *Gelidium* and *Gracilaria*, which is of a somewhat lower quality, is widely used in printing, brewing, canning and preserving, gel chroma-

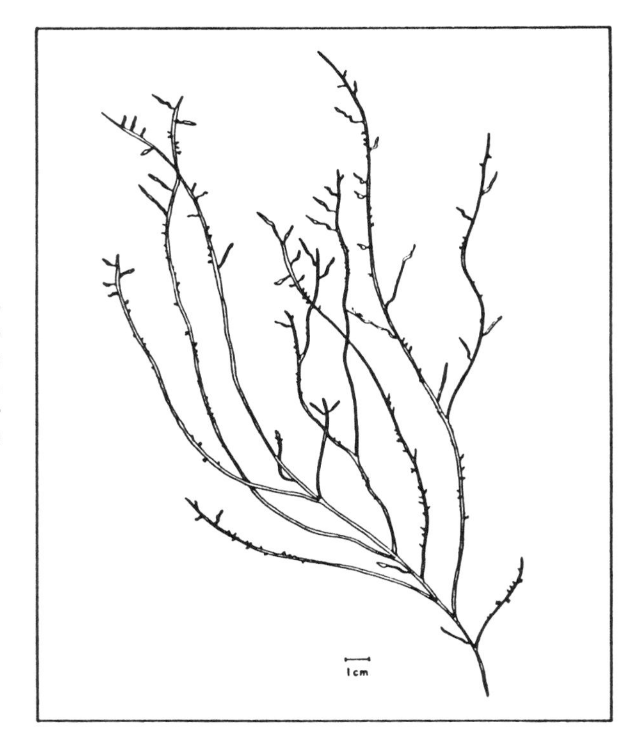

Figure 14–6. *Gracilaria verrucosa*, a red algae from which phycocolloids are extracted. (Source: J. E. Hansen, J. E. Packard, and W. T. Doyle, "Mariculture of Red Seaweeds," California Sea Grant College Program, Report No. T-CSGCP-002)

tography, and the making of dental impressions. Japan is the leading producer of agar, but there is some production in the U.S., with considerable increases in the last 20 years.

Carrageenan is actually a collection of phycocolloids from agar-producing plants. It may be extracted from numerous species of red algae and some brown algae, but its characteristics may change based on the plant from which it was extracted.

The U.S. leads the world in carrageenan production. About 75% of the carrageenan produced in the U.S. is used in the food industry, as water gels and as dairy protein applications. The water gel uses include dessert gels, pet foods, jellies, barbecue sauces, imitation milks and creamers, whipped toppings, and puddings. The dairy applications include ice cream, egg nog, evaporated milk, infant formulas, and custards. Carrageenan is also found in the hamburger patties produced by some hamburger chains.

Waste Water Treatment

Because of their ability to remove wastes from the water in which they grow, many aquatic plants are used in waste water treatment lagoons for the purpose of cleaning the water. The idea behind the lagoons is that the organic matter provides nutrients for unicellular algae to grow, the algae provide oxygen for bacteria, and the bacteria break down the organic matter so it can be used by the algae.

The only problem with the system is that the algae must be removed from the system periodically or the system will have the same amount of organic matter, in the form of algae, that it had with the wastes. The treatment facilities are not very cost-effective because the collection of the algae is expensive. Even though some of the costs can be recovered if the collected algae is sold as mulch, cattle feed, or fertilizer, a more cost-effective system would be beneficial.

Some waste water treatment facilities have solved the problem somewhat by using higher plants such as duckweed and water hyacinth. These plants are just as effective as the algae in removing wastes, and they are much easier to harvest.

Ornamentals

Plants grown for their aesthetic value include those grown for ornamental gardens, dried flower arrangements, and use in aquaria. One of the most common of these aquacrops is the cattail *Typha latifolia* or *T. angustifolia*. The cattail is used as an ornamental plant in many aquatic gardens and sometimes as a dried flower for arrangements. Arrowheads, several species of the genus *Sagittaria*, are also used in ornamental gardens, as are many other species found naturally in the U.S.

The cattail and the arrowhead also have edible parts, although they are not cultured for food as a general rule.

Fuel Sources

With fossil fuels in abundant supply and prices stable, very little attention is paid to other means of producing fuel. The fuel produced by these sources is not cost-efficient. However, when the supplies run low or prices increase because of trade problems, alternative sources of fuel may be necessary. Two types of aquatic plants have shown promise in the production of fuel. These are the water hyacinth and some algae species.

The water hyacinth has been used in experiments to produce methane gas, a clean, economical fuel. The method of using the water hyacinths from waste water treatment facilities might make both operations more economically feasible.

Some species of algae have the ability to break down sea water into hydrogen and oxygen. The hydrogen is a waste product which could be used as fuel. Hydrogen fuel, used in jet airplanes, is another clean, economical fuel.

SUMMARY

Although most of the aquaculture production in the U.S. has focused on animal production, there is good potential for plant aquaculture as well. All of the environmental factors are in place for significant plant aquaculture production. Freshwater supplies are abundant in many areas. Access to saltwater and brackish water is adequate in the areas where production would occur. Suitable climates are available for many aquaculture plants. Proven culture techniques have been established by aquafarmers in the U.S. and other countries. Table 14-1 contains the most common aquatic plants currently cultured.

Although the possibilities for increased production of aquatic plants exist, several factors have inhibited the growth of plant aquaculture. For the most part, Americans are not accustomed to including aquatic plants in their diets. Except for in Hawaii, aquatic plants historically have not been standard table fare. Also, the methods of producing aquatic plants for commercial purposes are unknown to most aquafarmers.

Table 14-1

Cultured Aquatic Plants

Common Name	Scientific Name	Uses	Location
Brown algae	*Undaria pinnatifida* *Macrocystis pyrifera*	Food, mulch, fertilizer	Saltwater
Red algae	*Porphyra* spp. *Gelidium* spp. *Gracilaria* spp.	Food, mulch, fertilizer, phycocolloids	Saltwater
Green algae	*Monostroma* *Enteromorpha*	Food, mulch, fertilizer	Saltwater, freshwater
Watercress	*Nasturtium officinale*	Food	Freshwater
Chinese waterchestnut	*Eleocharis dulcis*	Food	Freshwater
Water spinach	*Ipomoea reptans*	Animal feed	Freshwater
Arrowhead	*Sagittaria* spp.	Ornamental	Freshwater
Cattail	*Typha latifolia* *T. angustifolia*	Ornamental	Freshwater
Duckweed	*Lemna* spp. *Spirodela* spp. *Wolffia* spp. *Wolffiella* spp.	Waste water treatment, animal feed	Freshwater
Water hyacinth	*Eichhornia crassipes*	Waste water treatment	Freshwater

Many species of aquatic plants can be cultured. The uses for these plants include food for human consumption, animal feeds, waste water treatment, scientific culture material, food additives, and fuel.

When aquafarmers become aware of the possibility of polyculture as a means of increasing income from existing facilities, they will begin to take notice. Some promotion to inform the public of the nutritional benefits of aquacrops should also help increase production. As improved techniques for the culture of freshwater aquacrops are developed and markets are expanded, plant aquaculture in the U.S. should grow significantly.

QUESTIONS AND PROBLEMS FOR DISCUSSION

1. Compare the start-up costs for plant aquaculture to those for animal aquaculture.
2. What are the most commonly grown saltwater and brackish water plant aquacrops?
3. What are the two most commonly grown freshwater food plants?
4. Define "photosynthesis."
5. Define "respiration."
6. What gives plants their green color?
7. Where do most aquatic plants get their nutrients?
8. Describe the nutrient qualities of the various algae and freshwater food plants.
9. What are the three common habitat concerns of aquatic plants?
10. What is sexual reproduction in plants?
11. Describe the reproductive process of algae.
12. Describe the reproductive process of the Chinese waterchestnut.
13. Where does brown algae get its olive brown color?
14. Describe the spore collection process for brown algae.
15. How do aquafarmers obtain seedstock for the production of red algae?
16. Which algae is the most expensive seaweed cultured in Japan?
17. What are the two primary habitat considerations for watercress?
18. Describe the commercial propagation procedure for watercress.
19. What special nutrient requirements do Chinese waterchestnuts have?
20. What are some of the specialty uses of aquatic plants?

Appendices

Appendix A

TABLE OF EQUIVALENTS FOR AQUACULTURE

1 inch	=	2.54 centimeters
1 foot	=	30.48 centimeters
	=	0.348 meter
	=	12 inches
1 cubic foot	=	7.481 gallons
	=	62.4 pounds of water
	=	28,354.6 grams of water
1 yard	=	36 inches
1 cubic yard	=	27 cubic feet
	=	0.76 cubic meter
1 acre-foot	=	1 acre of surface area covered by 1 foot of water
	=	43,560 cubic feet
	=	2,718,144 pounds of water
	=	325,851 gallons of water

1 gallon	=	8.34 pounds of water
	=	3,800 cubic centimeters
1 quart	=	950 cubic centimeters
	=	950 grams of water
1 pound	=	453.6 grams
	=	16 ounces
1 ounce	=	28.35 grams
1 meter	=	39.37 inches
1 cubic meter	=	35.135 cubic feet
1 ppm requires:		2.72 pounds per acre-foot
		0.0038 gram per gallon
		0.0283 gram per cubic foot
		0.0000623 pound per cubic foot
		1.0 milligram per liter
		8.34 pounds per million gallons of water
1 ppt requires:		2,718 pounds per acre-foot
		3.80 grams per gallon
		28.30 grams per cubic foot

Appendix B

REGIONAL AQUACULTURE CENTERS

The U.S. Department of Agriculture has established five regional aquaculture centers. Each region is comprised of several states. A contact person is designated in each state to work with the regional centers. For additional information, contact the nearest center. The centers are listed below.

Northeastern Regional Aquaculture Center
Research Building, 201-B
Southeastern Massachusetts University
North Dartmouth, MA 02747

Western Regional Aquaculture Center
Division of Aquaculture
College of Fisheries
University of Washington
Seattle, WA 98195

Center for Tropical and Subtropical Aquaculture
The Oceanic Institute
Makapuu Point
Waimanalo, HI 96795

North Central Regional Aquaculture Center
Fisheries and Wildlife Department
Michigan State University
East Lansing, MI 48824

Southern Regional Aquaculture Center
P.O. Box 197
Stoneville, MS 38776

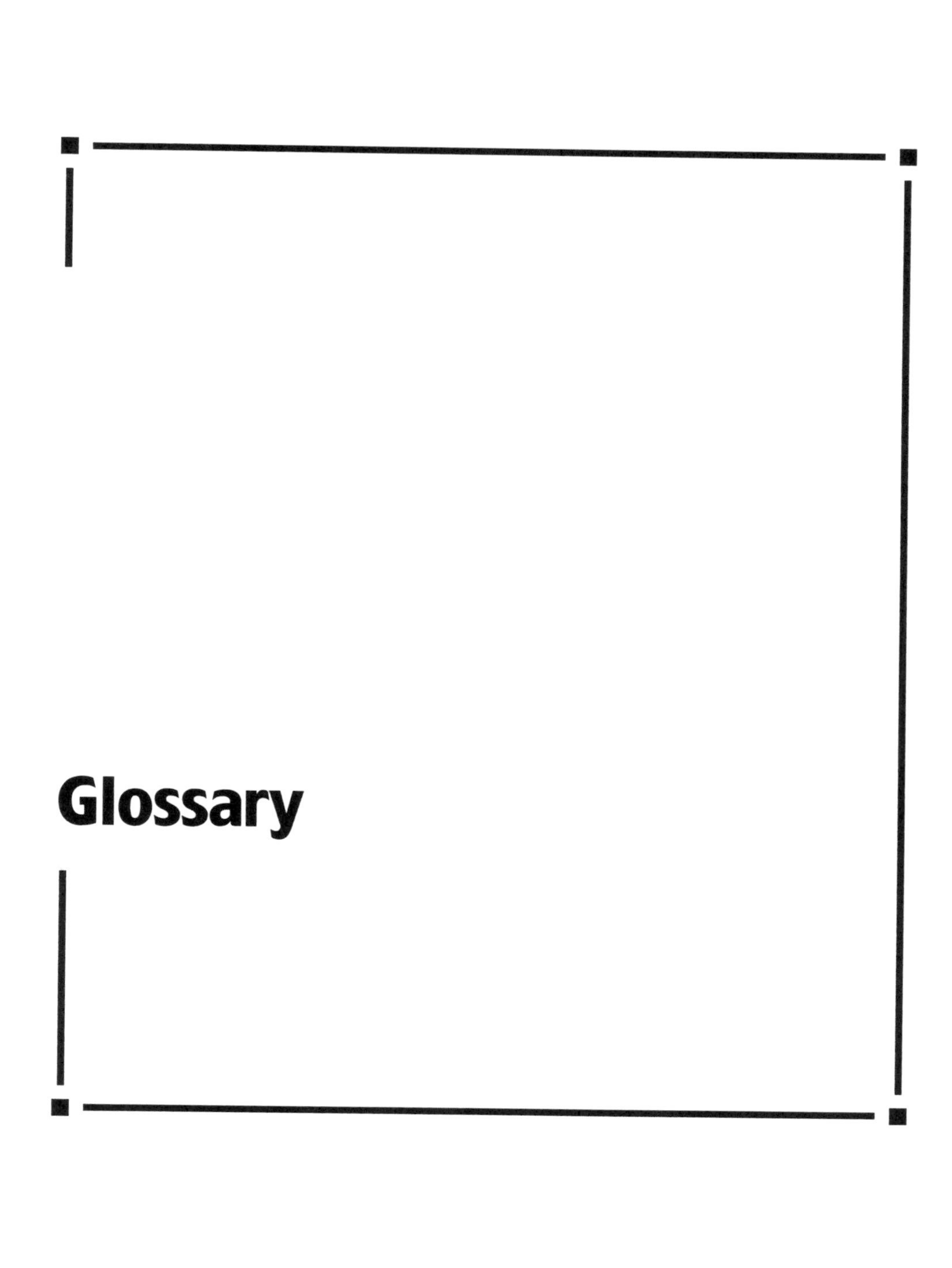

Glossary

GLOSSARY

A

Acute disease — a severe disease that lasts only a short time.

Aeration — the process of adding dissolved oxygen to water.

Alkalinity — a measure of the pH buffering capacity, usually of water in aquaculture.

Amino acids — compounds from which proteins are built; essential compounds in the diets of fish and other aquatic animals.

Ammonia — a gas formed of nitrogen and hydrogen by the decomposition of organic matter in aquaculture water.

Anadromous fish — types of fish that live in saltwater but spawn in freshwater.

Antibiotic — a chemical used to treat and control disease; may kill bacteria and other organisms.

Apprentice — an individual who works for another person to learn the skills of an occupation.

Aquabusiness — an enterprise or a business that provides for the needs of aquaculture production (such as feed and equipment) or markets aquatic commodities (such as processing).

Aquacrop — an aquatic crop.

Aquafarming — the culturing of aquatic crops.

Aquatic — living or growing on or in water.

B

Benthos—the organisms that live on or in the sediment on the bottom of a water facility.

Biofilter—a water filtration device that uses living organisms (usually bacteria) to convert harmful substances into less harmful substances.

Bivalve molluscs—animals with two sides of a shell hinged together by a ligament.

Brackish water—water near the shoreline of oceans that is a mixture of saltwater and freshwater.

Broodfish—sexually mature fish kept for reproduction purposes.

C

Capital—the wealth or property needed for establishing aquafarms or businesses; money or the access to money.

Carnivore—an animal that eats flesh; known as carnivorous.

Catadromous fish—types of fish that live in freshwater but spawn in saltwater.

Chronic disease—a disease that recurs or lasts a long time.

Consumer—a person or an organization that consumes or uses a product or service; often thought of as a consumer of food.

Corm—an enlarged, fleshy, bulb-like base or underground stem of a plant.

D

Decapod—an animal with 10 legs, such as a crawfish.

Detritus—the debris from plants and animals in water.

Dissolved oxygen (DO)—the oxygen that is dissolved in water and available for animal life; measured as parts per million (ppm).

Dorsal—the top side of the body of a fish.

Drawdown—the process of lowering the water level in a pond.

Dress—to prepare fish for the consumer by cleaning and eviscerating.

E

Ectothermic—referring to the body temperature of fish, which adjusts to that of the water.

Effluent—the water discharged from a fish farm, an industry, or another source.

Egg—the female sex cell.

Entrepreneur—someone who owns an aquafarm or business.

Estuary—the brackish water formed where a freshwater stream flows into an ocean.

Evaporation — the process of changing from a liquid to a vapor and passing into the air.

Eviscerate — to remove the internal organs of a fish in processing; to gut.

Extensive production — the growing of aquacrops in facilities where the supply of natural food is sufficient and less management is required.

Eyed-egg — a fish egg in which the embryo is developing; an egg with two black spots showing the eyes of the fish.

F

Fee-lake — a recreational fishing facility where individuals pay to fish.

Fertilization — the union of the sperm with the egg.

Fillet — a boneless piece of fish; usually cut from the sides along the backbone.

Fingerling — fish past the fry stage that are used in stocking growout facilities; often 1 inch to 8 or 10 inches long.

Flashing — the behavior of a fish in rubbing on plants or objects, turning sideways, and twisting.

Flatland — land with 3% or less slope.

Food fish — fish that are produced for human consumption.

Freeboard — the distance between the level or surface of the water in a pond and the top of the levee.

Fry — the stage of fish growth between hatching and the time a length of 1 inch is reached.

G

Gamete — a mature sexual reproduction cell; egg or sperm.

Gastropods — a class of molluscs having a single shell that is often coiled; examples include snails and slugs.

Genetics — the study of heredity.

H

Habitat — the usual environment of a plant or an animal.

Hemorrhage — a discharge of blood by an animal.

Herbivore — an animal that eats plants; known as herbivorous.

Holding — temporarily storing live fish; procedures must be followed to prevent stress.

Hybrids — fish or other animals or plants resulting from crosses of different species.

Hydroponics — the cultivation of terrestrial plants without soil in a water solution.

I

Infectious diseases—diseases that can be transferred from one organism to another; caused by pathogens.

Intensive production—the growing of aquacrops in higher population densities than could be supported in the natural environment.

Invertebrate—an organism without a spinal cord; has a hard outer skeleton.

L

Larvae—the early stage of an animal that is greatly different from the adult form.

Lateral—the side of the body.

Lesion—a wound or an injury on a fish or other animal.

Levee—an earthen dike used to enclose water.

Live car—a special type of seine attached to a harvesting seine used to impound (store) fish in a pond.

M

Marine—from or of the ocean or sea.

Mating—the interaction between the male and female which results in fertilization of the eggs.

Milt (semen)—the fluid produced by male fish that contains sperm.

Mollusc (or mollusk)—a member of the phylum Mollusca, containing all shellfish other than the crustaceans; includes snails, clams, oysters, and many others.

Monoculture—the cultivation of a single species of aquacrop in a water facility.

N

Niche market—a market that focuses on the unique needs or interests of a particular group or subpopulation.

Nitrate—a form of nitrogen found in aquaculture water that is converted from nitrite and is less toxic than ammonia and nitrite.

Nitrite—a form of nitrogen found in aquaculture water and other places and is less toxic to fish than ammonia; converted by bacteria from ammonia.

Nitrogen cycle—a natural process that occurs in ponds as the nitrogen in the organic matter is converted to ammonia, nitrite, and nitrate forms.

Noninfectious disease—a disease that cannot be transferred from one organism to another; not caused by a pathogen.

Nutrient—a substance that provides nourishment for plants and animals.

O

Off-flavor — a flavor that varies from that considered normal for fish, such as a muddy taste.

Organism — a life form made up of separate organs that carry on functions that are mutually dependent.

P

Parasite — an animal or a plant that lives on or in another organism.

Pathogen — an organism that causes disease, such as a bacterium or a virus.

Permeability — the quality of soil that allows water to pass through it; highly permeable soils do not hold water well and should not be used for ponds.

pH — an index of the acidity of a substance, with 7.0 being neutral and numbers below 7.0 becoming increasingly acid and those above 7.0 becoming increasingly alkaline (basic).

Photosynthesis — the process whereby plants, in the presence of light, convert carbon dioxide and other elements into simple sugar and produce oxygen.

Phytoplankton — microscopic plants that grow in water; microscopic aquatic plants.

Pigment — the coloring in the cells of animals and plants.

Piping — the gasping of fish at the surface of the water, usually for oxygen.

Plankton — microscopic plants or animals that grow in the water.

Planktonic — referring to organisms that grow suspended in water.

Polyculture — the cultivation of more than one species of organism in a growout facility.

Predator — an animal that hunts and kills other animals.

Prophylactic — taking steps to protect aquacrops against disease, such as adding treatments to water while fish are hauled.

R

Raceway — an aquaculture water facility that uses flowing water.

Ranching — obtaining wild stock for culturing to market size; releasing an aquacrop into the wild so it can mature and then be harvested at a later date.

Regulatory agency — an agency of government that establishes laws or regulations that govern aquaculture or other endeavors.

Respiration — the process used by plants and animals to transform chemical energy into other kinds of energy; oxygen is used and carbon dioxide is produced by the process.

Runoff — excess precipitation that does not infiltrate the soil; forms streams, ponds, and lakes.

S

Sac fry (alevins) — newly hatched fry that have a yolk sac attached for nourishment.

Salinity — the amount of salt in water; water with a high salinity has a high salt content.

Secchi disc — a device used to measure the transparency of water.

Sediment — matter (soil particles, feces, and uneaten feed) that settles to the bottom of water in aquaculture facilities.

Seed (seedstock) — the fry, fingerlings, or larval forms of aquacrops that are used to stock growout facilities.

Seine — a long net used for harvesting fish.

Sluice gate — a water regulatory device in a pond dam or other structure.

Spawning — the releasing of eggs from the female.

Species — the scientific class of animals or plants of the narrowest classification; capable of interbreeding; sharing a common name.

Sperm — the male sex cell, found in semen.

Spillway — the structure water passes over in leaving a pond; designed to serve in emergencies to regulate the level of the water in a pond or other water structure.

Stocker — fish larger than a fingerling but smaller than a food fish; usually more than 8 or 10 inches long.

Stress — conditions in water that cause fish to expend energy and experience physical strain or weakening; makes fish susceptible to disease.

Stripping — the process of artificially removing matured eggs from a female fish by exerting pressure on specific parts of the abdomen.

Supersaturation — a condition whereby there is more than the normal amount of a gas in water.

T

Taxonomy — the classification of plants and animals into groups.

Tide — the periodic rise and fall of the water in oceans and other large bodies of water caused by the attraction of the moon and sun; occurs about every 12 hours.

Topography — the lay of land; hills, valleys, and other features of land.

Topping — selectively harvesting only the fish that have reached a certain size.

Turbidity — cloudy or muddy water; caused by plankton or suspended soil particles.

V

Ventral — the underside of the body; the belly area.

Venturi drain — the water outlet or standpipe constructed in the middle of a tank or other water structure.

Vertebrate — an organism with a segmented spinal column and inner skeleton.

Viscera — the internal organs of the body; removed in processing.

W

Water chemistry — the composition of water; includes pH and minerals.

Watershed — the land around a pond or stream that provides water from precipitation or other sources.

Water quality — the suitability of water for aquaculture.

Water table — level below the surface of the earth where the ground is saturated with water.

Wetland — land that is covered with standing water much of the year.

Z

Zooplankton — microscopic animals that grow in water; microscopic aquatic animals.

Zygote — the cell produced by the union of two gametes.

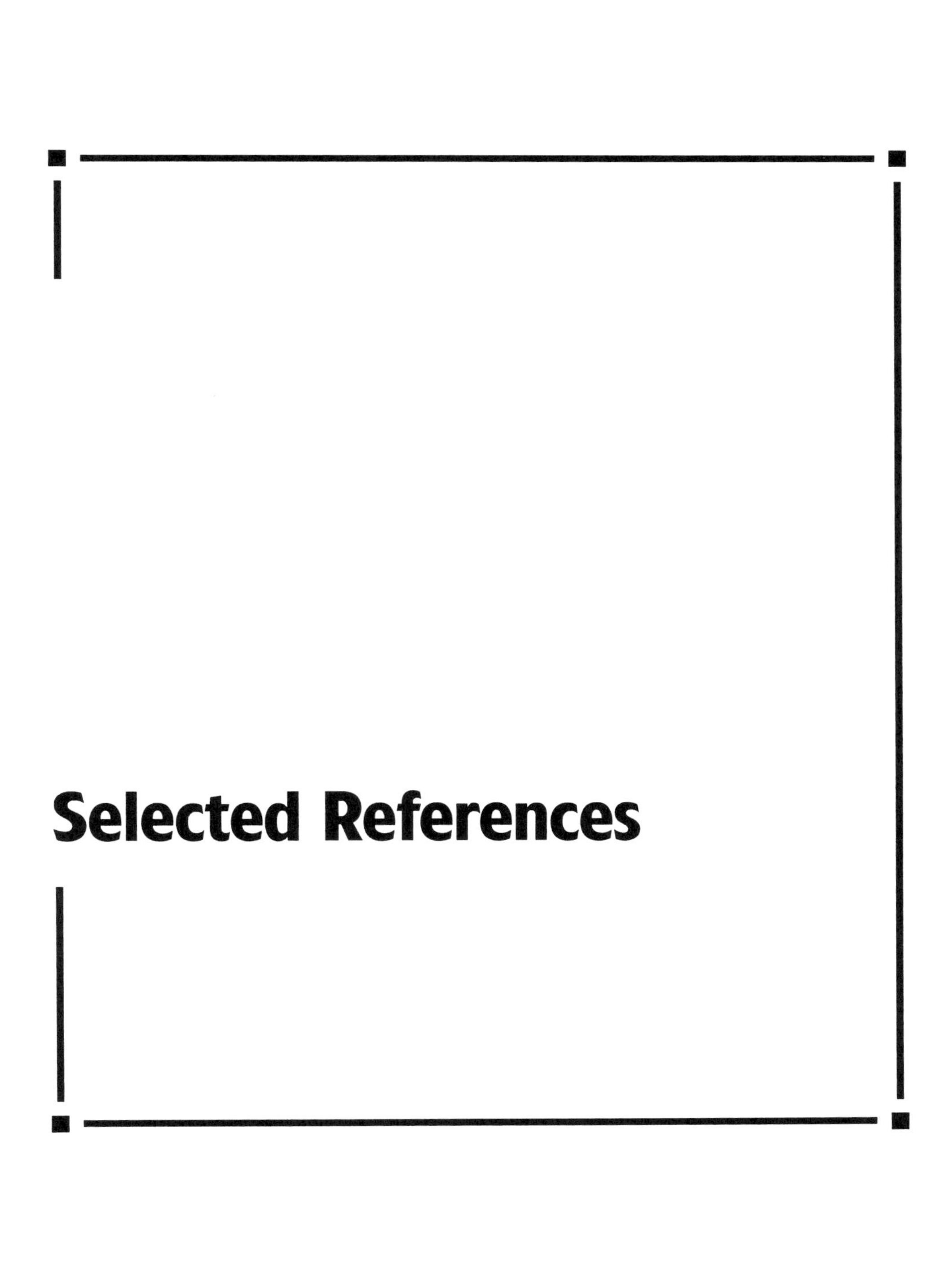

Selected References

SELECTED REFERENCES

Aquaculture: Opportunities for Appalachia. Washington: Appalachian Regional Commission, 1990.

"Aquaculture Resources Guide." Alexandria, Virginia: The Council on Agricultural Education, n.d.

"Aquaculture — Situation and Outlook Report." Washington: U.S. Department of Agriculture, March, 1991.

Avault, James W., Jr. "The Bare Bones of Crawfish Farming." *Aquaculture Magazine* (September - October, 1980), pp. 42–43.

Bardach, John E., John H. Ryther, and William O. McLarney. *Aquaculture: The Farming and Husbandry of Freshwater and Marine Organisms*. New York: John Wiley and Sons, 1974.

Bauer, Larry L. "Crawfish Production in South Carolina: Estimated Costs and Profit Potential." Clemson, South Carolina: Clemson University, August, 1984.

Bayer, Robert, and Juanita Bayer. *Lobsters Inside-Out*. Bar Harbor, Maine: Acadia Press, 1989.

Belusz, Larry. "Fish Farming Techniques." Columbia: University of Missouri, 1990.

Bliss, Dorothy E. *Internal Anatomy and Physiological Regulation*. New York: Academic Press, 1983.

Brown, E. E., and J. B. Gratzek. *Fish Farming Handbook*. Westport, Connecticut: AVI Publishing Company, Inc., 1980.

Brunson, Martin L. "Evaluation of Rice Varieties for Double Cropping Crawfish and Rice in Southwest Louisiana." Baton Rouge: Louisiana Agricultural Experiment Station, Bulletin 812, July, 1989.

Cardeilhac, Paul. "Nutritional Aspects of Hatchling Alligator Productions." *Today's Aquaculturist* (December, 1990), pp. 1–2.

Cobb, J. Stanley, and Bruce F. Phillips. *The Biology and Management of Lobsters*. New York: Academic Press, 1980.

Dupree, Harry K., and Jay V. Hunter, ed. "Third Report to the Fish Farmers." Washington, D.C.: U.S. Fish and Wildlife Service, 1984.

Emmens, Cliff W. *Tropical Fish: A Complete Introduction*. Neptune City, New Jersey: T. F. H. Publications, Inc., 1987.

Everhart, W. Harry, and William D. Youngs. *Principles of Fishery Science*, 2nd ed. London: Cornell University Press, 1989.

Hansen, Judith E., Julie E. Packard, and W. T. Doyle. "Mariculture of Red Seaweeds." La Jolla, California: California Sea Grant College Program, Report No. T-CSGCP-002, 1981.

Hedeen, Robert A. *The Oyster*. Centreville, Maryland: Tidewater Publishers, 1986.

Hinshaw, Jeffrey M. "Trout Production: Feeds and Feeding Methods." Raleigh: North Carolina State University, Southern Regional Aquaculture Center Publication 223, 1990.

Hinshaw, Jeffrey M. "Trout Production: Handling Eggs and Fry." Raleigh: North Carolina State University, Southern Regional Aquaculture Center Publication 220, January, 1990.

Hodson, Ronald G. "Food Fish Production of Hybrid Striped Bass," Raleigh: North Carolina State University, n.d.

Huet, Michael. *Textbook of Fish Culture*. Farnham, Surrey, England: Fishing News Books Ltd., 1986.

Jensen, Gary L., Joseph D. Bankston, and John W. Jensen. "Pond Aeration—Types and Uses of Aeration Equipment." Baton Rouge: Louisiana Cooperative Extension Service, Southern Regional Aquaculture Center Publication 371, 1989.

Jensen, John W. "Watershed Fish Production Ponds." Auburn: Alabama Cooperative Extension Service, Southern Regional Aquaculture Center Publication 102, 1989.

Kerby, James H., L. Curry Woods III, and Melvin T. Huish. "Pond Culture of Striped Bass X White Bass Hybrids," *Journal of World Mariculture Society* (1983), pp. 613–623.

Kight, Troy. "Hybrid Striped Bass—Pros and Cons," *Research Highlights*, Mississippi State: Mississippi Agricultural and Forestry Experiment Station, February, 1990.

Korringa, P. *Farming Marine Organisms Low in the Food Chain*. New York: Elsevier Scientific Publishing Company, 1976.

Krul, Syd. "Mahimahi Reproduction at the Waikiki Aquarium," *Today's Aquaculturist* (January, 1990),, pp. 1–2.

Lee, Jasper S. *Commercial Catfish Farming*, 3rd ed. Danville, Illinois: Interstate Publishers, Inc., 1991.

Masser, Michael P. "Cage Culture—Cage Construction and Placement." Frankfort: Kentucky State University, Southern Regional Aquaculture Center Publication No. 162, 1988.

Masser, Michael P. "Cage Culture—Handling and Feeding Caged Fish." Frankfort: Kentucky State University, Southern Regional Aquaculture Center Publication No. 164, 1988.

Masser, Michael P. "Cage Culture—Species Suitable for Cage Culture," Frankfort: Kentucky State University, Southern Regional Aquaculture Center Publication No. 163, 1988.

McHugh, John J., Steven K. Fukuda, and Kenneth Y. Takeda. *Hawaii Watercress Production.* Honolulu: University of Hawaii, College of Tropical Agriculture and Human Resources, 1987.

McLarney, William. *The Freshwater Aquaculture Book.* Point Roberts, Washington: Hartley & Marks, 1987.

Milne, P. H. *Fish and Shellfish Farming in Coastal Waters.* Farnham, Surrey, England: Fishing News Books Ltd., 1979.

Murphy, Tim R., and James L. Shelton. "Aquatic Weed Management." Athens: University of Georgia, Southern Regional Aquaculture Center Publication 361, April, 1989.

Nelson, Joseph S. *Fishes of the World*, 2nd ed. New York: John Wiley and Sons, 1984.

"Planning for Commercial Aquaculture." Blacksburg: Virginia Tech Publication 420, 1985.

Riemer, Donald N. *Introduction to Freshwater Vegetation.* Westport, Connecticut: AVI Publishing Company, Inc., 1984.

Silva, Juan L., James O. Hearnsberger, Fay Hagan, and Gale R. Ammerman. "A Summary of Processing Research on Freshwater Prawns at Mississippi State University." Mississippi State: Mississippi Agricultural and Forestry Experiment Station, Bulletin 961, August, 1989.

Smith, Theodore I. J. "Aquaculture of Striped Bass and Its Hybrids in North America." *Aquaculture Magazine* (January - February, 1990), pp. 40–49.

Tucker, C. S., and Edwin H. Robinson. *Channel Catfish Farming Handbook.* New York: Van Nostrand Reinhold, 1990.

Waite, Stephen W., Bruce C. Kinnett, and Andrew J. Roberts. "The Illinois Aquaculture Industry: Its Status and Potential." Springfield: Illinois Department of Agriculture, 1986.

Walker, Susan S. *Aquaculture.* Stillwater, Oklahoma: Mid-American Vocational Curriculum Consortium, Inc., 1990.

Wellborn, Thomas L. "Catfish Farmers Handbook." Mississippi State: Mississippi Cooperative Extension Service, Publication No. 1549, 1989.

Wellborn, Thomas L. "Construction of Levee-Type Ponds for Fish Production." Gainesville: University of Florida, Southern Regional Aquaculture Center Publication No. 101, 1988.

Wellborn, Thomas L. "Site Selection of Levee-Type Fish Production Ponds." Gainesville: University of Florida, Southern Regional Aquaculture Center Publication No. 100, 1988.

"World Shrimp Farming — 1989." San Diego: *Aquaculture Digest*, January, 1990.

Index

INDEX

B

C

F

N

O

Q

R

S

T

U

V

W

X

Y

Z